ADVANCED PRACTICAL ORGANIC CHEMISTRY

ADVANCED PRACTICAL ORGANIC CHEMISTRY

By

Amit Arora

DISCOVERY PUBLISHING HOUSE PVT. LTD.

NEW DELHI-110 002

First Published – 2009

Reprinted – 2025

ISBN: 978-81-8356-444-1

Advance Practical Organic Chemistry

Published by:

DISCOVERY PUBLISHING HOUSE
4383/4B, Ansari Road, Darya Ganj
New Delhi-110 002 (India)
Phone: +91-11-23279245; 23253475; 43596065
Mobile: +91 9811179893 / +91 9871656464
E-mail: discoverybooksindia@gmail.com
orderdphbooks@gmail.com
namitwasan9@gmail.com
web: www.discoverypublishinggroup.com

Printed at:
Infinity Imaging Systems
Delhi (INDIA)

Contents

Preface *ix*

1. **Introduction** **1**

Branched Chains • Rings • Simple Hydrocarbons • Isomers • Functional Groups • Molecules • Hydrocarbons • Carbon Rings • Carbon, Hydrogen and Oxygen • Isomerism • Adding Nitrogen • Other Atoms

2. **Functional Groups** **23**

Functional Groups: Aliphatic and Aromatic • Taxonomy of Functional Groups • Taxonomy: Primary, Secondary, Tertiary and Quaternary • Bonding of Intermolecule • Reactions and Properties

3. **Organic Synthesis Reagents** **41**

Mechanism Reaction • Transformation Regents • Phase Transfer Catalysts' Application • Yeast for Bakers • Reactions: Produced Form • Study of Crown Ethers • Study of Phase Transfer Catalysts • Reactions of Prevost and Woodward

4. **The Structure** 57

Making Structures • The Taxonomy • Bonding and Structure • Hybridised Centres and Bonds • Cycloalkanes and Alkanes • Hybridisation and Covalent Bonding

5. **Reactions of Organic Names** 83

Reaction of Barton • Baeyer-Villiger Oxidation • Reaction of Chichibabin • Rearrangement of Favorskii • Addition Reaction of Michael

6. **Different Elements** 111

Thiols Chemistry • Preparation of Ethers, Epoxides and Thioethers • Ether, Epoxides and Thioethers: Properties • Study of Amines and Nitriles • Amines' Properties • Nitrilis Chemistry • Formation of Alcohols, Phenols and Thiols • Making of Phenols • Alcohols and Phenols: Properties • Alcohol's Reaction • Phenol's Reaction

7. **Process of Reaction** 147

Reactions of Enolate • Process of Alkyl Halides • Substitution of Nucleophilic • Process of Elimination • Derivatives and Meaning of Carboxylic Acid • Substitution of Nucleophilic • Process of Reactivity

8. **Process of Oxidation and Reduction** 189

Alkenes: Oxidation and Reduction • Alkenes Hydroboration • Alkynes: Electrophilic Addition • Alkynes Reduction • Terminal Alkynes: Alkylation • Dienes, which are Conjugated

• Process of Aromaticity • Aromatic Compounds: Properties and Formation • Benzene's Electrophilic Substitution • Formation of Mono-substituted Bonzenes • Mono-substituted Aromatic Rings: Electrophilic Addition • Alkynes and Alkenes • Characteristics of Alkynes and Alkenes • Symmetrical Alkenes: Electrophilic Addition • Unsymmetrical Alkenes: Electrophilic Addition

9. Reaction and Mechanism **243**

Formation of Enolates • Reaction between Acid and Base • Strength of Acid • Strength of Base • Acids and Bases of Lewis

10. Significance of Nucleophilics **275**

Reactions in Enolate Ion • Reduction and Oxidation • Nucleophilic Additions • Nucleophilics in Operation

Bibliography 295

Index 303

Preface

Organic Chemistry is the study of the properties of the compounds of carbon that are organic. All the carbon compounds, except a few inorganic compounds, are organic. Organic Chemistry is that branch of chemistry, which deals with the structure, properties and reactions of compounds that contain carbon. It is a highly creative science. Chemists in general, and organic chemists in particular, can create new molecules, never before proposed, which may have important properties for the betterment of the human experience. However, the original definition of Organic Chemistry came from the misconception that organic compounds were always related to the process of life. Organic Chemistry received a boost, when it was realised that these compounds could be treated in ways, similar to inorganic compounds and could be created in laboratory by means, other than 'Vital Force'.

The history of Organic Chemistry continues with the discovery of petroleum and its separation into fractions, according to boiling ranges. Many individuals consider Organic Chemistry to be very complicated and unrelated to daily life. Though, the study of Organic Chemistry may be complex, it is very important in everyday life.

Although many books on the subject are available in the market, yet, there is a dearth. Hence this humble effort. Hopefully it will prove to be beneficial for all concerned readers. Positive and enlightening feedback is solicited

Introduction

Organic chemistry is the branch of chemistry in which covalent carbon compounds and their reactions are studied. A wide variety of classes of compounds such as vitamins, drugs, natural and synthetic fibres, as well as carbohydrates, peptides and fats consist of organic molecules. Organic chemists determine the structures of organic molecules, study their various reactions, and develop procedures for the synthesis of organic substances.

Organic chemistry is the study of the properties of the compounds of carbon that are organic. All carbon compounds except for a few inorganic carbon compounds are organic. Inorganic carbon compounds include the oxides of carbon, the bicarbonates and carbonates of metal ions, the metal cyanides, and a few others.

Organic chemistry is the most important branch of chemistry — but of course it would be nothing without the many other areas of chemistry — in fact all branches of chemistry should not be viewed in isolation, even though they may often be taught in isolation.

Organic chemistry is all around us, life is based on organic chemistry, the clothes we wear, the drugs we take, the cars we drive and the fuel that propels them, wood, paper, plastics and paints.

Organic chemistry is the study of compounds containing carbon the ability of carbon to form as many as 4 strong bonds to many other atoms, e.g., carbon, hydrogen, oxygen, nitrogen, halogens, sulphur, phosphorus ensures a virtual infinite number of possible compounds the constituent atoms and their exact combination determines the chemical and physical properties of compounds and hence, their suitability for applications.

To understand life as we know it, we must first understand a little bit of organic chemistry. Organic molecules contain both carbon and hydrogen. Though many organic chemicals also contain other elements, it is the carbon-hydrogen bond that defines them as organic. Organic chemistry defines life. Just as there are millions of different types of living organisms on this planet, there are millions of different organic molecules, each with different chemical and physical properties. There are organic chemicals that make up your hair, your skin, your fingernails, and so on. The diversity of organic chemicals is due to the versatility of the carbon atom. Why is carbon such a special element? Let's look at its chemistry in a little more detail.

Carbon (C) appears in the second row of the periodic table and has four bonding electrons in its valence shell. Similar to other non-metals, carbon needs eight electrons to satisfy its valence shell. Carbon, therefore, forms four bonds with other atoms (each bond consisting of one of carbon's electrons and one of the bonding atom's electrons). Every valence electron participates in bonding, thus a carbon atom's bonds will be distributed evenly over the atom's surface. These bonds form a tetrahedron (a pyramid with a spike at the top), as illustrated below:

Carbon forms 4 bonds

Organic chemicals gets their diversity from many different ways carbon can bond to other atoms. The simplest organic chemicals, called *hydrocarbons,* contain only carbon and hydrogen atoms; the simplest hydrocarbon (called *methane*) contains a single carbon atom bonded to four hydrogen atoms:

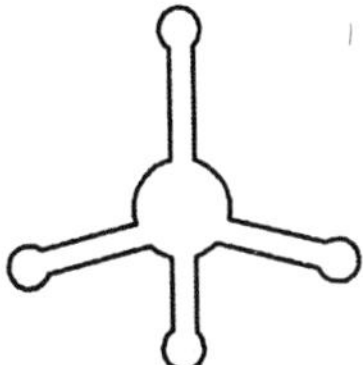

Methane: A carbon atom bonded to 4 hydrogen atoms

But carbon can bond to other carbon atoms in addition to hydrogen, as illustrated in the molecule ethane below:

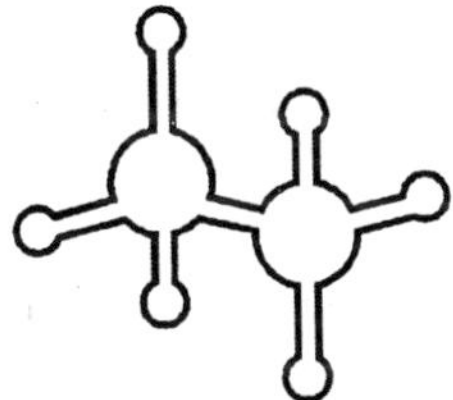

Ethane: A carbon-carbon bond

In fact, the uniqueness of carbon comes from the fact that it can bond to itself in many different ways. Carbon atoms can form long chains:

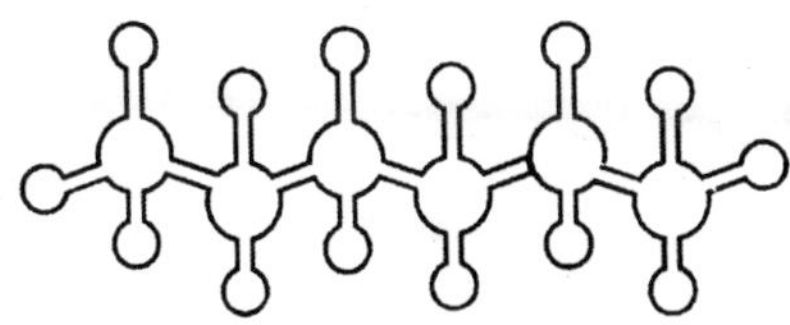

Hexane: A 6-carbon chain

Branched Chains

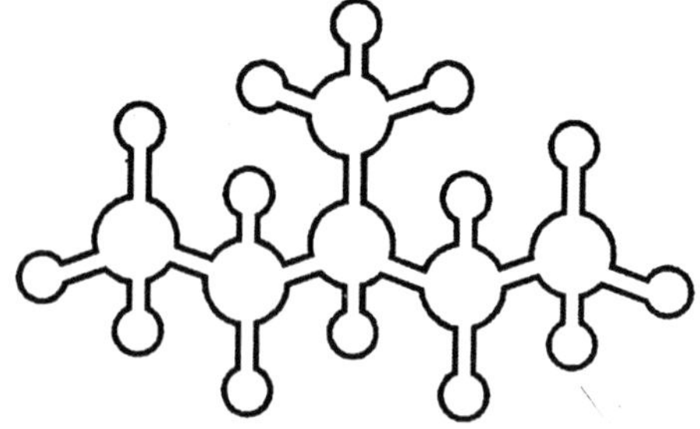

Isohexane: A branched-carbon chain

Rings

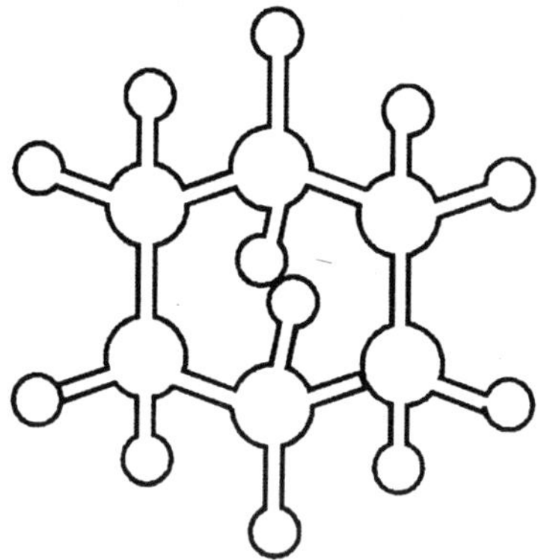

Cyclohexane: A ringed hydrocarbon

They appears to be almost no limit to the number of different structures that carbon can form. To add to the complexity of organic chemistry, neighbouring carbon atoms can form double and triple bonds in addition to single carbon-carbon bonds:

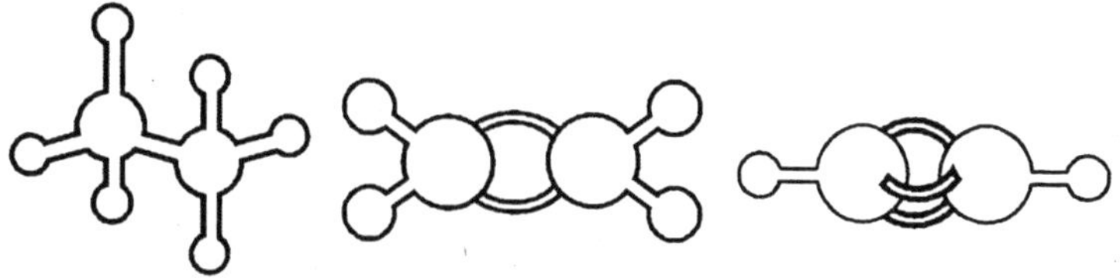

Single bonding Double bonding Triple bonding

Keep in mind that each carbon atom forms four bonds. As the number of bonds between any two carbon atoms increases, the number of hydrogen atoms in the molecule decreases.

Simple Hydrocarbons

The simplest hydrocarbons are those that contain only carbon and hydrogen. These simple hydrocarbons come in three varieties depending on the type of carbon-carbon bonds that occur in the molecule. Alkanes are the first class of simple hydrocarbons and contain only carbon-carbon single bonds.

The alkanes are named by combining a prefix that describes the number of carbon atoms in the molecule with the root ending "ane". The names and prefixes for the first ten alkanes are given in the following table:

Carbon Atoms	*Prefix Name*	*Alkane Formula*	*Chemical Formula*	*Structural*
1.	Meth-	Methane	CH_4	CH_4
2.	Eth-	Ethane	C_2H_6	CH_3CH_3
3.	Prop-	Propane	C_3H_8	$CH_3CH_2CH_3$
4.	But-	Butane	C_4H_{10}	$CH_3CH_2CH_2CH_3$
5.	Pent-	Pentane	C_5H_{12}	$CH_3CH_2CH_2CH_2CH_3$
6.	Hex-	Hexane	C_6H_{14}	$CH_3CH_2CH_2CH_2CH_2CH_3$
7.	Hept-	Heptane	C_7H_{16}	$CH_3CH_2CH_2CH_2CH_2CH_2CH_3$
8.	Oct-	Octane	C_8H_{18}	$CH_3CH_2CH_2CH_2CH_2CH_2CH_2CH_3$
9.	Non-	Nonane	C_9H_{20}	$CH_3CH_2CH_2CH_2CH_2CH_2CH_2CH_2CH_3$
10.	Dec-	Decane	$C_{10}H_{22}$	$CH_3CH_2CH_2CH_2CH_2CH_2CH_2CH_2CH_2CH_3$

The chemical formula for any alkane is given by the expression C_nH_{2n+2}. The structural formula, shown for the first five alkanes in the table, shows each carbon atom and the elements that are attached to it. This structural formula is important when we begin to discuss more complex hydrocarbons. The simple alkanes share many properties in common. All enter into combustion reactions with oxygen to produce carbon dioxide and water vapour. In other words, many alkanes are flammable. This makes them good fuels. For example, methane is the principle component of natural gas, and butane is common lighter fluid.

$$CH_4 + 2O_2 \rightarrow CO_2 + 2H_2O$$

The combustion of methane

The second class of simple hydrocarbons, the alkenes, consists of molecules that contain at least one double-bonded carbon pair. Alkenes follow the same naming convention used for alkanes. A prefix (to describe the number of carbon atoms) is combined with the ending "ene" to denote an alkene. Ethene, for example is the two-carbon molecule that contains one double bond. The chemical formula for the simple alkenes follows the expression C_nH_{2n}. Because one of the carbon pairs is double bonded, simple alkenes have two fewer hydrogen atoms than alkanes.

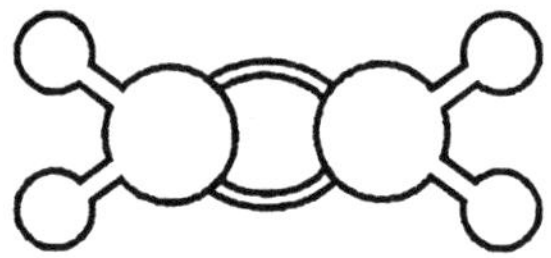

Ethene

Alkynes are the third class of simple hydrocarbons and are molecules that contain at least one triple-bonded carbon pair. Like the alkanes and alkenes, alkynes are named by combining a prefix with the ending "yne" to denote the triple bond. The chemical formula for the simple alkynes follows the expression C_nH_{2n-2}.

Ethyne

Isomers

Because carbon can bond in so many different ways, a single molecule can have different bonding configurations. Consider the two molecules illustrated here:

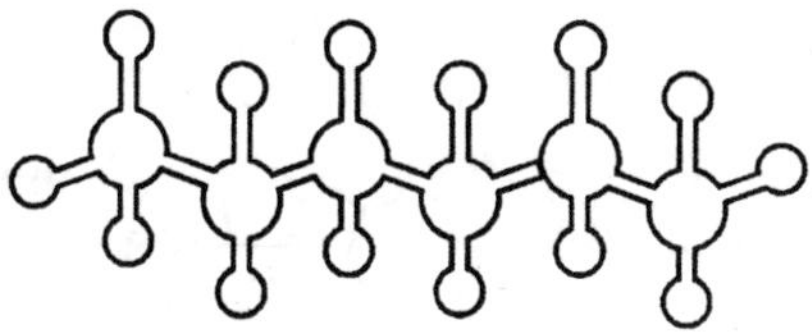

C_6H_{14} $CH_3CH_2CH_2CH_2CH_2CH_3$

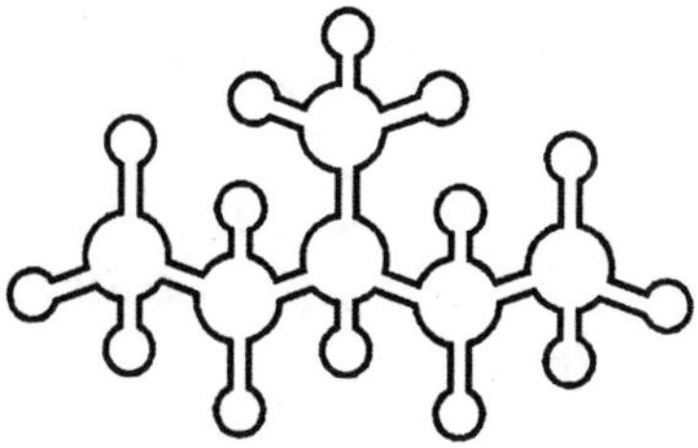

CH_3
|
C_6H_{14} $CH_3CH_2CH_2CH_2CH_3$

Both molecules have identical chemical formulas; however, their structural formulas (and thus some chemical properties) are different. These two molecules are called *isomers*. Isomers are molecules that have the same chemical formula but different structural formulas.

Functional Groups

In addition to carbon and hydrogen, hydrocarbons can also contain other elements. In fact, many common groups of atoms can occur within organic molecules, these groups of atoms are called *functional groups*. One good example is the hydroxyl functional group. The hydroxyl group consists of a single oxygen atom bound to a single hydrogen atom (-OH).

The group of hydrocarbons that contain a hydroxyl functional group is called *alcohols*. The alcohols are named in a similar fashion to the simple hydrocarbons, a prefix is attached to a root ending (in this case "anol") that designates the alcohol.

The existence of the functional group completely changes the chemical properties of the molecule. Ethane, the two-carbon alkane, is a gas at room temperature; ethanol, the two-carbon alcohol, is a liquid.

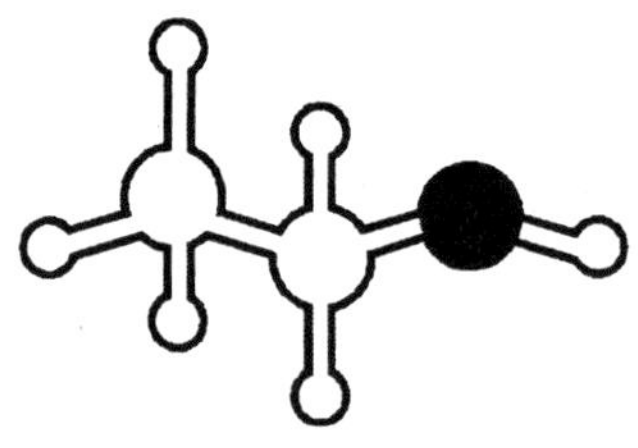

Ethanol

Ethanol, common drinking alcohol, is the active ingredient in "alcoholic" beverages such as beer and wine.

Molecules

All substances are made up of molecules which are collections of atoms. All the molecules in existence are made up of about a hundred different kinds of atoms.

For example, a water molecule is composed of two atoms of hydrogen and one atom of oxygen. We write its formula as H_2O.

A molecule of sulphuric acid contains two atoms of hydrogen, one atom of sulphur and four atoms of oxygen. Its formula is H_2SO_4.

These are simple molecules containing only a few atoms. Most inorganic molecules are small. Below are a few common inorganic substances with their formulas:

Name of Substance	*Formula*
Carbon Dioxide	CO_2
Salt	NaCl
Nitric Acid	HNO_3
Laughing Gas	N_2O
Ammonia	NH_3
Saltpetre (used in gunpowder)	KNO_3
Carbon Monoxide	CO
Potassium Permanganate (used in labs)	$KMnO_4$
Calcium Carbonate (chalk)	$CaCO_3$

All of these molecules have less than a dozen atoms.

The symbols Ca, K, Mn, Na and Cl stand for calcium, potassium, manganese, sodium and chlorine, respectively.

Molecules with Carbon

Most atoms are only capable of forming small molecules. However, one or two can form larger molecules.

By far and away the best atom for making large molecules with, is carbon. Carbon can make molecules that have tens, hundreds, thousands even millions of atoms. The huge number of possible combinations means that there are more carbon compounds that those of all the other elements put together.

A single carbon atom is capable of combining with up to four other atoms. We say it has a valency of 4. Sometimes a carbon atom will combine with fewer atoms.

The carbon atom is one of the few that will combine with itself.

In other words, carbon combines with other carbon atoms.

This means that carbon atoms can form chains and rings onto which other atoms can be attached.

This leads to a huge number of different compounds. Organic chemistry is essentially the chemistry of carbon.

Carbon compounds are classified according to how the carbon atoms are arranged and what other groups of atoms are attached.

Hydrocarbons

The simplest organic compounds are made up of only carbon and hydrogen atoms only. Even these run into thousands! Compounds of carbon and hydrogen only are called *Hydrocarbons.*

Alkanes

The simplest hydrocarbon is methane, CH_4. This is the simplest member of a series of hydrocarbons. Each successive member of the series has one more carbon atom than the preceding member. This is shown in the table below:

Formula	*Structure*	*Name / Uses*
CH_4	H | H—C—H | H	Methane — gas used for cooking.
C_2H_6	H H | | H—C—C—H | | H H	Ethane — in chemical industry.
C_3H_8	H H H | | | H—C—C—C—H | | | H H H	Propane — heating fuel.
C_4H_{10}	H H H H | | | | H—C—C—C—C—H | | | | H H H H	Butane — lighter/camping fuel.

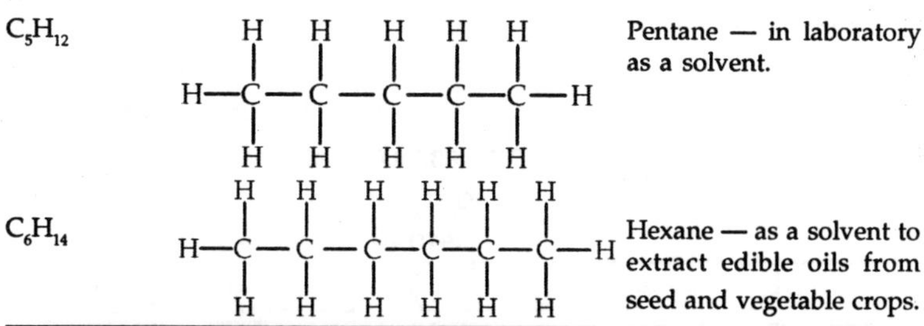

As the reader can see, there is a series of these compounds with this general formula:

$$C_nH_{2n+2}$$

This series of compounds are called *alkanes*. The lighter ones are gases and used as fuels. The middle ones (7 carbons to 12 carbons) are liquids used in petrol (gasoline). The higher ones are waxy solids. Candle wax is a mixture of alkanes.

After Butane, the names of these compounds are from the Greek for the number of carbon atoms followed by the suffix -ane. So, Decane would have the formula $C_{10}H_{22}$.

Polythene is a very large alkane with millions of atoms in a single molecule. Apart from being flammable, alkanes are stable compounds found underground.

In the alkanes, all four of the carbon valency bonds are taken up with links to different atoms. These types of bonds are called *single bonds* and are generally stable and resistant to attack by other chemicals. Alkanes contain the maximum number of hydrogen atoms possible. They are said to be saturated.

The alkanes are not the only hydrocarbons.

Alkenes

Another series of compounds is called the *alkenes*. These have a general formula:

$$C_nH_{2n}$$

Alkenes have fewer hydrogen atoms than the alkanes. The extra valencies left over occur as double bonds between a pair of Carbon atoms. The double bonds are more reactive than single bonds making the alkenes chemically more reactive.

The simplest alkenes are listed in the table below:

<table>
<tr><th>Formula</th><th>Structure</th><th>Name / Uses</th></tr>
<tr><td>C_2H_4</td><td>H—C=C—H
    | |
    H H</td><td>Ethene — used as an industrial starter chemical.</td></tr>
<tr><td>C_3H_6</td><td>   H
    |
H—C—C=C—H
    | | |
    H H H</td><td>Propene — raw material for the production of poly-propylene.</td></tr>
<tr><td>C_4H_8</td><td>   H H
    | |
H—C—C—C=C—H
    | | | |
    H H H H</td><td>Butene — commonly used as a co-polymer.</td></tr>
<tr><td>C_5H_{10}</td><td>   H H H
    | | |
H—C—C—C—C=C—H
    | | | | |
    H H H H H</td><td>Pentene — used in making of plastics polyethylene.</td></tr>
<tr><td>C_6H_{12}</td><td>   H H H H
    | | | |
H—C—C—C—C—C=C—H
    | | | | | |
    H H H H H H</td><td>Hexene — as a comonomer in production of polyethene.</td></tr>
</table>

These compounds are named in a similar manner to the alkanes except that the suffix is -ene.

Alkynes

A third series are the alkynes. These have the following formula:

$$C_nH_{2n-2}$$

Alkynes have two carbon atoms joined by a triple bond. This is highly reactive making these compounds unstable:

<table>
<tr><th>Formula</th><th>Structure</th><th>Name / Uses</th></tr>
<tr><td>C_2H_2</td><td>H—C≡C—H</td><td>Ethyne — better known as acetylene which is used for welding under-water.</td></tr>
<tr><td>C_3H_4</td><td>H
|
H—C—C≡C—H
|
H</td><td>Propyne — used as a rocket fuel.</td></tr>
<tr><td>C_4H_6</td><td>H H
| |
H—C—C—C≡C—H
| |
H H</td><td>Butyne — used in electroplating.</td></tr>
<tr><td>C_5H_8</td><td>H H H
| | |
H—C—C—C—C≡C—H
| | |
H H H</td><td>Pentyne — used in chemical industry.</td></tr>
<tr><td>C_6H_{10}</td><td>H H H H
| | | |
H—C—C—C—C—C≡C—H
| | | |
H H H H</td><td>Hexyne — in petro-leum products.</td></tr>
</table>

These highly reactive substances have many industrial uses. Again the naming of these compounds is similar to the alkanes except that the suffix is -yne.

Carbon Rings

Alkanes, alkenes and alkynes all contain Carbon atoms in linear chains. There are also hydrocarbons arranged in rings. When rings are combined with chains, the number of hydrocarbons is virtually infinite. And we are still using only two types of atoms (carbon and hydrogen). We will now add a third.

Formula	Structure	Name / Uses
C_6H_{12}		Cyclohexane — a saturated hydrocarbon with the atoms arranged in a hexagonal ring. In organic chemistry, the presence of hydrogen atoms is often assumed and this compound can be represented by a hexagonal ring:
C_6H_6		Benzene — an industrial solvent. The Benzine Ring is one of the most important structures in organic chemistry. In reality, its alternate double and single bonds are "spread around" the ring so that the molecule is symmetrical. This structure is represented by a hexagon with a circle:
C_7H_8		Toluene — an important solvent and starter chemical. Using the Benzine Ring, this molecule can also be depicted as:

Contd...

Formula	*Structure*	*Name / Uses*
$C_{10}H_8$	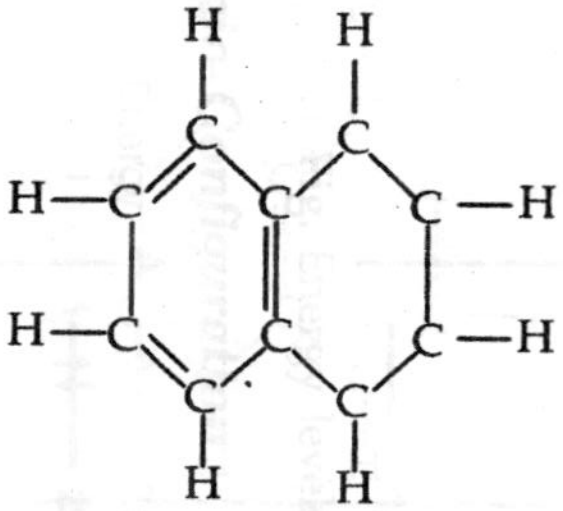	Naphthalene — used in moth balls. This can be depicted as two fused Benzine Rings: 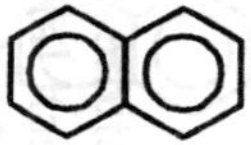

Carbon, Hydrogen and Oxygen

When oxygen atoms are added, the variety of compounds grows enormously. In the table below, each row discusses a series of compounds:

In the above examples, each molecule has a single functional group. It is possible to have two or more functional groups on a molecule. These can be the same group (as in Oxalic Acid — a poison found in rhubarb leaves — which has two fatty acid groups) or different (as in Hydroxymethanoic Acid — which has a hydroxyl group and a fatty acid group):

```
          H
          ||
H—O— C— C—O—H
               ||
               O
```

(COOH)[2]: Oxalic Acid

```
          H
          |
H—O—C— C—O—H
          |    ||
          H   O
```

$CH_2OHCOOH$: Hydroxymethanoic Acid

The most famous compounds containing carbon, hydrogen and oxygen are the Carbohydrates. An example is the common sugar, Sucrose ($C_{12}H_{22}O_{11}$).

This shows how varied and complex even simple organic compounds can be. Sucrose has a pair of rings: one hexagonal, the other pentagonal. Each ring contains an oxygen atom. The rings are joined by an oxygen (Ether) link. The entire compound contains several Hydroxyl (OH) groups.

Sucrose

Formula	*Name*		
$C_nH_{2n+1}OH$	Alcohols	Alcohols have the OH (hydroxyl) group in the molecule. A group of atoms that gives an organic series its distinctive character is called a *functional group.*	CH_3OH Methanol—wood, alcohol. C_2H_5OH Ethanol—drinking alcohol. C_6H_5OH Phenolcarbolic acid - used as disinfectant.
$(C_nH_{2n+1})_2O$	Ethers	Ethers have an O atom attached to two hydrocarbon chains (or rings).	$(CH_3)_2O$ Dimethyl Ethera—gas. $(C_2H_5)_2O$ Diethyl Ethera—liquid— used as an anaesthetic).

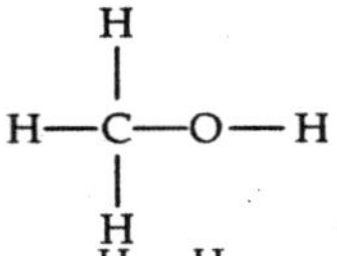

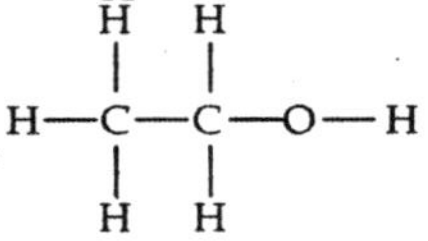

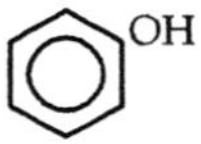

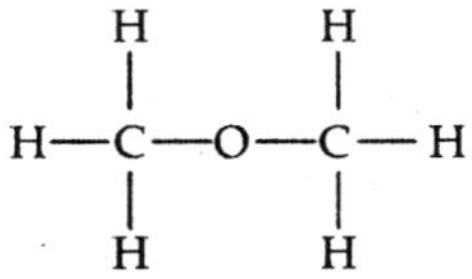

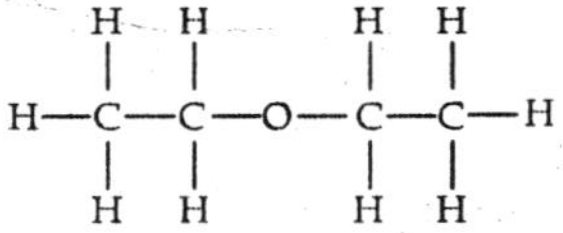

Contd...

Formula	*Name*			
$(C_nH_{2n+1})_2CO$	Ketones	Ketones have a CO group attached to two hydrocarbon chains (or rings).	CH_3COCH_3 Dimethyl Ketone (also known as *acetone*)—nail-varnish remover.	$\begin{array}{ccccccccc} & & H & & & & H & & \\ & & \vert & & & & \vert & & \\ H & — & C & — & C & — & C & — & H \\ & & \vert & & \Vert & & \vert & & \\ & & H & & O & & H & & \end{array}$
$C_nH_{2n+1}CHO$	Aldehydes	Aldehydes have a CHO group attached to a hydrocarbon chain (or ring).	HCHO Formaldehyde — preservative in labs.	$\begin{array}{ccccccc} & & H & & & & \\ & & \vert & & & & \\ H & — & C & — & C & — & H \\ & & \vert & & \Vert & & \\ & & H & & O & & \end{array}$
			CH_3CHO Acetaldehyde	$\begin{array}{ccccccccc} & & H & & H & & & & \\ & & \vert & & \vert & & & & \\ H & — & C & — & C & — & C & — & H \\ & & \vert & & \vert & & \Vert & & \\ & & H & & H & & O & & \end{array}$
$C_nH_{2n+1}CO_2H$	Fatty Acids	Fatty Acids contain the CO_2H (or COOH) group attached to a	HCO_2H Formic Acidin—ant bites and stinging nettles.	$\begin{array}{ccccccc} H & — & C & — & O & — & H \\ & & \Vert & & & & \\ & & O & & & & \end{array}$

Contd...

Formula	*Name*			
		hydrocarbon chain or ring.	CH_3CO_2H Acetic Acidvinegar	H | H—C—C—O—H | ‖ H O
			$C_2H_5CO_2H$ Butyric Acidthe — rancid butter smell.	H H | | H—C—C—C—O—H | | ‖ H H O
RCO_2R' (R, R' are Hydrocarbon chains or rings).	Esters	Esters are similar to Fatty Acids except that the H in the COOH group is another hydrocarbon chain. They are usually very sweet smelling liquids used in perfumes.	$CH_3CO_2CH_3$ Methyl— Methoateessence of pear drops.	H H | | H—C—C—O—C—H | ‖ | H O H

Isomerism

An interesting phenomenon with organic molecules is called *isomerism*. Let us look at two compounds introduced earlier. Dimethyl Ether: $(CH_3)_2O$ and Ethanol: C_2H_5OH.

The first is a gas which will knock you out if inhaled. The second is common alcohol drunk in spirits. The two molecules are shown below:

```
     H       H
     |       |
  H—C—O—C—H
     |       |
     H       H
```

Dimethyl Ether

```
     H   H
     |   |
  H—C—C—O—H
     |   |
     H   H
```

Ethanol

Notice that both compounds contain 2 carbon atoms, 6 hydrogen atoms and 1 oxygen atom.

Even though the atoms are the same, they are arranged differently. This yields two different compounds with the same number of atoms. These compounds are isomers and the phenomenon is called *Isomerism*.

In this example, the two molecules have different functional groups. They are structural isomers. Other types of isomers exist.

Isomerism increases the number of organic compounds. The more carbon atoms in a compound, the more ways of arranging the atoms and the larger number of isomers.

Adding Nitrogen

Many very important organic compounds contain nitrogen. This produces more series of compounds.

A famous compound containing nitrogen is Trinitro Toluene [$C_6H_2CH_3(NO_2)_3$ — usually abbreviated to TNT]. This is an artificially made explosive. Its structure is shown below:

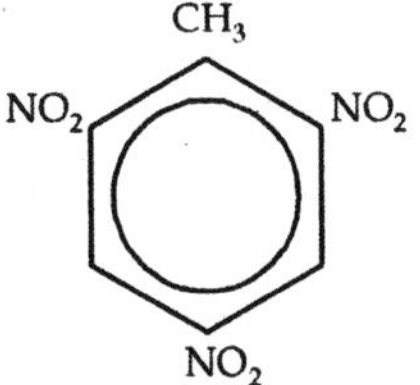

Trinitro Toluene (TNT)

There are six isomers of this compound as the three NO_2 groups can be placed in six different arrangements on the ring. These are known as *positional isomers*.

Other Atoms

The vast majority of organic compounds contain carbon, hydrogen, oxygen and nitrogen. Other types of atoms can be included to form even more compounds. These can contain atoms like phosphorus, sulphur (e.g. thiamine, vitamin B_1), magnesium (e.g. chlorophyll) and iron (e.g. haemoglobin). As can be imagined, these additions increase the number of compounds. Apart from the naturally occurring organic compounds, millions more can be synthesised. These can include atoms like Chlorine (used in pesticides). Examples of organic compounds containing Chlorine are shown below.

There is no difference between the same substance extracted from living organisms and made in a laboratory.

Formula	*Structure*	*Name / Uses*
$CHCl_3$	H Cl—C—Cl Cl	Chloroform — a human-made anaesthetic.
$C_{14}H_9Cl_5$	H Cl— —C— —Cl CCl_3	Dichloro Diphenyl Tri-chloro Methane — DDT an insecticide.

Functional Groups

Functional Groups: Aliphatic and Aromatic

Aliphatic Functional Groups

An aliphatic functional group is one in which there is no aromatic ring directly attached to the functional group:

(a) (b)

Fig. (a) Aliphatic Ketone; (b) Aliphatic Ester.

Aromatic Functional Groups

An aromatic functional group is one in which an aromatic ring is directly attached to the functional group:

(a) (b)

Fig. (a) Aromatic Carboxylic Acid; (b) Aromatic Ketone.

In case of esters and amides, the functional groups are defined as aromatic or aliphatic depending on whether the aryl group is directly attached to the carbonyl end of the functional group, i.e., Ar-CO-X. If the aromatic ring is attached to the heteroatom instead, then the ester or amide is classified as an aliphatic amide:

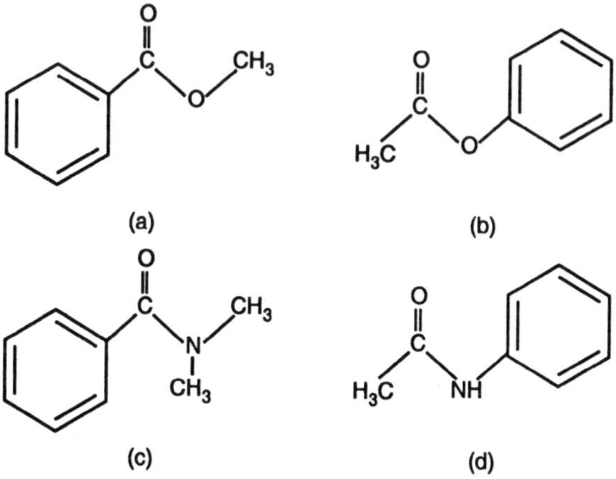

(a) (b)

(c) (d)

Fig. (a) Aromatic ester; (b) aliphatic ester; (c) aromatic amide; (d) aliphatic amide.

Taxonomy of Functional Groups

General Rules

Various nomenclature rules for alkanes hold true for molecules containing a functional group, but some extra rules are required to define the type of functional group present and its position within the molecule. Important rules are as follows:

(i) The main (or parent) chain must include the carbon containing functional group, and so may not necessarily be the longest chain;

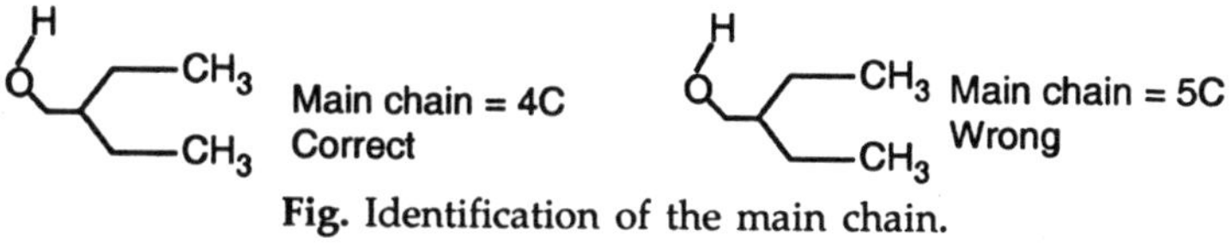

Main chain = 4C Correct

Main chain = 5C Wrong

Fig. Identification of the main chain.

(ii) The presence of some functional groups is indicated by replacing -ane for the parent alkane chain with a suffixes depending on the functional group present, e.g.,

Functional Group	*Suffix*	*Functional Group*	*Suffix*
alkene	-ene	alkyne	-yne
alcohol	-anol	aldehyde	-anal
ketone	-anone	carboxylic acid	-anoic acid
acid chloride	-anoyl chloride	amine	-ylamine.

The example given in figure above is a butanol.

(iii) Numbering of carbon atoms must start from the end of the main chain nearest to the functional group. Therefore, the numbering should place the alcohol at position 1 and not position 4. (Lowest position number to the carbon containing the functional group):

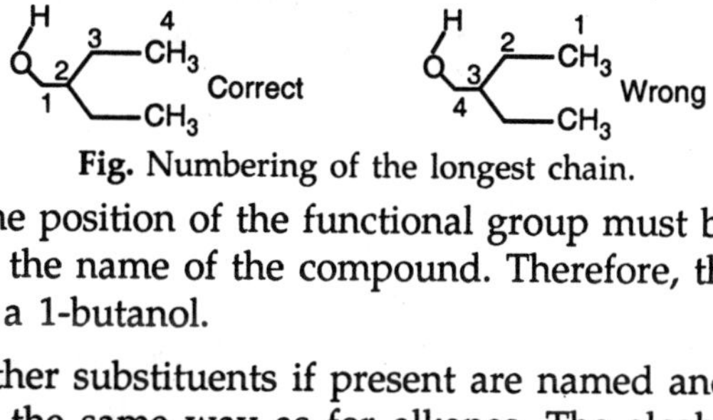

Correct   Wrong

Fig. Numbering of the longest chain.

(iv) The position of the functional group must be defined in the name of the compound. Therefore, the alcohol is a 1-butanol.

(v) Other substituents if present are named and ordered in the same way as for alkanes. The alcohol has an ethyl group at position 3 and so the full name for the structure is 3-ethyl-1-butanol.

Some other rules are needed to deal with a specific situation. For example, if the functional group is at equal distance from either end of the main chain, the numbering starts from the end of the chain nearest to any substituents. For example, the alcohol is 2-methyl-3-pentanol and not 4-methyl-3-pentanol. (Lowest number rule):

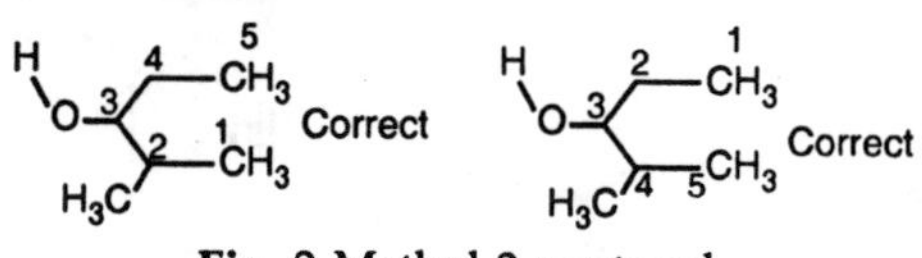

Correct   Correct

Fig. 2-Methyl-3-pentanol.

Alkenes and Alkynes

The names of alkenes and alkynes contain the suffixes -ene and -yne, respectively. With some alkenes it is necessary to define the stereochemistry of the double bond:

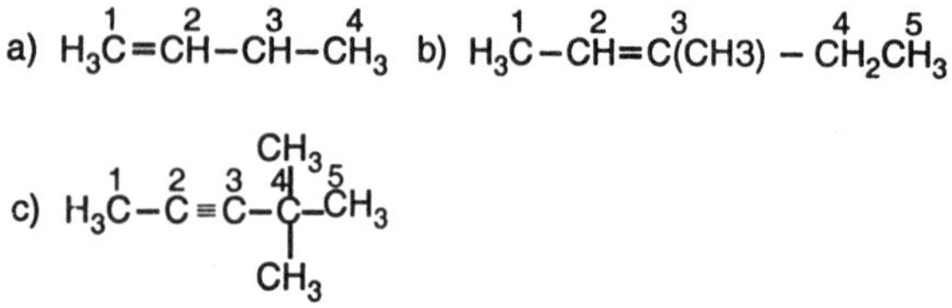

a) $H_3\overset{1}{C}{=}\overset{2}{C}H{-}\overset{3}{C}H{-}\overset{4}{C}H_3$ b) $H_3\overset{1}{C}{-}\overset{2}{C}H{=}\overset{3}{C}(CH3) - \overset{4}{C}H_2\overset{5}{C}H_3$

c) $H_3\overset{1}{C}{-}\overset{2}{C}{\equiv}\overset{3}{C}{-}\overset{4}{C}(CH_3)_2{-}\overset{5}{C}H_3$

Fig. (a) 2-Butene; (b) 3-methyl-2-pentene; (c) 4,4-dimethyl-2-pentyne.

Aromatics

The well-known aromatic structure is benzene. If an alkane chain is linked to a benzene molecule, then the alkane chain is generally considered to be an alkyl substituent of the benzene ring. However, if the alkane chain contains more than six carbons, then the benzene molecule is considered to be a phenyl substituent of the alkane chain:

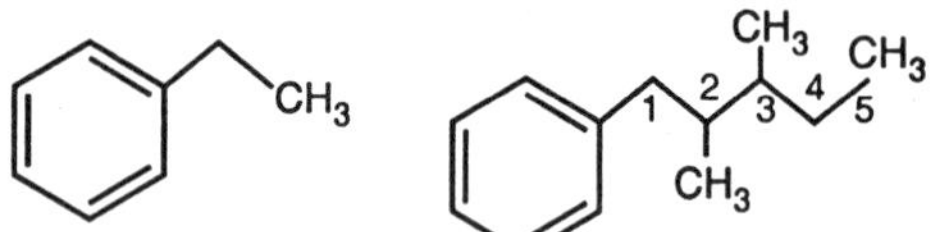

Fig. (a) Ethylbenzene; (b) 3-phenyl-2,3-dimethylpentane.

A benzyl group is made up of an *aromatic ring* and a *methylene group*.

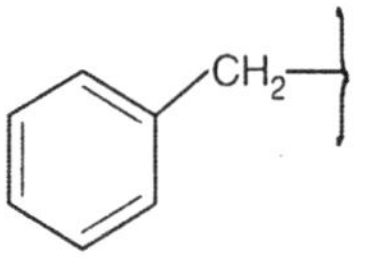

Fig. Benzyl group.

Benzene is not the only parent name that can be used for aromatic compounds:

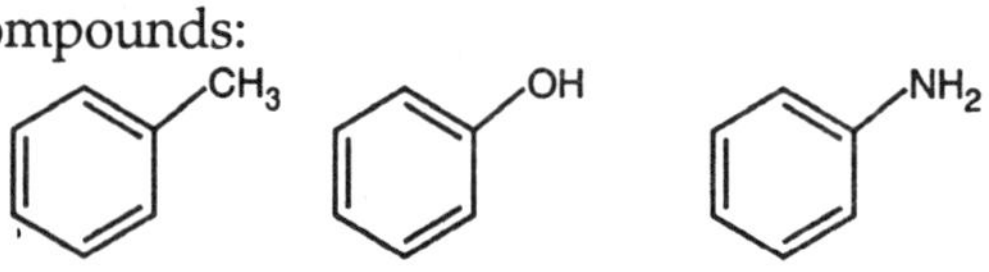

(a) (b) (c)

(d) (e) (f)

Fig. (a) Toluene; (b) phenol; (c) aniline; (d) benzoic acid; (e) benzalde-hyde; (f) acetophenone.

In case of distributed aromatic rings, the position of substituents has to be defined by numbering around the ring, in such a way that the substituents are positioned at the lowest numbers possible, for example, the structure is 1,3-dichlorobenzene and not 1,5-dichlorobenzene:

Correct Wrong

Fig. 1,3-Dichlorobenzene.

Alternatively, the terms *ortho, meta,* and *para* can be used. These terms define the relative position of one substituent to another. Thus, 1,3-dichlorobenzene can also be called *meta-dichlorobenzene*. This can be shortened to *m-dichlorobenzene*. The examples in figure given below shows how different parent names can be used. The substituent which defines the parent name is given as position 1. For example, if the parent name is toluene, the methyl group must be at position 1.

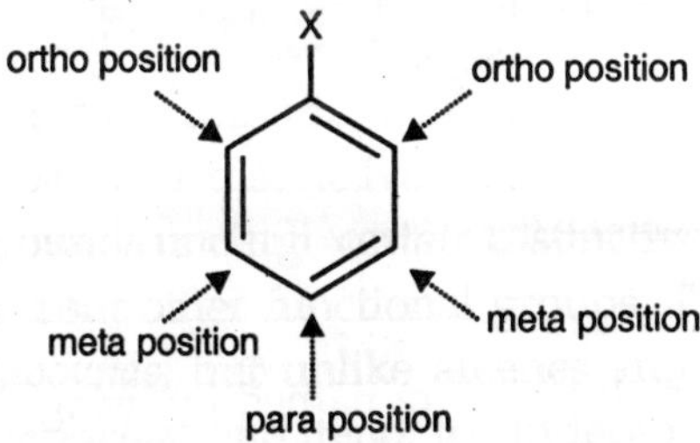

Fig. *Ortho, meta* and *para* positions of an aromatic ring.

When more than two substituents are present on the aromatic ring, the *ortho, meta, para* nomenclature is no longer valid and numbering has to be used. In such a case the relevant substituent has to be placed at position 1 if the parent name is toluene, aniline, *etc.*

If the parent name is benzene, the numbering is done in such a way that the lowest possible numbers are used. In the example shown, any other numbering would result in the substituents having higher numbers.

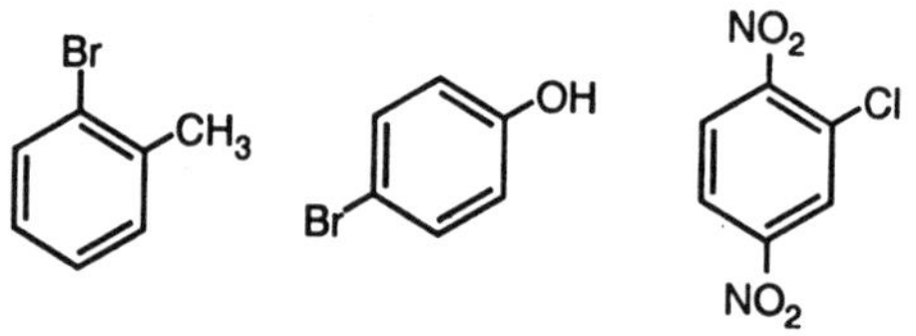

(a) (b) (c)

Fig. (a) 2-Bromotoleune or o-bromotolucne; (b) 4-bromophenol or p-bromophenol; (c) 3-chloroaniline or m-chloroaniline.

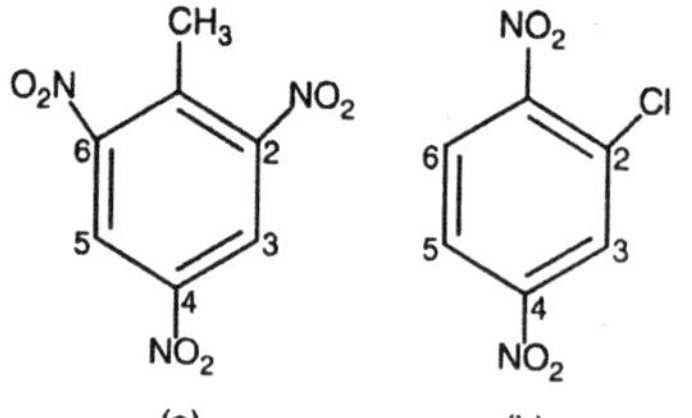

(a) (b)

Fig. (a) 2.4.6-Trinitrotoluene; (b) 2-chloro-1,4-dinitrobenzene.

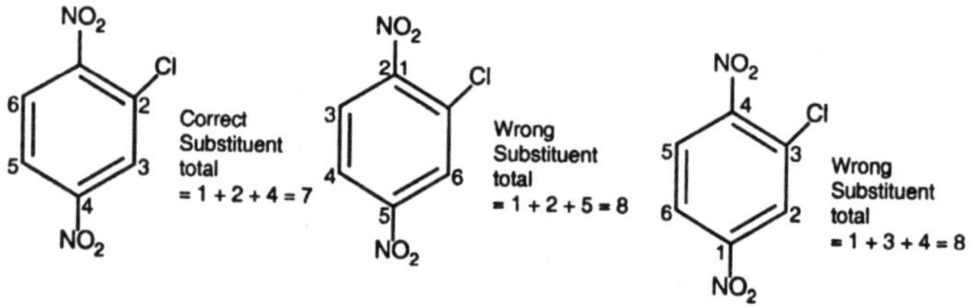

Correct Substituent total = 1 + 2 + 4 = 7

Wrong Substituent total = 1 + 2 + 5 = 8

Wrong Substituent total = 1 + 3 + 4 = 8

Fig. Possible numbering systems if tri-substituted aromatic ring.

Alcohols

Alcohols or alkanols are named by using the suffix -anol. The general rules discussed earlier are used to name alcohols.

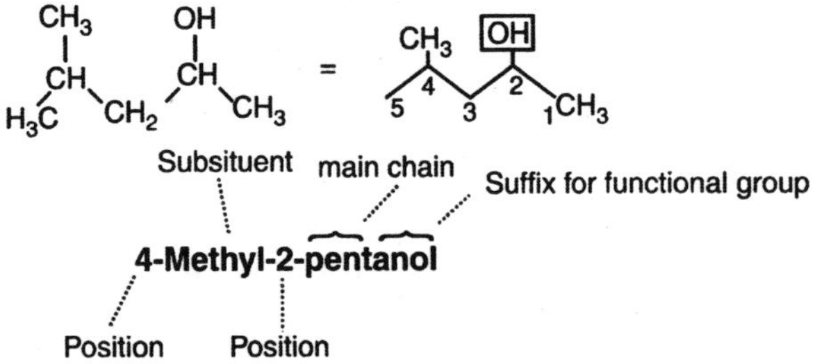

Fig. 4-Methyl-2-pentanol.

Ethers and Alkyl Halides

For the nomenclature for ethers and alkyl halides the functional group is considered to be a substituent of the main alkane chain. The functional group is numbered and named as a substituent:

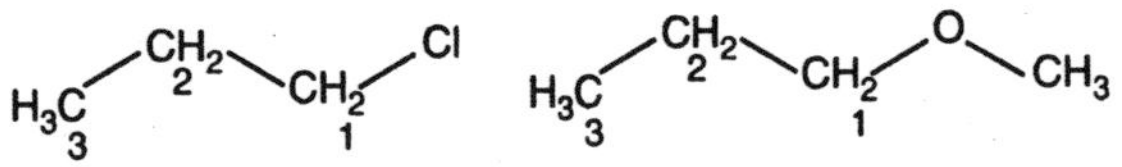

Fig. (a) 1-Chloropropane: (b) 1-methoxypropane.

In ethers we have two alkyl groups on either side of the oxygen. The larger alkyl group is the parent alkane. The smaller alkyl group along with the oxygen is the substituent and is called an *alkoxy group*.

Aldehydes and Ketones

The suffix for an aldehyde (or alkanal) is *-anal,* and the suffix for a ketone (or alkanone) is *-anone*. The main chain must include the functional group and the numbering is such that the functional group is at the lowest number possible.

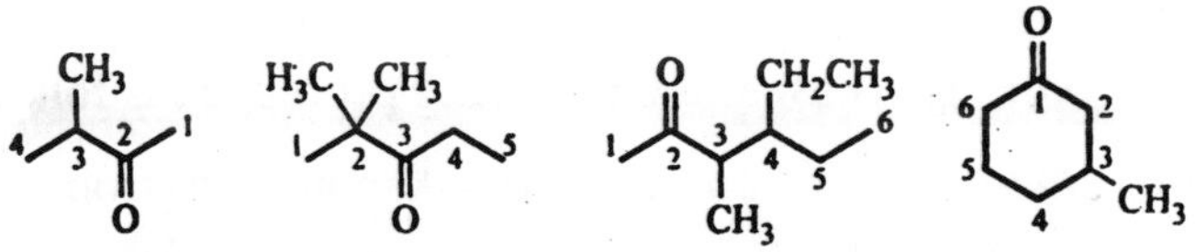

Fig. (a) 3-Methyl-2-butanone; (b) 2,2-dimelhyl-3-pentanone; (c) 4-ethyl-3-methyl-2-hexanone; (d) 3-methylcyclohexanone.

If the functional group is in the centre of the main chain, the numbering is done in such a way that other substituents have the lowest numbers possible, (*e.g.*, 2,2-dimethyl-3-pentanone and not 4,4-dimethyl-3-pentanone):

3-Methyl-2-butanone can in fact be simplified to 3-methylbutanone because there is only one possible place for the ketone functional group in this molecule. In case the carbonyl C = O group is at the end of the chain; it would be an aldehyde and not a ketone. Numbering is also not necessary in locating an aldehyde group since it can only be at the end of a chain:

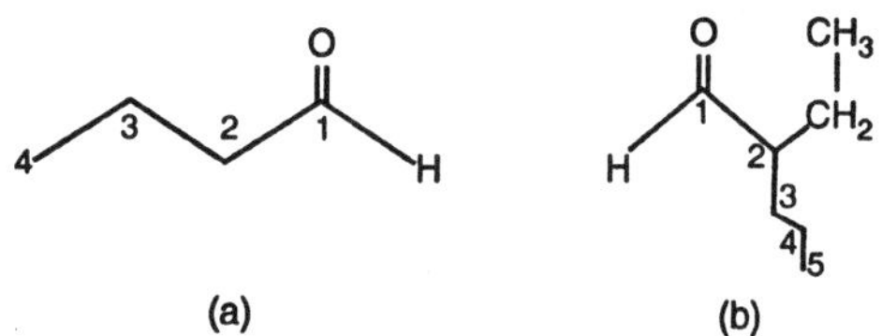

Fig. (a) Butanal; (b) 2-ethylpentanal.

Carboxylic Acids and Acid Chlorides

Carboxylic acids and acid chlorides can be identified by adding the suffix *–anoic acid* and *–anoyl chloride,* respectively. Both these functional groups are always at the end of the main chain and need not be numbered:

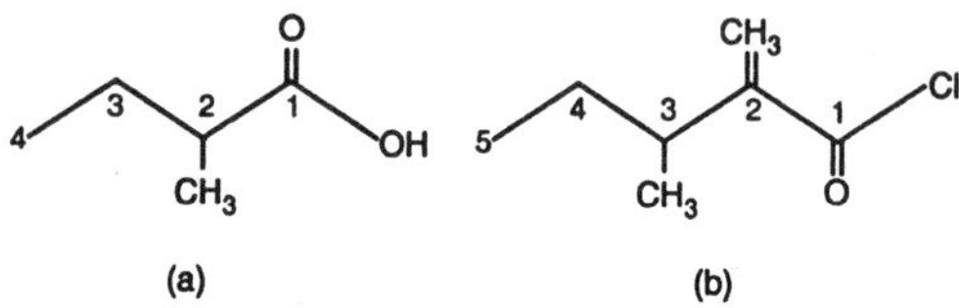

Fig. (a) 2-Methylbutanoic; (b) 2,3-dimethylpentanoyl chloride.

Esters

For naming an ester, the following procedure is followed:

(i) Identify the carboxylic acid (alkanoic acid) from which it was derived.

(ii) Change the name to an alkanoate rather than an alkanoic acid.

(iii) Identify the alcohol from which the ester was derived and consider this as an alkyl substituent.

(iv) The name becomes an alkyl alkanoate.

For example, the ester is derived from ethanoic acid and methanol. The ester would be an *alkyl ethanoate* since it is derived from ethanoic acid.

The alkyl group comes from methanol and is a methyl group. Therefore, the full name is *methyl ethanoate*. (Note that there is a space between both parts of the name).

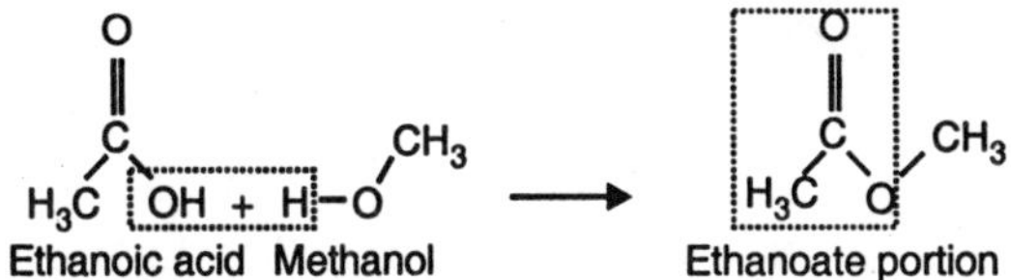

Fig. Ester formation.

Amides

Amides are the derivatives of the carboxylic acids. In amides the carboxylic acid is linked with ammonia or an amine. Like esters, the parent carboxylic acid is identified. This is then termed an alkanamide and includes the nitrogen atom. For example, linking ethanoic acid with ammonia gives ethanamide:

$$H_3C{-}C({=}O){-}OH + NH_3 \xrightarrow{-H_2O} H_3C{-}C({=}O){-}NH_2$$

Fig. Formation of ethanamide.

If the carboxylic acid is linked with an amine, then the amide will have alkyl groups on the nitrogen. These are considered as *alkyl substituents* and come at the beginning of the name.

The symbol N is used to show that the substituents are on the nitrogen and not some other part of the alkanamide skeleton. For example, the structure in the following figure is named N-ethylethanamide:

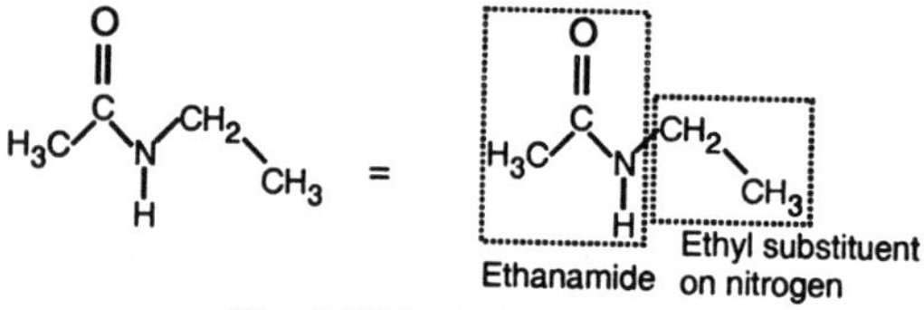

Fig. N-Ethylethanamide.

Amines

For naming amines the main part (or root) of the name is an alkane and the amino group is considered to be a substituent.

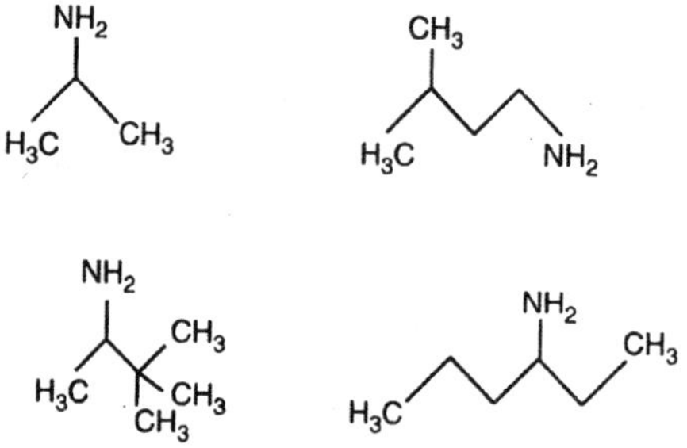

Fig. (a) 2-Aminopropane; (b) 1-amino-3-methylbutane; (c) 2-amino-3,3-dimethylbutane; (d) 3-aminohexane.

Simple amines are also named by placing the suffix -*ylamine* after the main part of the name.

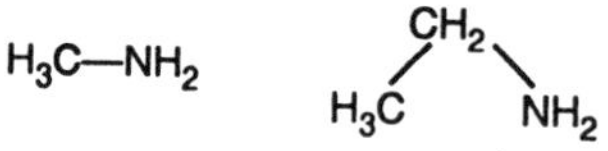

$H_3C—NH_2$

(a) (b)

Fig. (a) Methylamine; (b) ethylamine.

Amines containing more than one alkyl group attached are named by selecting the longest carbon chain attached to the nitrogen. In the example, that is an ethane chain and so this molecule is an aminoethane (N,N-dimethylaminoethane):

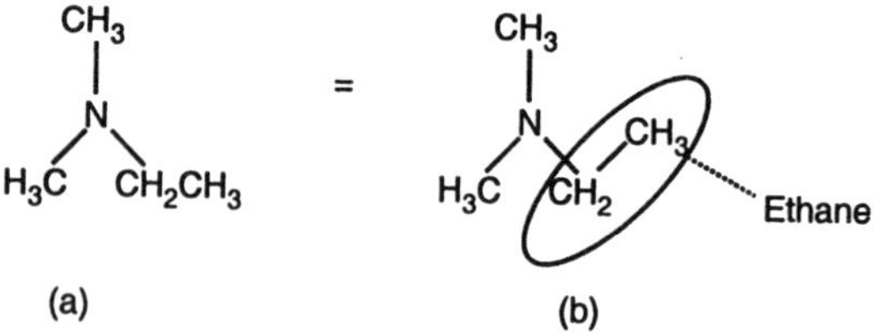

(a) (b)

Fig. N,N-Dimethylaminoethane.

Some simple secondary and tertiary amines have common names:

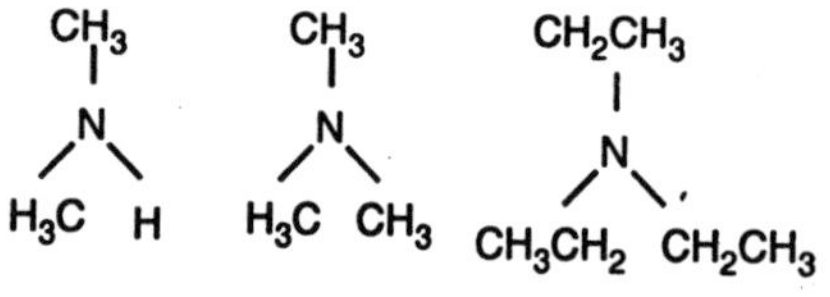

$$\underset{\text{(a)}}{H_3C-\overset{\overset{CH_3}{|}}{N}-H} \qquad \underset{\text{(b)}}{H_3C-\overset{\overset{CH_3}{|}}{N}-CH_3} \qquad \underset{\text{(c)}}{CH_3CH_2-\overset{\overset{CH_2CH_3}{|}}{N}-CH_2CH_3}$$

Fig. (a) Dimethylamine; (b) trimethylamine; (c) triethylamine.

Thiols and Thioethers

For naming thiols we add the suffix *-thoil* to the name of the parent alkane. Thioethers are named like ethers using the prefix *alkylthio-,* for example, 1 -(methylthio) propane. Simple thioethers can be named by identifying the thioether as a sulphide and prefixing this term with the alkyl substituents, for example, dimethyl sulphide.

$$\underset{\text{(a)}}{CH_2CH_3 - SH} \quad \underset{\text{(b)}}{H_3C - S - CH_3} \quad \underset{\text{(c)}}{H_3C-S-CH_2CH_2CH_3}$$

Fig. (a) Ethanethiol; (b) dimethylsulphide;
(c) 1–(methylthiopropane).

Taxonomy: Primary, Secondary, Tertiary and Quaternary

Definition

The primary (1°), secondary (2°), tertiary (3°) and quaternary (4°) nomenclature is used in a number of situations: to define a carbon centre, or to define functional groups like *alcohols, halides, amines* and *amides.* Identifying functional groups in this way can be important because the properties and reactivities of these groups vary depending on whether they are *primary, secondary, tertiary* or *quaternary.*

Carbon Centres

The easiest ways of determining if a carbon centre is 1°, 2°, 3°, or 4° is to count the number of bonds leading from that carbon centre to another carbon atom.

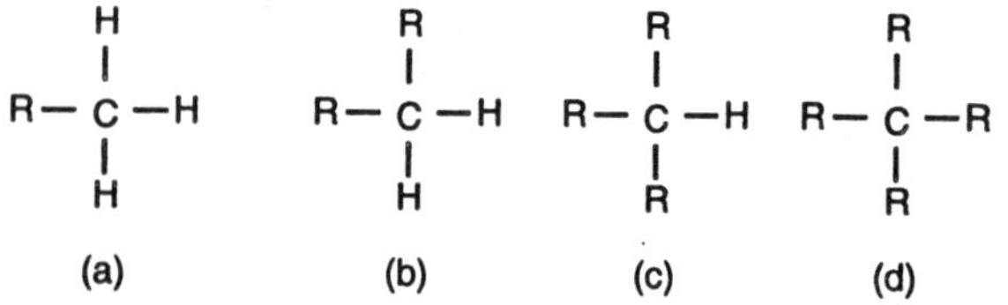

Fig. Carbon centres; (a) primary; (b) secondary; (c) tertiary; (d) quaternary.

A *methyl group* (CH_3) is a primary carbon centre (attached to one carbon), a *methylene group* (CH_2) is a secondary carbon centre (attached to the other carbons), a *methine group* (CH) is a tertiary carbon centre (attached to three other carbons) and a carbon centre with four alkyl substituents (C) is a quaternary carbon centre (attached to four other carbons):

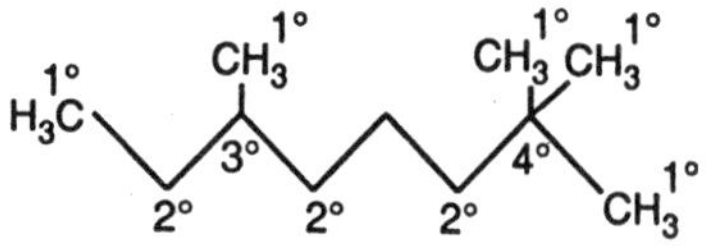

Fig. Primary, secondary, tertiary, and quaternary carbon centres.

Amines and Amides

Amines and amides are classified as primary, secondary, tertiary, or quaternary depending on the number of bonds from nitrogen to carbon. Note that a quaternary amine is positively charged and is therefore known as a *quaternary ammonium ion.* It is not possible to get a quaternary amide.

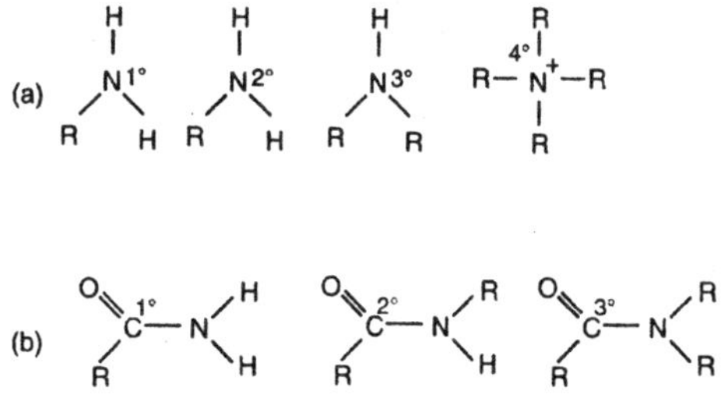

Fig. (a) Amines; (b) amides.

Alcohols and Alkyl Halides

Alcohols and alkyl halides can be classified as primary, secondary, or tertiary depending on the carbon to which the alcohol or halide is attached and it ignores the bond to the functional group. Thus, quaternary alcohols or alkyl halides are not possible.

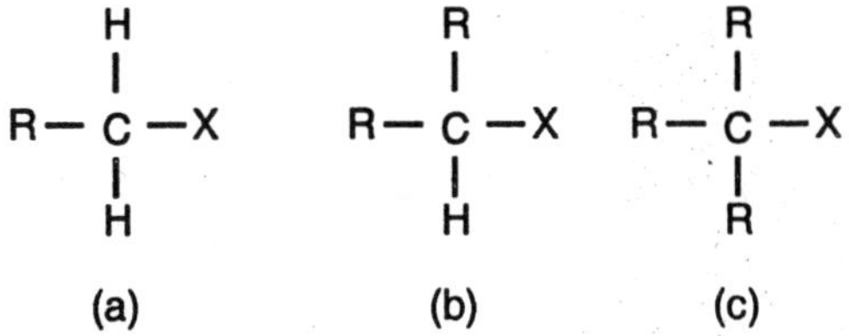

Fig. Alcohols and alkyl halides; (a) primary; (b) secondary; (c) tertiary.

For example, which illustrate different types of alcohols and alkyl halides:

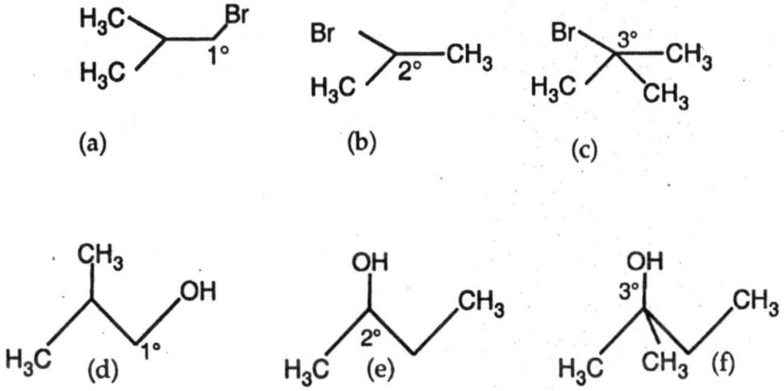

Fig. (a) 1° alkyl bromide; (b) 2° alkyl bromide; (c) 3° alkyl bromide; (d) 1° alcohol; (e) 2° alcohol; (f) 3° alcohol.

Bonding of Intermolecule

Definition

Intermolecular bonding refers to the bonding interaction that occurs between different molecules. This can take the form of ionic bonding, hydrogen bonding, dipole-dipole interactions or van der Waals interactions.

These bonding forces are weaker than the covalent bonds, but they do have an important influence on the physical and biological properties of a compound.

Ionic Bonding

Ionic bonding occurs between molecules which have opposite charges and it involves an electrostatic interaction between the two opposite charges, the functional groups that most easily ionise are amines and carboxylic acids:

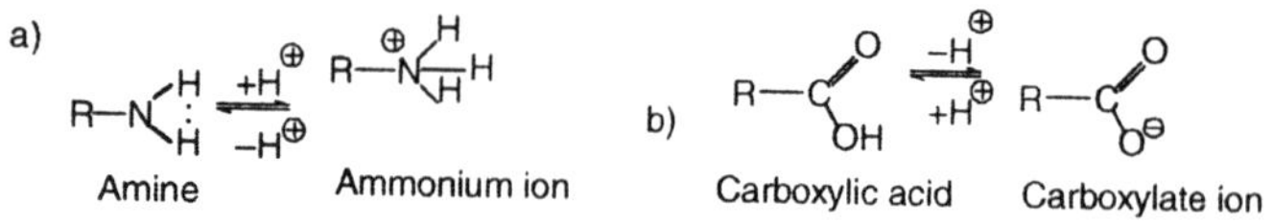

a) Amine $\underset{-H^{\oplus}}{\overset{+H^{\oplus}}{\rightleftharpoons}}$ Ammonium ion   b) Carboxylic acid $\underset{+H^{\oplus}}{\overset{-H^{\oplus}}{\rightleftharpoons}}$ Carboxylate ion

Fig. (a) Ionisation of an amine: (b) ionisation of a carboxylic acid.

Ionic bonding can occur between a molecule containing an ammonium ion and another molecule containing a carboxylate ion. Some important naturally occurring molecules that contain both groups arc the *amino acids.* Both these functional groups are ionised to form a structure called *zwitterion* (a neutral molecule bearing both a positive an a negative charge) and intermolecular ionic bonding can occur:

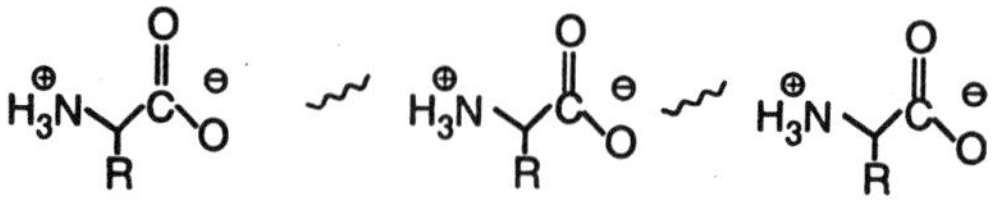

Fig. Intermolecular ionic bonding of amino acids.

Hydrogen Bonding

Hydrogen bonding can occur if molecules have a hydrogen atom attached to a heteroatom like nitrogen or oxygen. The common functional groups that can participate in hydrogen bonding are *alcohols, phenols, carboxylic acids, amides* and *amines.* Hydrogen bonding is possible because of the polar nature of the N–H or O–H bond. Nitrogen and oxygen are more electronegative than hydrogen. Thus, the heteroatom gains a slightly negative charge and the hydrogen gains a slightly positive charge. Hydrogen bonding involves the partially charged hydrogen of one molecule (the H bond donor) interacting with the partially charged heteroatom of another molecule (the H bond acceptor):

Fig. Intermolecular hydrogen bonding between alcohols.

Dipole-dipole Interactions

Dipole-dipole interactions can occur between polarised bonds other than N–H or O–H bonds. The most likely functional groups that can interact in this way are those containing a carbonyl group (C = O), the electrons in the carbonyl bond are polarised towards the more electronegative oxygen such that the oxygen acquires a partial negative charge and the carbon acquires a partial positive charge. This results in a dipole moment that can be represented by the arrow shown in figure given below.

The arrow points to the negative end of the dipole moment. Molecules containing dipole moments can align themselves with each other in such a way that the dipole moments are pointing in opposite directions.

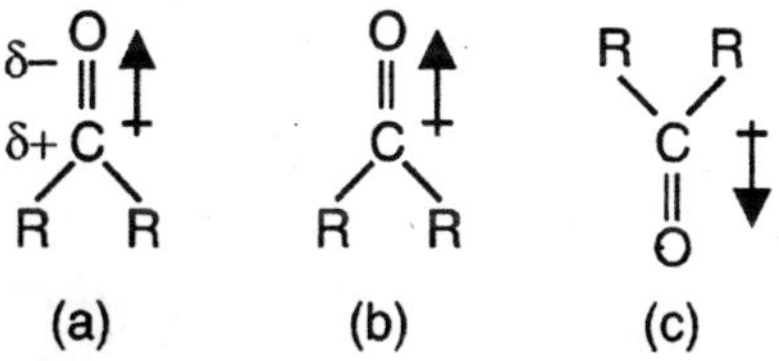

Fig. (a) Dipole moment of a ketone; (b) intermolecular dipole-dipole interaction between ketones.

Van der Waals Interactions

Van der Waals interactions are the weakest of the intermolecular bonding forces and involve the transient existence of partial charges in a molecule. Electrons are continually moving in an unpredictable fashion around any molecule.

At any given moment of time, there is a slight excess of electrons in part of the molecule and a slight deficit in another part. Although the charges are very weak and fluctuate around the molecule, they are sufficiently strong to permit a weak interaction between molecules, where regions of opposite charge in different molecules attract each other.

Alkane molecules can interact in this way and the strength of the interaction increases with the size of the alkane molecule, van der Waals interactions are also important for alkanes, alkynes and aromatic rings.

The types of molecules involved in this form of intermolecular bonding are 'fatty' molecules that do not dissolve easily in water and such molecules are called *hydrophobic* (water-hating). Hydrophobic molecules can dissolve in non-polar, hydrophobic solvents because of van der Waals interactions and so this form of intermolecular bonding is also called a *hydrophobic interaction.*

Reactions and Properties

Properties

The chemical and physical properties of an organic compound depend on the sort of intermolecular bonding forces present, which in turn depends on the functional group present. A molecule like methane has a low boiling point and is a gas at room temperature because its molecules are bound together by weak van der Waals forces (fig. a). In contrast, methanol is a liquid at room temperature because hydrogen bonding is possible between the alcoholic functional groups (fig. b).

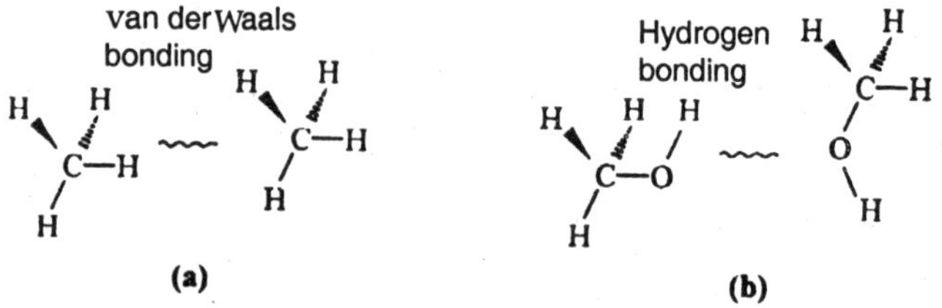

van der Waals bonding

(a)

Hydrogen bonding

(b)

Fig. (a) Intermolecular van der Waals (methane); (b) intermolecular hydrogen bonding (methanol).

The polarity of molecules depends on the functional groups present in the molecule. A molecule will be polar and have a dipole moment if it has a polar functional groups like an *alcohol, amine* or *ketone.* Polarity also determines solubility in different solvents. Polar molecules dissolve in polar solvents like water or alcohols, whereas non-polar molecules dissolve in non-polar solvents like ether and chloroform. Polar molecules that can dissolve in water are called *hydrophilic* (water-loving) while non-polar molecules are called *hydrophobic* (water-hating).

Generally, the presence of a polar functional group determines the physical properties of the molecule. But this is not always true. If a molecule has a polar group like a carboxylic acid, but has a long hydrophobic alkane chain, then the molecule will be hydrophobic.

Reactions

Most of the organic reactions occur at functional groups and are characteristic of that functional group. However, the reactivity of the functional group is affected by stereo-electronic effects. For example, a functional group may be surrounded by bulky groups that hinder the approach of a reagent and slow down the rate of reaction. This is called as *steric shielding.*

Electronic effects can also influence the rate of a reaction. Neighbouring groups can influence the reactivity of a functional group if they are *electron-withdrawing* or *electron-donating* and influence the electronic density within the functional group. *Conjugation* and *aromaticity* also effect the reactivity of functional groups. For example, an aromatic ketone reacts at a different rate from an aliphatic ketone. The aromatic ring is in conjugation with the carbonyl group and this increases the stability of the overall system, making it less reactive.

Organic Synthesis Reagents

Mechanism Reaction

Here are given two Reactions :

(i) $CH_3CH = CH_2 + H_2 \xrightarrow{\text{Pd, Pt or Ni}}$

(ii) $CH_3CH = CH_2 + H_2 \xrightarrow[\text{(Wilkinsons catalyst)}]{RhCl[(Ph_3P)_3]}$

In terms of the catalysts used, The two reactions classified. Shown the steps in the mechanism of reaction :

Product in both the reactions is $CH_3CH_2CH_3$. Both reactions are catalytic hydrogenations (addition of H_2). (i) is heterogeneous catalytic hydrogenation and (ii) is homogeneous catalytic hydrogenation.

$$RhL_3Cl \underset{-L}{\overset{H_2}{\rightleftharpoons}} \underset{(A)}{RhL_2(H)_2Cl} \overset{>C=C<}{\rightleftharpoons} \underset{(B)}{RhL_2(H)_2Cl(\leftarrow >C=C<)} \rightleftharpoons \underset{(C)}{RhL_2(H)Cl(CH_2CH_2H)}$$

In step-1 an H_2 adds to the rhodium complex and one Ph_3P ligand (L) is lost, resulting in a five coordinate rhodium complex, A (L = Ph_3P). In this oxidative addition, the Rh changes oxidation state from + 1 to + 3. In step-2 the alkene reacts with A to form a π complex, B, which undergoes rearrangement (step-3) of an H to one of the C's of the double bond, the other C forming a σ bond to the Rh (C).

In the last step, a second H is transferred to the other C, and the alkane is lost with simultaneous regeneration of the metal catalyst—

$$C \xrightleftharpoons{+L} Rh(Ph_3P)_3Cl + HCH_2CH_2H$$

The catalysis depends on the ability of rhodium, a transition metal to form a π complex with

$>C=C<$

The reagents for carrying out the following transformations, indicated:

(a) $CH_3(CH_2)_7CH=CH(CH_2)_7CH_3 \xrightarrow{?} CH_3(CH_2)_7CH(OH)-CH(OH)(CH_2)_7CH_3$

(b) cyclopentene $\xrightarrow{?}$ cis-1,2-cyclopentanediol

(c) cyclohexanone $\xrightarrow{?}$ cyclohexane-1,2-dione

(d) $CHCO_2C_2H_5$ (cyclohexylidene) $\xrightarrow{?, AC_2O}$ $CHCO_2C_2H_5$, OAC

(e) $\xrightarrow[C_5H_5N]{?}$

(f) $\xrightarrow[H_2O]{?}$

(a) OsO_4 (b) OsO_4 (c) SeO_2 (d) SeO_2 (e) SeO_2 (f) OsO_4

Transformation Regents

(1) Cyclohexenone $\xrightarrow[H^+]{?}$ 3-methylcyclohexanone (Me)

1,4 addition

(2) $CH_3CH_2OH \xrightarrow[\text{Triethyl amine}]{?} CH_3CH_2OSi(CH_3)_3 + Et_3\overset{+}{N}HCl^{\ominus}$

(1) $(CH_3)_2CuLi$ (Lithium dimethyl cuprate).

(2) $(CH_3)_3SiCl$ (Trimethylsilyl chloride).

Rearrangement of Lobry de Bruyn-van Ekenstein

(1) $C_6H_5—COOH + H_2\dot{O}_2 + HOOC_6H_5 \xrightarrow{?}$

$C_6H_5CO—O—O—OCC_6H_5 + C_6H_{11}–NH\overset{O}{\overset{\|}{C}}NHC_6H_{11}$

Dibenzoyl peroxide Dicyclohexyl urea

(2) n Nucleotides $\xrightarrow{?}$ (Nucleotides)$_n$ + RNA

(3) α-Picoline (CH_3) $\xrightarrow{?}$ α-Picolinic acid (COOH)

α-Picoline α-Picolinic acid

Reducing sugars on treatment with dicyclohexyl carbodiimide (DCC) undergo rearrangement to form a mixture of other reducing sugars *e.g.*, glucose forms a mixture of fructose and mannose.

Glucose		Fructose	Mannose
CHO		CH₂OH	CHO
HCOH		CO	HOCH
HOCH	$\xrightarrow{DCC}$	HOCH	HOCH
HOCH		H—COH	H—COH
HCOH		H—COH	HCOH
CH_2OH		CH_2OH	CH_2OH

Glucose Fructose Mannose

Suitable Mechanism for the Reaction :

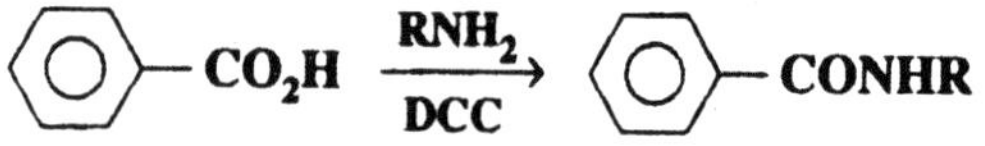

$$C_6H_5-CO_2H \xrightarrow[\text{DCC}]{RNH_2} C_6H_5-CONHR$$

(i) 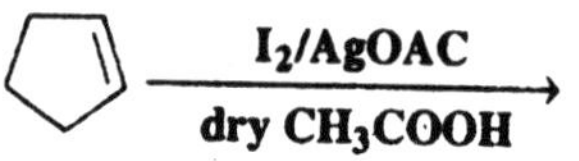 Cyclopentene $\xrightarrow[\text{dry } CH_3COOH]{I_2/AgOAC}$ (ii) 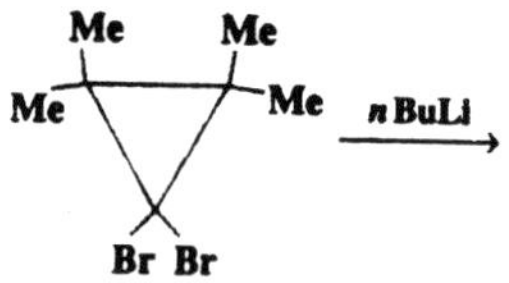 Me, Me, Me, Me, Br Br $\xrightarrow{n\text{BuLi}}$

DCC is dicyclohexylcarbodiimide.

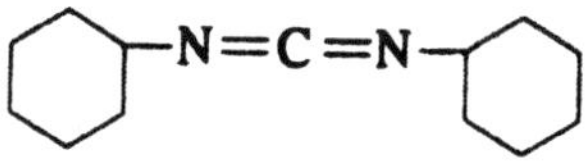

$$C_6H_{11}-N=C=N-C_6H_{11}$$

A carboxylic acid cannot be directly converted to an acid amide by the action of an amine. Hence, the need of a dehydrating agent like DCC. The acid is converted to a compound with a better leaving group.

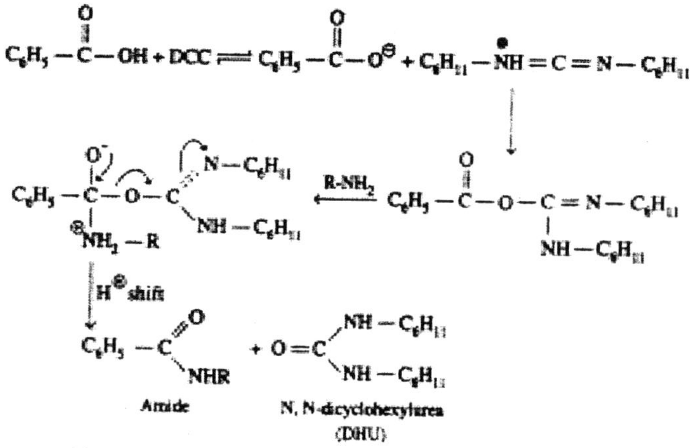

$$C_6H_5-\overset{O}{\overset{\|}{C}}-OH + DCC \rightleftharpoons C_6H_5-\overset{O}{\overset{\|}{C}}-O^{\ominus} + C_6H_{11}-\overset{\bullet}{N}H=C=N-C_6H_{11}$$

$$\downarrow$$

$$C_6H_5-\overset{O}{\overset{\|}{C}}-O-\underset{NH-C_6H_{11}}{\underset{|}{C}}=N-C_6H_{11}$$

$$\xleftarrow{R\text{-}NH_2} \quad C_6H_5-\overset{O^-}{\underset{^{\oplus}NH_2-R}{C}}-O-C(=N-C_6H_{11})(NH-C_6H_{11})$$

$$\downarrow H^{\oplus} \text{ shift}$$

$$C_6H_5-C(=O)-NHR + O=C(NH-C_6H_{11})_2$$

Amide N, N-dicyclohexylurea (DHU)

Difference in behaviour of RMgX and R_2CuLi:

Since the C—to—Mg bond has more ionic character than the C—to—Cu bond, its R group is more like R : $^-$ and is much more reactive.

A set of reagents is listed under column A and their uses under column B. The reagent with its most appropriate use—

Column A	Column B
(1) $(Ph_3P)_3RhCl$	(a) Hydroboration
(2) 18-crown-6	(b) Epoxidation
(3) $Bu_4N^+ Br^-$	(c) Alkylation
(4) Disiamyl borane	(d) Hydrogenation
(5) DDQ	(e) Dehydration
(6) DCC	(f) Dehydrogenation
(7) CH_3CO_3H	(g) Cation complexing
(8) LDA	(h) Phase transfer catalysis

(1)—(d), (2)—(g), (3)—(h), (4)—(a), (5)—(f), (6)—(e), (7)—(b), (8)—(c).

Formulas for A through D in the following reactions

$$RCH_2-\overset{|}{C}=O + (i\text{-}Pr_2)N^- Li^+ \longrightarrow A + B$$

Lithium di-isopropylamide

$$A + R'X \longrightarrow C + D$$

LDA rather than NaOEt is used as the base in the abvoe reaction:

A is carbanion enolate, $[R\ddot{C}H^{-}-\overset{|}{C}=\ddot{O}: \longleftrightarrow RCH=\overset{|}{C}-\ddot{O}:^-]$; B is $(i\text{-}Pr)_2NH$; C is alkylated product, $RCHR'\overset{|}{C}=O$; D is the o-alkylated product, an enol ether, $RCH=\overset{|}{C}-OR'$; LDA is a bulky base capable of abstracting H^+ but incapable of displacing X^- from R'X and thereby nullifying the alkylation whereas NaOEt would react with R'X.

Structure for A, B and C

$$Cl(CH_2)_3I \longrightarrow A \xrightarrow{BuLi} B \xrightarrow[HgCl_2,\ MeOH]{H_2O} C$$

(a) A is [1,3-dithiane ring: S S; H, $(CH_2)_3Cl$] (I $^-$ is a better leaving group)

B is [1,3-dithiane ring: S S; $\ddot{-}$, $(CH_2)_3Cl$], C is [cyclobutane ring]=O

cyclobutanone

Tributylstannane $(C_4H_9)_3SnH$ reduces an alkyl halide to the corresponding alkane by a free radical mechanism. The initiator is an azo compound,

$$(CH_3)_2C(CN)—N = N—C(CN)(CH_3)_2$$

which breaks down to N_2 and a radical. A possible mechanism:

The initiator gives the carbon radical $(CH_3)_2C(CN)$, (Radical) which abstracts an H from the tin compound.

$$\text{Radical} + (C_4H_9)_3Sn—H \longrightarrow (C_4H_9)_3Sn + \text{Radical—H}$$

The tributyltin radical abstracts a halogen atom from the alkyl halide and the chain is propagated as follows—

$$(C_4H_9)_3Sn\cdot + R—X \longrightarrow (C_4H_9)_3Sn—X + R\cdot$$

then

$$R\cdot + (C_4H_9)_3Sn—H \longrightarrow R—H + (C_4H_9)_3Sn\cdot$$

Matched:

(1) Wilkinsons catalyst	**(a) Cyclic polyethers**
(2) DDQ	**(b)** 
(3) Prevost reagent	**(c)** $[(CH_3)_2CHCH(CH_3)]_2BH \equiv Sia_2BH$
(4) Crown ethers	**(d)** $RhCl[(Ph_3P)_3]$
(5) 1, 3-dithiane	**(e)** $(i\text{-}Pr)_2N^- Li^+$
(6) Disiamyl borane	**(f) 2,3-dichloro- 5,6-dicyano-1, 4-benzoquinone**
(7) LDA	**(g) Mixture of silver salt of an acid and iodine.**

(1)—(d), (2)—(f), (3)—(g), (4)—(a), (5)—(b), (6)—(c), (7)—(e).

Phase Transfer Catalysts' Application

In water :

$$Na^+ : CN^- + n\text{-}Bu_4N^+Cl^- \longrightarrow n\text{-}Bu_4N^+ : CN^- + Na^+ Cl^-$$

In non-polar solvent :

$n\text{-}C_8H_{17}Cl + n\text{-}Bu_4N^+ : C^- \longrightarrow n\text{-}Bu_4N + Cl^- + n\text{-}C_8H_{17}CN$

$n\text{-}Bu_4N^+ Cl^-$ goes back into H_2O and the two steps repeat.

Activation of the C-terminus of an AA, followed by coupling with a second AA, is accomplished with the reagent, DCC, C_6H_{11}—N=C=N—C_6H_{11}. Give the structure of the product of reaction of DCC with RCOOH:

$$R{-}\overset{\overset{\displaystyle O}{\|}}{C}{-}OH + C_6H_{11}{-}N{=}C{=}N{-}C_6H_{11} \longrightarrow R{-}\overset{\overset{\displaystyle O}{\|}}{C}{-}O{-}\overset{\overset{\displaystyle NHC_6H_{11}}{|}}{C}{=}N{-}C_6H_{11}$$

(A)

The Merrifield solid-phase process for synthesising peptides :

The solid phase is beads of polystyrene whose mer is mainly $-CH_2-\overset{\overset{\displaystyle PH}{|}}{CH}-$. However, occasional benzene rings have p-CH_2Cl substituents projecting out to the surface. The solid phase may be indicated as. [P]—CH_2Cl (A), where [P] is the polystyrene backbone. The peptide chain is started at the C-terminus by bonding the BOC protected AA to the solid phase through benzyl ester formation, followed by removal of the BOC group.

$$[P]{-}CH_2Cl + {}^-OOCCHRNHBOC \xrightarrow{-Cl^-} [P]{-}C_6H_4CH_2OOCCHRNH\,BOC \xrightarrow{CF_3COOH} [P]{-}C_6H_4CH_2OOCCHRNH_2$$

A second AA (BOC-protected so that it will not self couple) is added, along with DCC. The two steps of addition of a BOC-protected AA and regeneration of NH_2 with CF_3COOH are repeated as many times as required. In between each step, the [P]-growing peptide chain is washed with suitable solvents to remove excess reagents and undesirable products. Thus, intermediates do not have to be isolated and purified and the yields are high. Reactant BOC-AA's, DCC and CF_3COOH are added through an automated system. Removal of the completed peptide from the polymer is accomplished with anhydrous HF, which also removes the last BOC group.

The Reactions Completed :

(i) 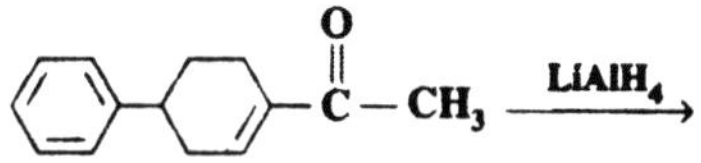 C_6H_5–(cyclohexenyl)–$\overset{O}{\overset{\|}{C}}$–$CH_3$ $\xrightarrow{LiAlH_4}$ (ii)  (cyclobutyl)–CH_2CN $\xrightarrow{LiAlH_4}$

(iii) 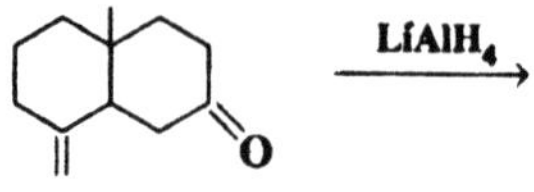 (bicyclic ketone) $\xrightarrow{LiAlH_4}$ (iv)  (bicyclic ketone) $\xrightarrow{LiAlH(o\text{-}t\text{-}Bu)_3}$

(i) 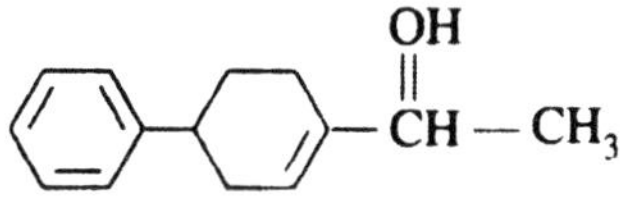 C_6H_5–(cyclohexenyl)–$\overset{OH}{\overset{|}{CH}}$–$CH_3$ (ii) 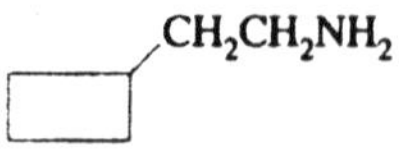 (cyclobutyl)–$CH_2CH_2NH_2$

(iii) 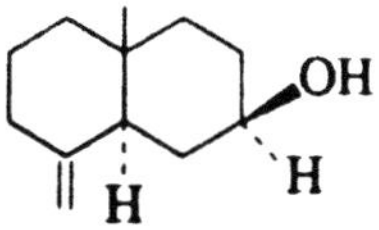 (bicyclic alcohol, OH, H, H) (iv) 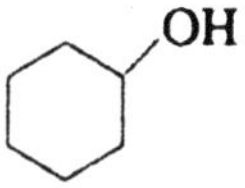 (cyclohexyl)–OH

SeO_2: Use

(i) 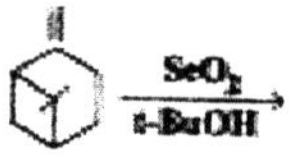 $\xrightarrow[t\text{-}BuOH]{SeO_2}$ (ii) 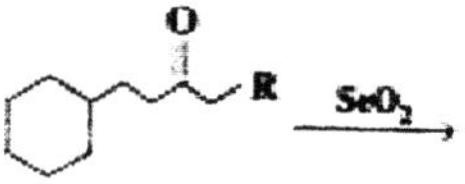 (cyclohexyl)–CH₂CH₂–CO–CH₂–R $\xrightarrow{SeO_2}$

(iii) 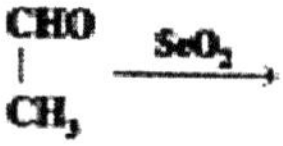 CHO–CH_3 $\xrightarrow{SeO_2}$ (iv) 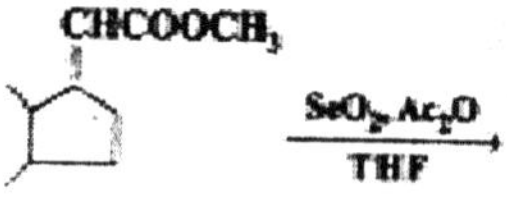 $CHCOOCH_3$ $\xrightarrow[THF]{SeO_2, Ac_2O}$

(v) 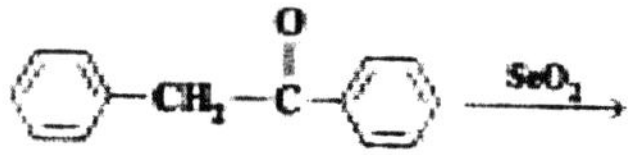 C_6H_5–CH_2–$\overset{O}{\overset{\|}{C}}$–$C_6H_5$ $\xrightarrow{SeO_2}$

SeO_2 is used as oxidising agent, dehydrogenating agent and carbonylating agent. Selenuous acid is formed when water reacts with SeO_2.

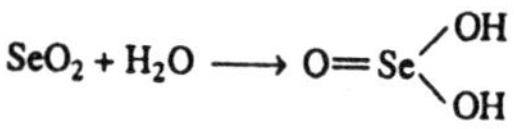

$$SeO_2 + H_2O \longrightarrow O{=}Se(OH)_2$$

(i) 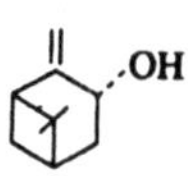 OH (ii) 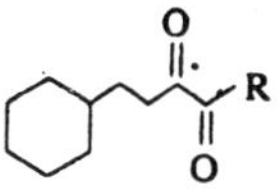 R

(iii) CHO–CHO (iv) 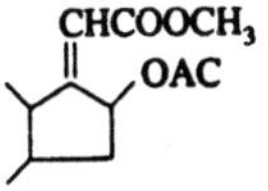 CHCOOCH$_3$ … OAC (v) 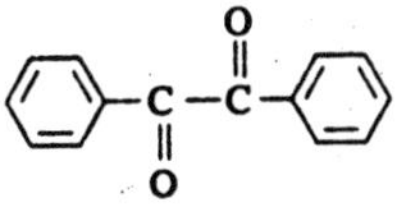 Ph–C(=O)–C(=O)–Ph

The following reactions:

(i) 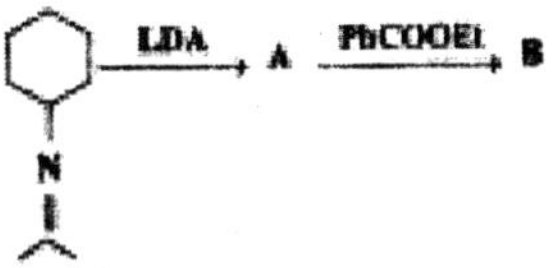 LDA → A, PhCOOEt → B

(ii) 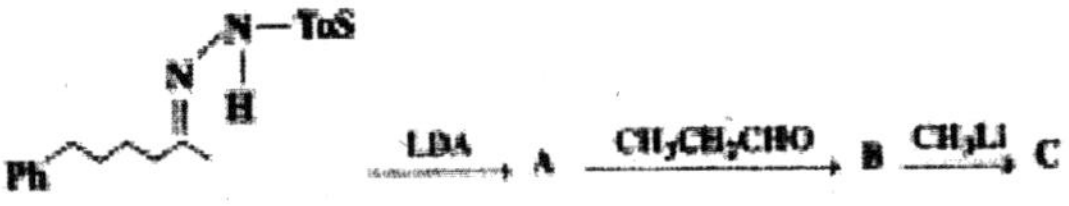 Ph … N–N(H)–TosS; LDA → A, CH_3CH_2CHO → B, CH_3Li → C

(TosS → toluene sulphonyl)

(iii) 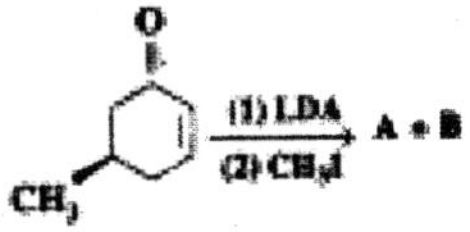 (1) LDA, (2) CH_3I → A + B

(iv) O … X + Me_3SiI ⟶

(v) R–C(=O)–CH(R)–OH, Me_3SiI →

(i) A — … N, H, $\overset{\ominus}{C}H_2\overset{\oplus}{Li}$, B — … N, O, H, Ph

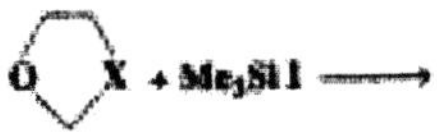

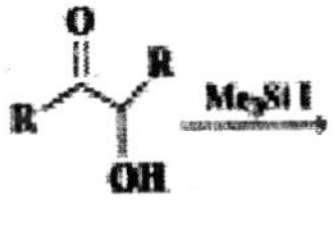

(ii) A — … N–TosS, N, Li, Li, B — Ph … N–NHTosS … OH, C — Ph … OH

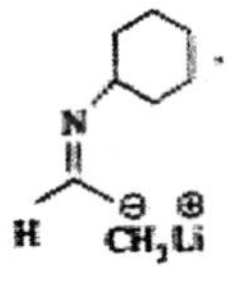

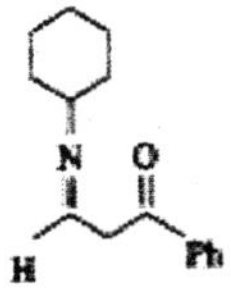

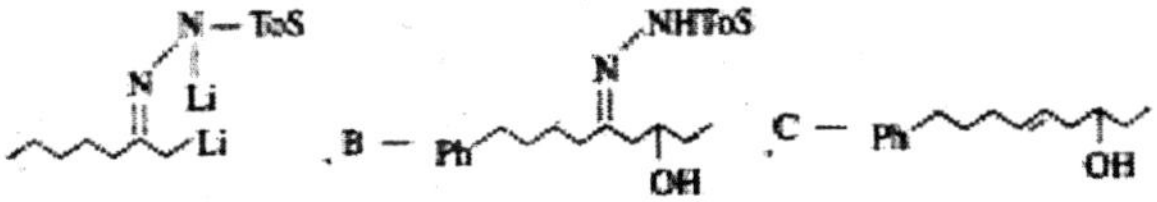

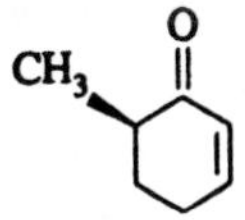

(iii) A — CH_3 … O (Minor), B — CH_3 … O (Major)

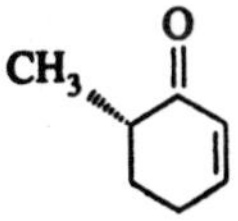

(iv)  Me_3SiO–X–I (v) 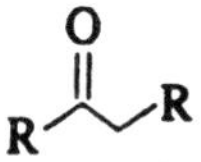 $RCOCH_2R$

Yeast for Bakers

(i) COOH ... N ... C_5H_{11} —Baker Yeast→ A (ii) CH_2OTS —Bu_3SnH→ A

(iii) $(CO)_3Cr$ —Baker Yeast→ A (iv) Br —Bu_3SnH→ A + B

(v) Ph O R O —Baker Yeast→ A

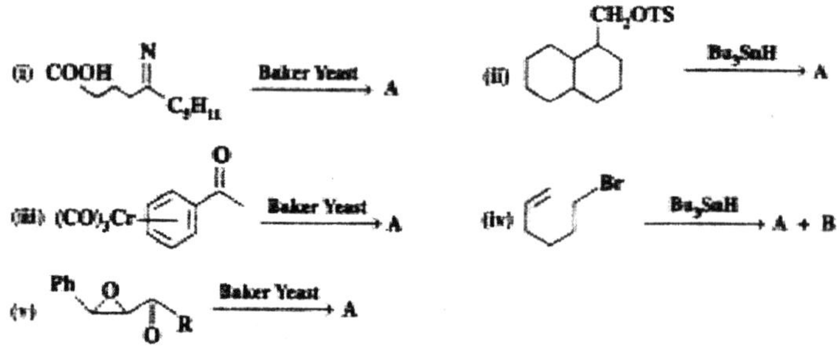

Saccharomyces cerevisiae (Baker Yeast) are micro organisms and are used for the reduction of carbonyl group to hydroxyl group and for reduction of double bond.

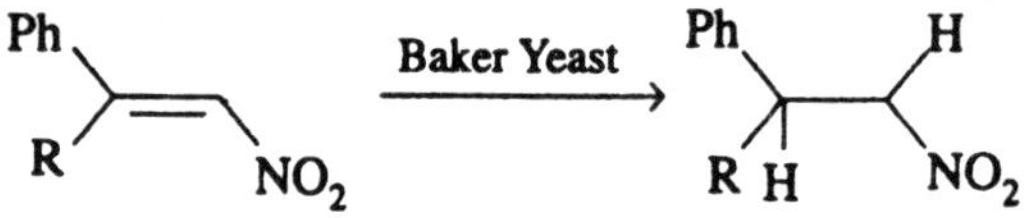

Ph, R, NO_2 —Baker Yeast→ Ph, H, R H, NO_2

Reactions: Produced Form

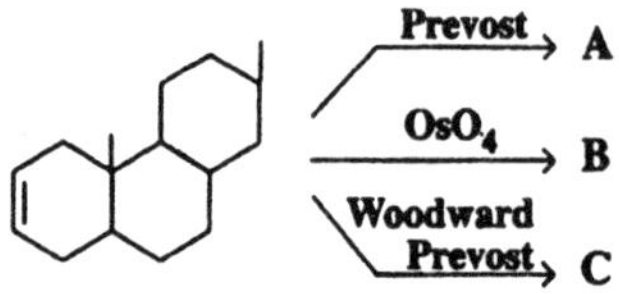

—Prevost→ A

—OsO_4→ B

—Woodward Prevost→ C

In Prevost

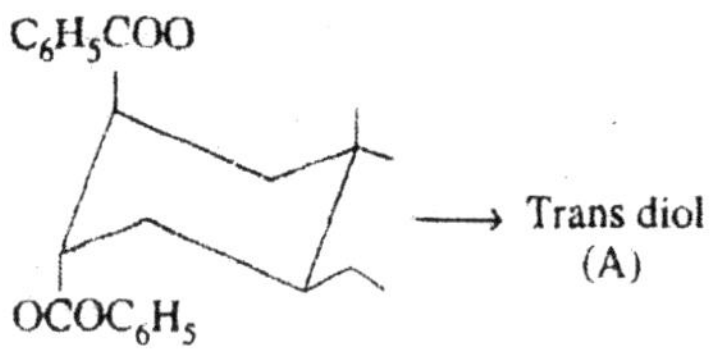

C_6H_5COO, $OCOC_6H_5$ ⟶ Trans diol (A)

With OsO_4

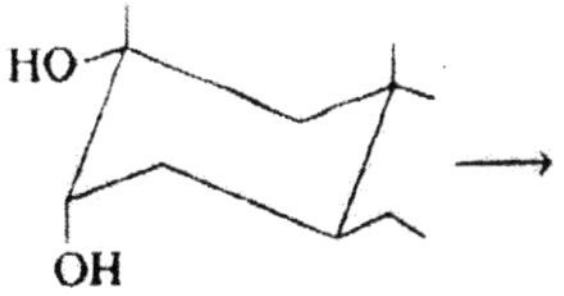

HO OH → Trans diol (B) (less hindered)

In Woodward-Prevost

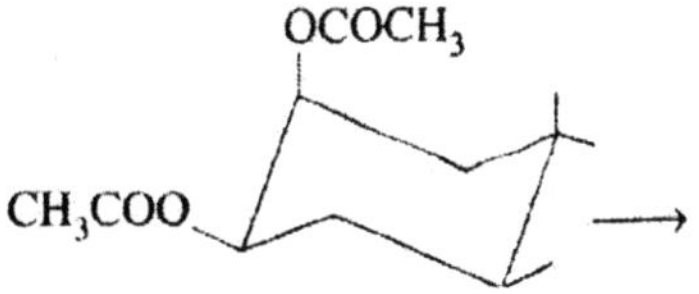

$OCOCH_3$

CH_3COO → Cis diol (C) (More hindered)

The Reactions Completed:

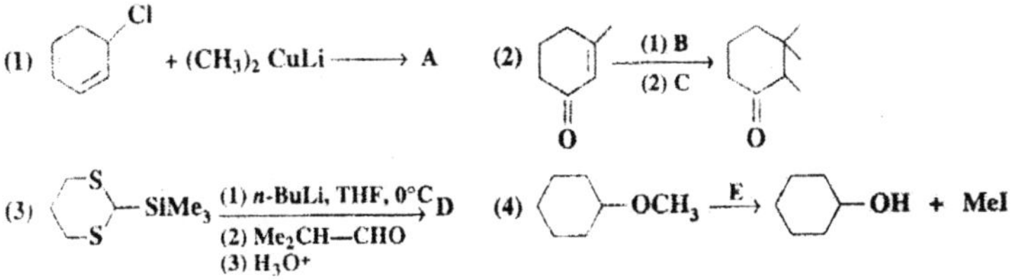

(1) Cl + $(CH_3)_2$ CuLi → A (2) O $\xrightarrow[(2)\ C]{(1)\ B}$ O

(3) S S —SiMe$_3$ $\xrightarrow[(2)\ Me_2CH—CHO \quad (3)\ H_3O^+]{(1)\ n\text{-BuLi, THF, }0°C}$ D (4) —OCH_3 $\xrightarrow{E}$ —OH + MeI

(1) A CH_3 (2) B $(CH_3)_2CuLi$
C CH_3I

(3) D 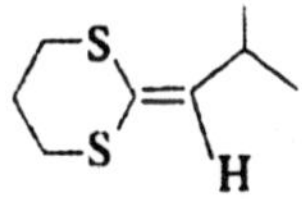 S S = H (4) E Me_3SiI.

Study of Crown Ethers

Crown ethers are cyclic polyethers and their structure permits a conformation with certain sized 'holes' in which cations can be trapped by coordination with the lone pair electrons on the oxygen atoms.

These are used as phase transfer catalysts. The cyclic polymers of ethylene glycol $(OCH_2CH_2)_n$ are named as X-crown-Y. X refers to total number of atoms in the ring and Y to the total number of oxygens in the ring.

The ability of a crown ether depends to complex a cation: Why the nucleophilic reactions

$$K^+CN^- + RCH_2X \xrightarrow[\text{benzene}]{\text{18-crown-6}} RCH_2CN + K^+X^-$$

$$C_6H_5CH_2Cl + KF \xrightarrow[\text{acetonitrile}]{\text{18-crown-6}} C_6H_5CH_2F + K^+Cl^-$$

under the reaction conditions give almost same quantitative yields ?

On the size of the cavity in the crown ether which can be tailored to allow for the selective binding of only certain cations, *i.e.*, whose ionic radius is best accommodated by the polyether. Thus, 18-crown‾6 shows a high affinity for K^+ (ionic diameter 2.66 Å) and 15-crown-5 for Na^+ (ionic diameter 1.80 Å).

The crown ether, 18-crown-6 acts as a phase transfer catalyst and gets the anion into the organic phase. On coordination with a metal cation the crown ethers convert the metal ion into a species with a hydrocarbon like exterior. The crown ether, 18-crown-6, *e.g.*, coordinates effectively with potassium ions as the cavity size is correct and because the six oxygens are ideally situated to donate their electron pairs to the central ion.

Crown ethers render many organic salts soluble in non-polar solvents. Therefore, salts like CH_3COOK, KCN can be transferred into aprotic solvents by using catalytic amounts of 18-crown-6. In the organic phase the relatively unsolvated anions of these salts bring about nucleophilic substitution reaction on an organic substate.

Study of Phase Transfer Catalysts

A compound whose addition to a two-phase organic water system helps to transfer a water soluble ionic reactant across the interface to the organic phase where a homogeneous reaction can take place is called a *phase transfer catalyst*. These catalysts enhance the rate of a reaction. A quaternary ammonium halide $R_4N^+X^-$ *e.g.*, tetrabutylammonium halide is phase transfer catalyst. It can cause the transfer of the nuclepphile for example CN^- as an ion pair (Q^+CN^-) into the organic phase; since the cation (Q^+) of the ion pair with its four bulky alkyl groups resembles a hydrocarbon even though it has a positive charge. It is thus lipophilic, *i.e.*, it prefers a non-polar environment to an aqueous one.

Reactions of Prevost and Woodward

Both the reactions are essentially the additions of iodine carboxylate (formed in situ) to an alkene, *i.e.*, the reaction of an alkene with iodine and silver salt. The Prevost procedure employs iodine and silver carboxylate under dry conditions. The initially formed transiodocarboxylate (b) from a cyclic iodonium ion (a) undergoes internal displacement to a common intermediate acylium ion (c). The formation of the diester (d) with retention of configuration provides an example of neighbouring group participation. The diester on subsequent hydrolysis gives a transglycol.

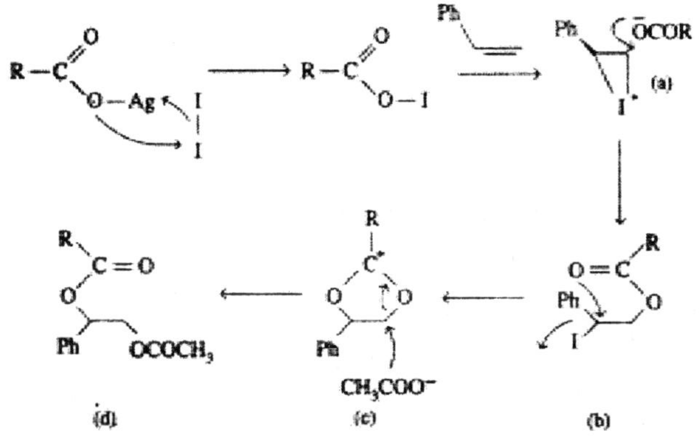

R—C(=O)—O—Ag + I—I → R—C(=O)—O—I; Ph-alkene → (a) Ph, $\bar{O}COR$, I^+

(b) O=C(R)—O, Ph, I → (c) R—C^+, O, O, Ph, CH_3COO^- → (d) R—C=O, O, Ph, $OCOCH_3$

(d) (c) (b)

Prevost Reaction

In the Woodward procedure, water is present which intercepts the intermediate (c) to give a cishydroxy ester (e) and a cisglycol on subsequent hydrolysis.

(c) R—^+C, O, O, Ph + H—Ö—H $\xrightarrow{-H^+}$ $\xrightarrow{H^+}$ R, C, OH, $H\overset{+}{O}$, O, Ph → (e) OH, OCOR, Ph

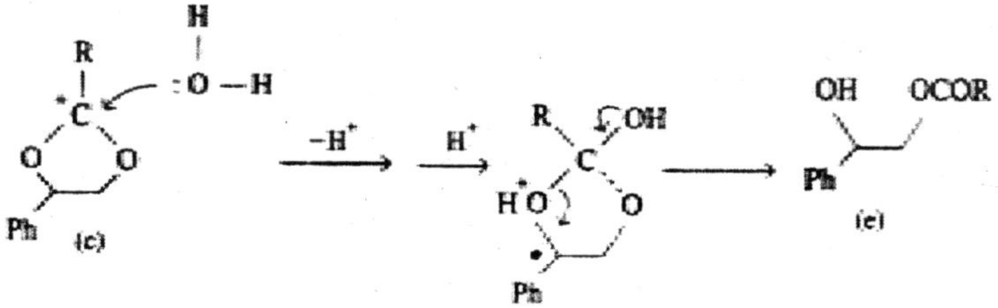

Umpolung: The reversal of polarity of the carbonyl carbon atom is termed umpolung (German : for polarity reversal). Normally the carbonyl carbon atom of an aldehyde (or a ketone) is partially positive, *i.e.*, electrophilic and therefore it reacts with nucleophiles. When the aldehyde is converted to a dithiane by reaction with 1,3-propanedithiol and reacted with butyl lithium the same carbon now becomes negatively charged to react with electrophiles. This reversed polarity of the carbonyl carbon is termed umpolung which increases the versatility of the carbonyl group in synthesis. The sulphur atoms stabilise carbanions because sulphur has the capacity to utilise 3d orbitals for bonding and to occur in valence states higher than 2.

$$R-\overset{\overset{O^{\delta-}}{\|}}{C^{\delta+}}-H \xrightarrow[(2)\ C_4H_9Li]{(1)\ HSCH_2CH_2CH_2SH,\ H^+} \text{(dithiane ring: S, S)}\ R-C{:}^{\ominus}$$

Umpolung

Dithianes: These are thioacetals which can be prepared by treating an aldehyde with 1,3-propanedithiol in the presence of trace of acid.

$$\underset{\text{Aldehyde}}{R\overset{\overset{O}{\|}}{C}H} + \underset{\text{1, 3-propanedithiol}}{HSCH_2CH_2CH_2SH} \xrightarrow{H^+} \underset{\text{Dithiane}}{\text{(S, S ring) } C(R)(H)}$$

A 1,3-dithiane is a weak proton acid ($pKa = 32$) which can be deprotonated by strong bases such as *n*-butyllithium. The resulting carbanion is stabilised by the electron withdrawing effect of the two sulphur atoms.

$$\underset{\text{1, 3-dithiane}}{\text{(S, S ring) } CH_2} + C_4H_9Li \longrightarrow \underset{\text{Dithiane anion}}{\text{(S, S ring) } \bar{C}H\ Li^+} + \underset{\text{Butane}}{C_4H_{10}} \uparrow$$

The reactions completed by writing the appropriate product:

(a) $C_6H_5CH = CHNO_2 \xrightarrow[C_6H_6]{H_2, [(C_6H_5)_3P]_3RhCl}$? (b) 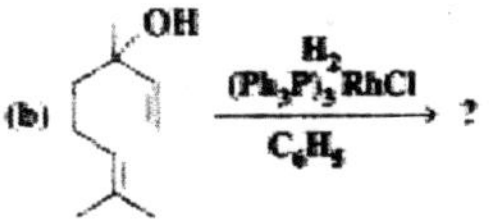 $\xrightarrow[C_6H_6]{H_2, (Ph_3P)_3RhCl}$?

w-nitrostyrene

(c) + SeO_2 ⟶ ? (d) 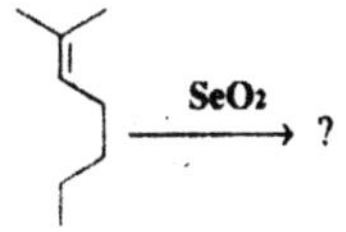 $\xrightarrow{SeO_2}$?

(a) $C_6H_5CH_2CH_2NO_2$ (Phenylnitroethene) (b)

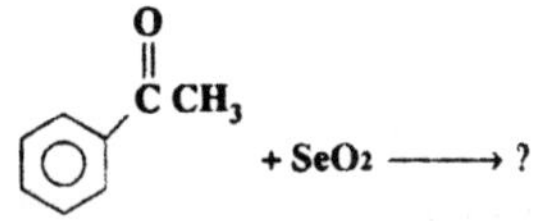

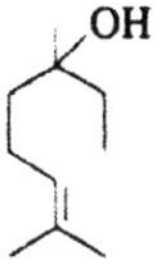

(c) 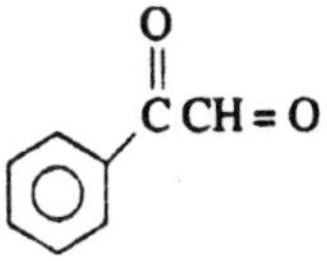 (d) 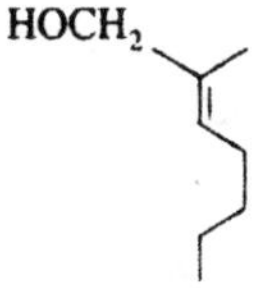 $HOCH_2$

Reaction completed :

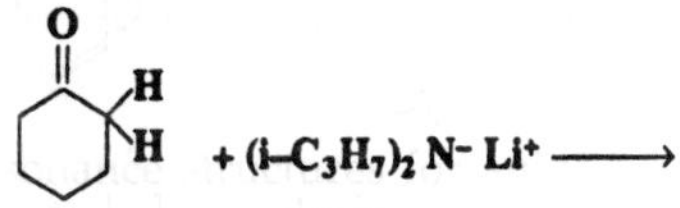

$+ (i\text{–}C_3H_7)_2N^- Li^+ \longrightarrow$

Cyclohexanone (LDA)

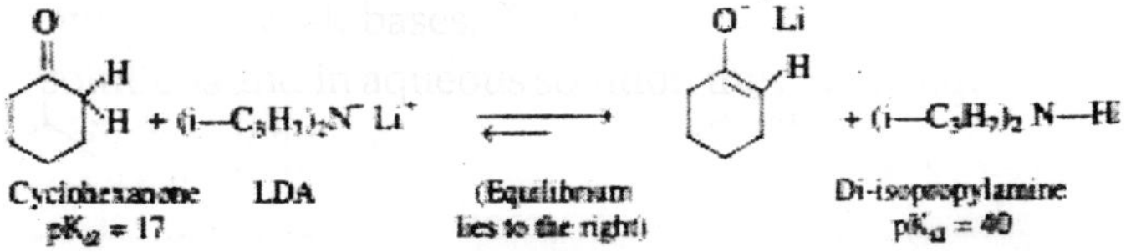

$+ (i\text{—}C_3H_7)_2N^- Li^+ \rightleftharpoons$ $O^- Li^+$ $+ (i\text{—}C_3H_7)_2N\text{—}H$

Cyclohexanone $pK_a = 17$ LDA (Equilibrium lies to the right) Di-isopropylamine $pK_a = 40$

Peterson's Synthesis :

In the Peterson's synthesis the b-hydroxysilanes are converted to alkenes in either acidic or basic solution.

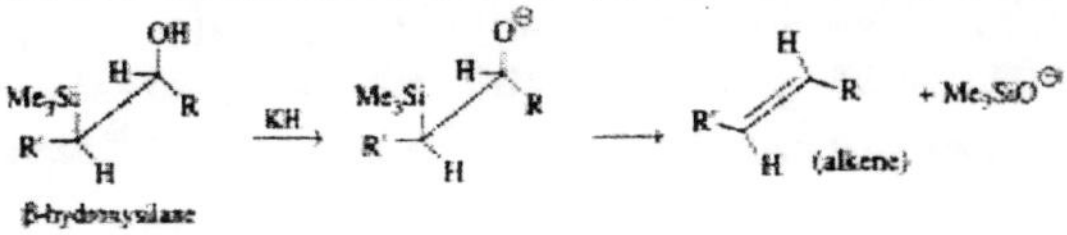

$\xrightarrow{KH}$ ⟶ (alkene) $+ Me_3SiO^{\ominus}$

β-hydroxysilane

The Structure

Making Structures

C–H Bond Omission

There are various ways to drawing structures of organic molecules. A molecule like ethane can be drawn showing every C–C and C–H bond. However, this becomes difficult particularly with more complex molecules, and it is much easier to miss out the C–H bonds:

$$\begin{array}{ccccccc} & & H & & H & & \\ & & | & & | & & \\ H & - & C & - & C & - & H \\ & & | & & | & & \\ & & H & & H & & \end{array} \quad = \quad H_3C{-}CH_3$$

Fig. Ethane.

Skeletal Drawings

A further simplification is generally used when only the carbon-carbon bonds are shown. This is a skeletal drawing of the molecule. In such drawings, it is understood that a carbon atom is present at every bond junction and that every carbon has sufficient hydrogens attached to make up four bonds:

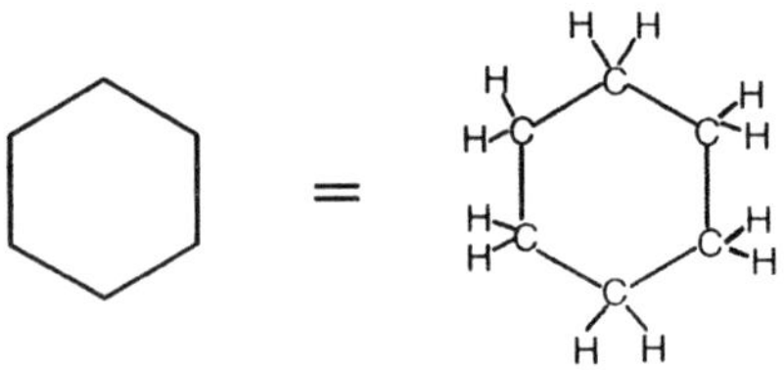

Fig. Skeletal drawing of cyclohexane.

Straight chain alkanes can also be represented by drawing the C–C bonds in a zigzag fashion:

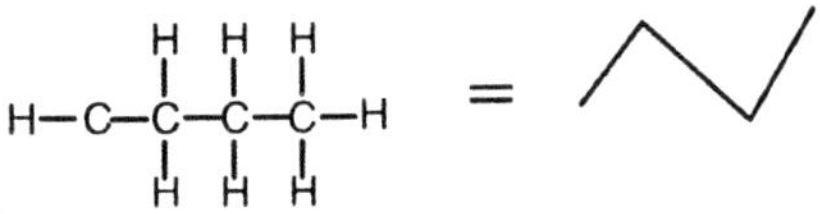

Fig. Skeletal drawing of butane.

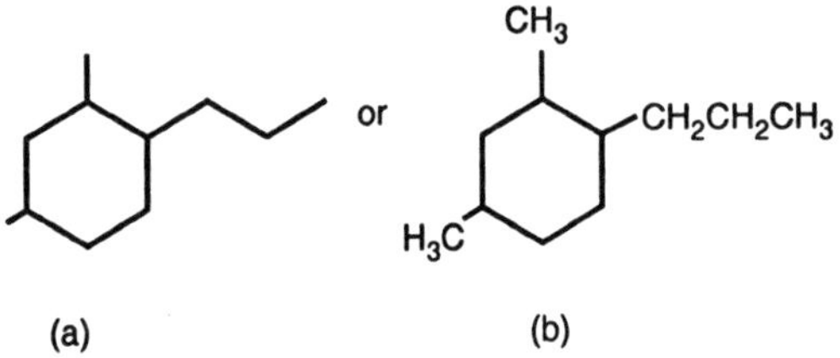

(a) (b)

Fig. Drawings of an alkyl substituted cyclohexane.

Alkyl Groups

Alkyl groups (C_nH_{2n+1}) are the substituents of a complex molecule. Simple alkyl groups can be indicated in skeletal form, or as CH_3, CH_2CH_3, $CH_2CH_2CH_3$, *etc.*

The CH_3 groups shown in figure given below, the structure in figure (a) is more correct than the structure in figure (b) because the bond shown is between the carbons.

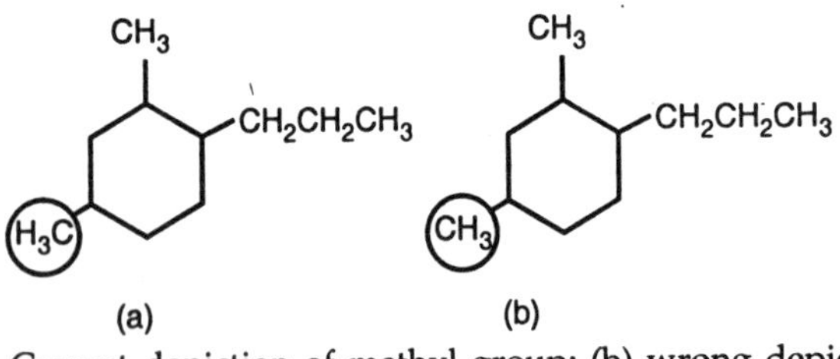

(a) (b)

Fig. (a) Correct depiction of methyl group; (b) wrong depiction of methyl group.

The Taxonomy

Simple Alkanes

To names simple alkanes the longest carbon chain is selected. The names of the simplest straight chain alkanes are shown in the following figure.

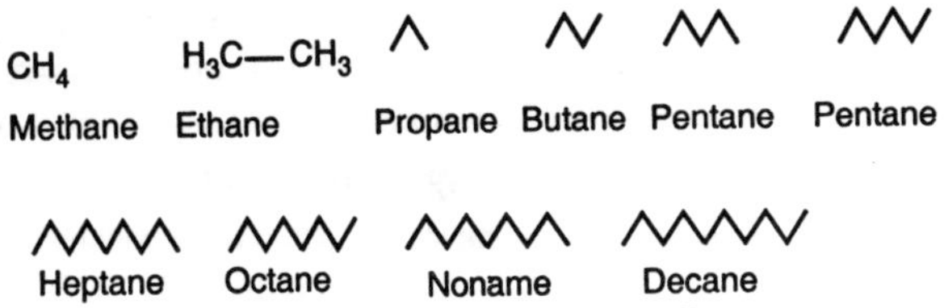

CH_4 Methane $H_3C—CH_3$ Ethane   Propane   Butane   Pentane   Pentane

Heptane   Octane   Noname   Decane

Fig. Nomenclature of simple alkanes.

Branched Alkanes

Branched alkanes are named by the following procedure:

(i) Identify the longest chain of carbon atoms. In the example shown below, the longest chain consists of five carbon atoms and a pentane chain:

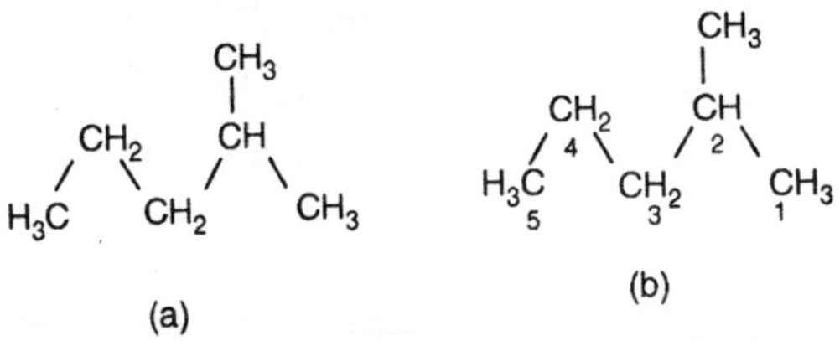

(a): CH_3, CH_2, CH, H_3C, CH_2, CH_3

(b): CH_3, CH_2 4, CH 2, H_3C 5, CH_2 3, CH_3 1

Fig. (a) identify the longest chain; (b) number the longest chain.

(ii) Number the longest chain of carbons, starting from the end nearest the branch point (fig. b).

(iii) Identify the carbon with the branching group (number 2 in fig. b)

(iv) Identify and name the branching group. (In this example it is CH_3. Branching groups (or substituents) are referred to as alkyl groups (C_nH_{2n+1}) rather than alkanes (C_nH_{2n+2}). Therefore, CH_3 is called *methyl* and not methane.).

(v) Name the structure by first identifying the substituent and its position in the chain, then naming the longest chain. The structure in above fig is called

2-methylpentane. Notice that the substituent and the main chain is one complete word, that is, 2-methylpentane rather than 2-methly pentane.

Multi-Branched Alkanes

In case there is more than one alkyl substituent present in the structure then the substituents are named in *alphabetical order*, numbering again from the end of the chain nearest the substituents. The structure in following figure is 4-ethyl-3-methyloctane and not 3-methyl-4-ethyloctane.

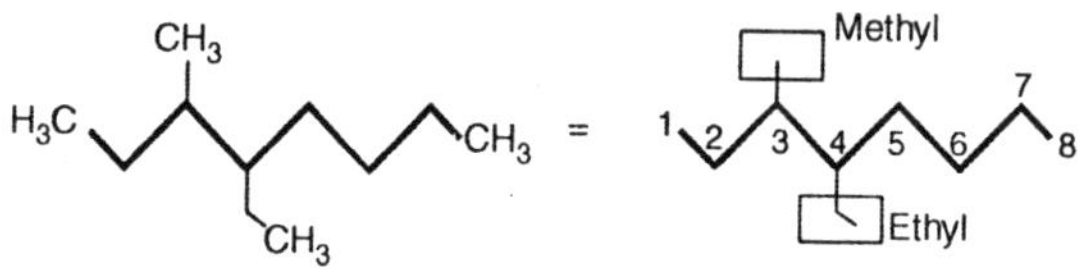

Fig. 4-Ethyl-3methyloctane.

If a structure has identical substituents, then the prefixes di-, tri-, tetra-, etc. are used to represent the number of substituents. For example, the structure in figure given below is called *2,2-dimethylpentane* and not 2-methyl-2-methylpentane.

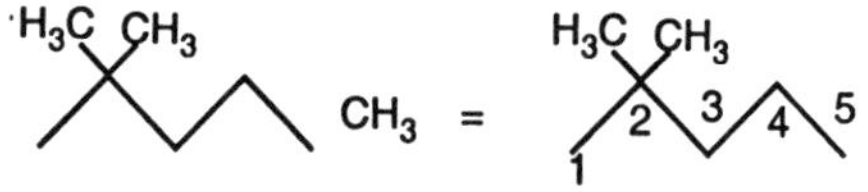

Fig. 2, 2-Dimethylpentane.

The prefixes di-, tri-, tetra-, etc., are used for identical substituents, but the order in which they are written is still dependent on the alphabetical order of the substituents themselves (i.e. ignore the di-, tri-, tetra-, *etc.*). For example, the structure in following figure is called *5-ethyl-2, 2-dimethyldecane* and not 2, 2-dimethyl-5-ehtyldecane.

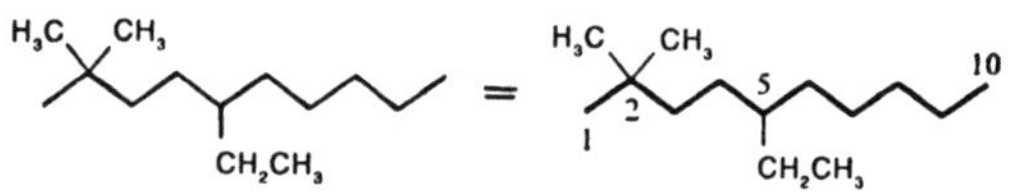

Fig. 5-Ethyl-2,2-dimcthyldecane.

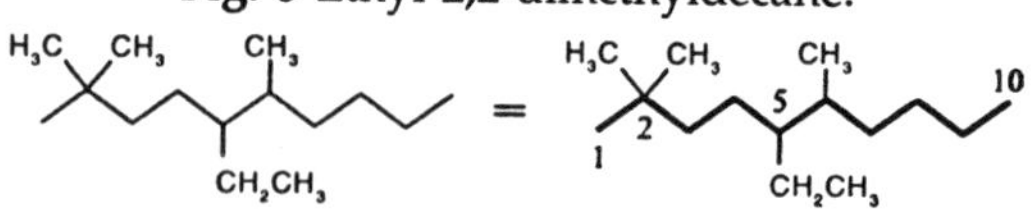

Fig. 5-Ethyl-2,2,6-trimethyldecane.

Identical substituents can be in different positions on the chain, but the same rules apply. For example, the structure in figure given above is called *5-ethyl-2,2,6-trimethyldecane*. In some structures, it is difficult to decide which end of the chain to number from. For example, two different substituents might be placed at equal distances from either end of the chain. In such a case, the group with alphabetical priority should be given the lowest numbering. For example, the structure in following figure (a) is 3-ethyl-5-methylheptane and not 5-ethyl-3-methylheptane. There is another rule that might take precedence over the above rule. The structure (fig. c) has ethyl and methyl groups equally placed from each end of the chain, but there are two methyl groups to one ethyl group. Numbering should be chosen such that the smallest total is obtained. In this example, the structure is called *5-ethyl-3,3-dimethylheptane* rather than 3-ethyl-5,5 dimethylheptane (fig. b) since 5+3+3 = 11 is less than 3 + 5 + 5=13.

(a) (b) (c)

Fig. (a) 3-Ethyl-5-methylheptane; (b) incorrect numbering; (c) 5-ethyl-3,3-dimethylheptane.

Cycloalkanes

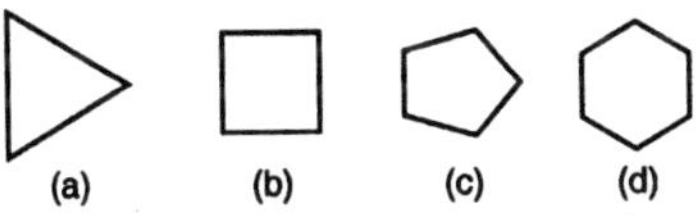

Fig. (a) Cyclopropane; (b) cyclobutane; (c) cyclopentane; (d) cyclohexane.

Cycloalkanes are simply named by identifying the number of carbons in the ring and prefixing the alkane name with cyclo.

Branched Cyclohexanes

Cycloalkanes made up of a cycloalkane moiety linked to an alkane moiety are generally named in such a way that the cycloalkane is the parent system and the alkane moiety is considered to be an alkyl substituent. Therefore, the structure in following figure (a) is methylcyclohexane and not cyclohexylmethane. In it there is no need to number the cycloalkane ring when only one substituent is present.

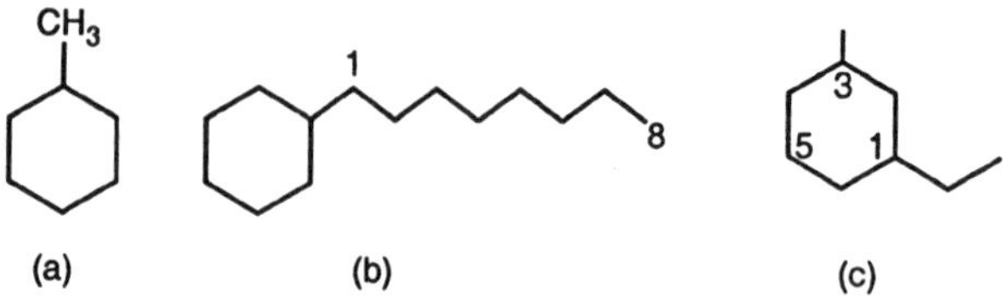

CH_3 1 8 3 5 1

(a) (b) (c)

Fig. (a) Methylcyclohexane; (b) 1-cyclohexyloctane; (c) l-ethyl-3 methylcyclohexane.

If the alkane moiety contains more carbon atoms than the ring, the alkane moiety becomes the parent system and the cycloalkane group becomes the substituent. For example, the structure in above figure (b) is called *1–cyclohexyloctane* and not octylcyclohexane. The numbering is necessary to identify the position of the cycloalkane on the alkane chain.

Multi-Branched Cycloalkanes

Branched cycloalkanes having different substituents are numbered such that the alkyl substituent having alphabetical priority is at position 1.

The numbering of the rest of the ring is then done such that the substituent positions add up to a minimum. For example, the structure in figure (c) is called *1-ethyl-3-methylcyclohexane* rather than l-methyl-3-ethylcyclohexane or l-ethyl-5-methylcyclohexane. The last name is incorrect since the total obtained by adding the substituent positions together is 5 + 1 = 6 which is higher than the total obtained from the correct name (i.e. 1 + 3 = 4).

Bonding and Structure

Atomic Structure of Carbon

Atomic Orbitals: Carbon has six electrons, i.e. $1s^2 2s^2 p_x^{\,1} p_y^{\,1}$ and is placed in second period of the periodic table. In it there are two shells of atomic orbitals available for these electrons. The first shell closest to the nucleus has a single 5 orbital–the 1*s* orbital. The second shell has a single *s* orbital (the 2*s* orbital) and three *p* orbitals (3 × 2*p*). Therefore, there are a total of five atomic orbitals into which these six electrons can fit. The *s* orbitals are spherical in shape with the 2*s* orbital being much larger than the 1*s* orbital. The *p* orbitals are dumb-bell-shaped and are aligned along the *x*, *y* and *z* axes. Therefore, they are assigned the $2p_x$, $2p_y$ and $2p_z$ atomic orbitals.

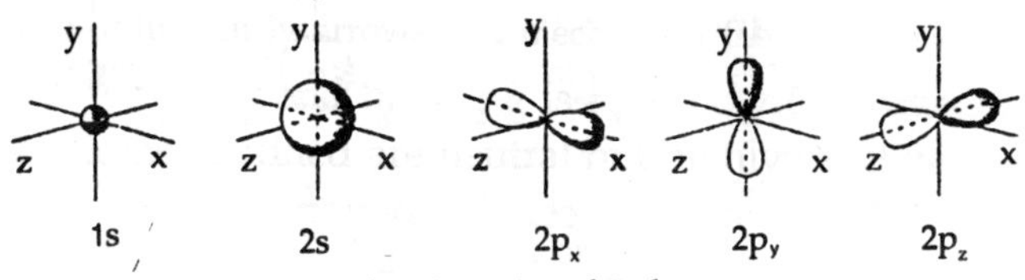

Fig. Atomic orbitals.

Energy Levels

Various atomic orbitals of carbon atom are not of equal energy. The 1*s* orbital has the lowest energy. The 2*s* orbital is next in energy and the 2*p* orbitals have the highest energies. The three 2*p* orbitals have the same energy, i.e. they are *degenerate*.

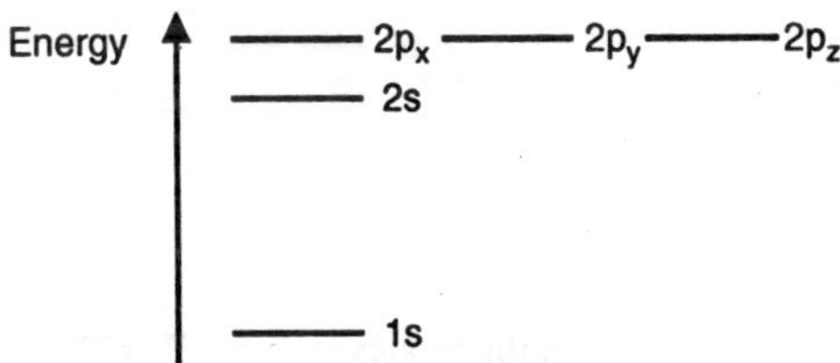

Fig. Energy levels of atomic orbitals.

Electronic Configuration

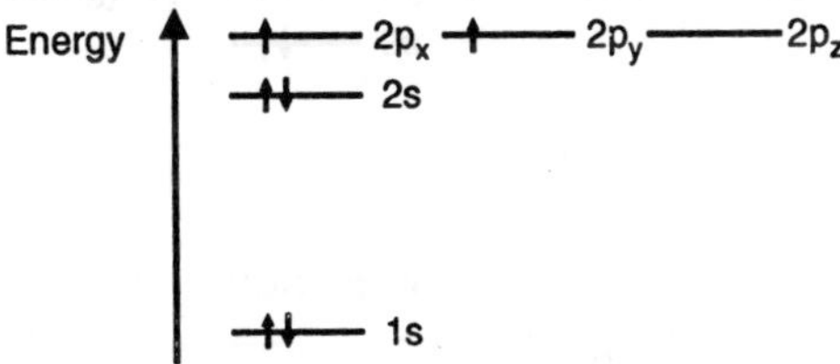

Fig. Electronic configuration for carbon.

According to the *Aufbau principle, Pauli exclusion principle* and *Hund's rule*, the electronic configuration of carbon is $1s^2 2s^2 2p_x^{\ 1} 2p_y^{\ 1}$:

The electronic configuration for carbon is $1s^2 2s^2 2p_x^{\ 1} 2p_y^{\ 1}$. The numbers in superscript refer to the numbers of electrons in each orbital.

Hybridised Centres and Bonds

σ and π Bonds

Identifying σ and π bonds in a molecule is quite simple if remember the following rules:

(i) All bonds in organic structures are either sigma (σ) or pi (π) bonds.

(ii) All single bonds are σ bonds.

(iii) All double bonds are made up of one σ bond and one π bond.

(iv) All triple bonds are made up of one σ bond and two π bonds.

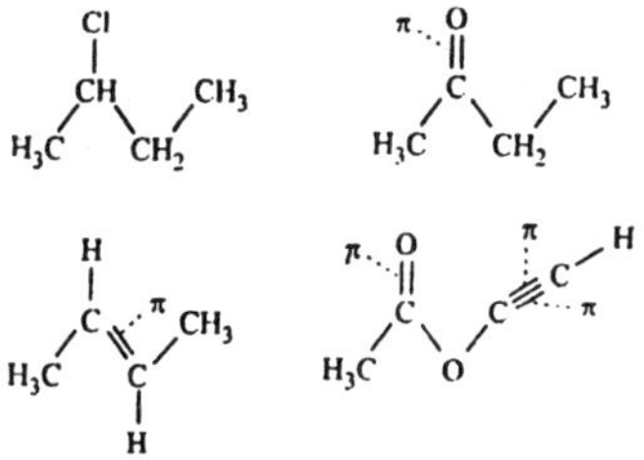

Fig. Examples–all the bonds shown are σ bonds except those labelled as p.

Hybridised Centres

All the atoms in an organic structure (except hydrogen) are either *sp, sp^2 or sp^3* hybridised:

To identify the type of hybridisation remember the following rules:

(i) All atoms joined by single bonds are sp^3 hybridised (except hydrogen).

(ii) Both the carbons involved in C=C (alkene) are sp^2 hybridised.

(iii) Both carbon and oxygen in carboxyl group (>C=0) must be sp^2 hybridised.

(iv) All aromatic carbons must be sp^2 hybridised.

(v) Both carbon atoms involved in a triple bond (C≡C) are *sp* hybridised (e.g. in alkynes).

(vi) Both atoms involved in a triple bond must be *sp* hybridised.

(vii) Hydrogen uses its 1*s* orbital in bonding and is not hybridised.

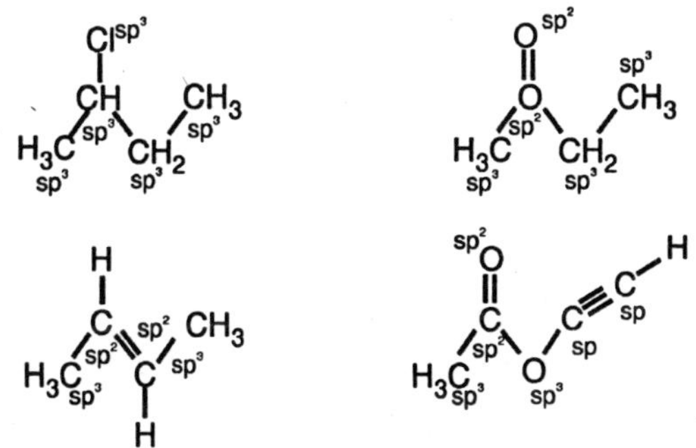

Fig. Example of *sp*, sp^2 and sp^3 hybridised centres.

Hybridisation is not Possible in Case of Hydrogen

Oxygen, nitrogen and halogen atoms can form hybridised orbitals. These hybridised orbitals may be involved in bonding or may be used to accommodate a lone pair of electrons.

Shape

The shape of a molecule and the functional group depends on the type of hybridisation of atoms involved in its formation, e.g:

1. Functional groups with sp^3 hybridisation are *tetrahedral.*
2. Functional groups having sp^2 hybridisation are *planar.*
3. Functional groups with *sp* hybridisation are *linear.*

Examples:

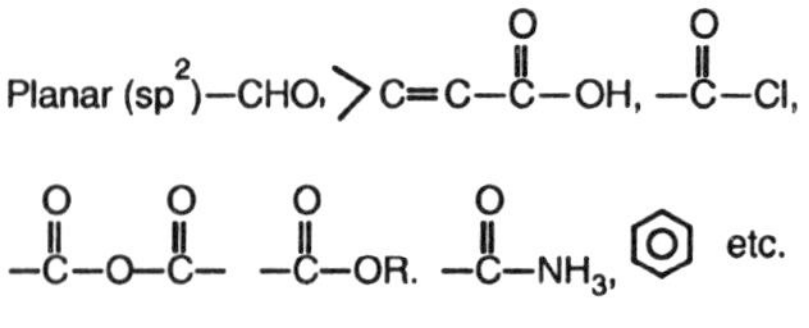

Planar (sp^2)—CHO, $>C=C-\overset{O}{\overset{\|}{C}}-OH$, $-\overset{O}{\overset{\|}{C}}-Cl$,

$-\overset{O}{\overset{\|}{C}}-O-\overset{O}{\overset{\|}{C}}-$ $-\overset{O}{\overset{\|}{C}}-OR$. $-\overset{O}{\overset{\|}{C}}-NH_3$, (benzene ring) etc.

Linear (sp)— C≡O. C≡N, etc.

Tetrahedral (sp^3)— R—OH. R—O.—R R—X
Carbon where R is an alkyl group.

Reactivity

Functional groups containing π-bonds are more reactive because π-bond is weaker and can be easily broken, e.g. aromatic rings, alkenes, alkynes, aldehydes and ketones, carboxylic acids, esters, amides, acid chlorides, acid anhydrides, nitriles, etc.

Cycloalkanes and Alkanes

Definitions

Alkanes: These are the open chain organic compounds having the general formula C_nH_{2n+2}. In them all the bonds are σ-bonds and so they are also called *saturated hydrocarbons*. All the carbon atoms in any alkane are sp^3 hybridised and so their shape in tetrahedral. Since C–C and C–H σ-bonds are strong. So the alkanes are unreactive to most of the chemical reagents.

Alkanes are also referred to as *straight chain* or *acyclic compounds* (Hydrocarbons).

Cycloalkanes

These are cyclic alkanes (hydrocarbons) and are also called *alicyclic compounds*. Their general formula in C_nH_{2n}. In these the carbon atoms are linked to form a rings of various sizes. The six membered carbon ring being most commonly found. Most of there cycloalkanes are unreactive to chemical reagents.

The cycloalkane with three or four carbon atom rings behave like alkanes (i.e., they are reactive). Their reactivity is due to the fact that such cyclic structures are highly strained.

Hybridisation and Covalent Bonding

Covalent Bonding

A covalent bond is formed between two atoms together in a molecular structure. It is formed when atomic orbitals overlap to produce a *molecular orbital*. For example, the formation of a hydrogen molecule (H_2) from two hydrogen atoms. Each hydrogen atom has a half-filled 1*s* atomic orbital and when the atoms approach each other, the atomic orbitals interact to produce two MOs (the number of resulting MOs must equal the number of original atomic orbitals):

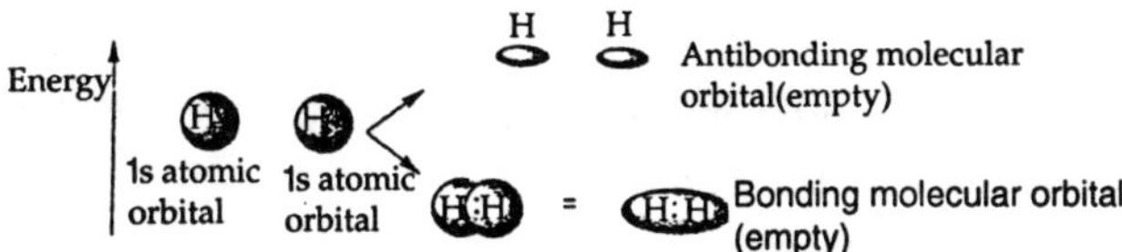

Fig. Molecular orbitals for hydrogen (H_2).

The MOs are of different energies. One is more stable than the combining atomic orbitals and is known as the *bonding MO*. The other is less stable and is known as *anti-bonding MO*. The bonding MO is like a rugby ball and is formed by the combination of the 1*s* atomic orbitals. As this is the more stable MO, the valence electrons (one from each hydrogen) enter this orbital and pair up. The anti-bonding MO is of higher energy and consists of two deformed spheres. This remains empty. As the electrons end up in a bonding MO that is more stable than the original atomic orbitals, energy is released and bond formation is favoured.

Sigma Bonds

σ-bonds have a circular cross-section and are formed by the head-on overlap of two atomic orbitals. This involves a strong interaction and thus sigma bonds are strong bonds.

Hybridisation

Atoms can form bonds with each other by sharing unpaired electrons such that each bond contains two electrons. A carbon atom has two unpaired electrons and so we expect carbon to form two bonds. However, carbon forms four bonds. How does a carbon atom form four bonds with only two unpaired electrons?

When a carbon atom forms bonds and is part of a molecular structure, it can 'mix' the *s* and *p* orbitals of its second shell (the valence shell). This is called *hybridisation* and it allows carbon to form the four bonds which are observed in its compounds.

There are three important ways in which the mixing process can occurs are:

- The 2*s* orbital is mixed with all three 2*p* orbitals. This is called *sp^3 hybridisation.*
- The 2*s* orbital is mixed with two of the 2*p* orbitals. This is called *sp^2 hybridisation.*
- The 2*s* orbital is mixed with one of the 2*p* orbitals. This is called as *sp hybridisation.*

The other types of hybridisation observed in some other compounds are:

(i) *dsp^2*: In it one $(n - 1)d$ orbital and one *ns* orbital and 2 *np* orbitals combine to form 4 hybridised orbitals.

(ii) *d^2sp^3* or *sp^3d^2*: The six hybridised orbitals are formed by mixing up of two $\lambda(n - 1)d$ orbitals, or two *nd* orbitals with one *ns* and three *np* orbitals.

sp^3 Hybridisation

Definition: The sp^3 hybridisation of carbon involves mixing up of the 2*s* orbitals with all three of the 2*p* orbitals to give a set of four sp^3 hybrid orbitals. The hybrid orbitals will each have the same energy but will be different in energy from the original atomic orbitals. That energy difference will reflect the mixing of the respective atomic orbitals. The energy of each hybrid orbital is greater than the original *s* orbital but less than the original *p* orbitals.

Electronic Configuration

The valence electrons for carbon can now be fitted into the sp^3 hybridised orbitals. There are a total of four electrons in the *2s* and *2p* orbitals of carbon. The *s* orbital was filled and two of the *p* orbitals were half filled. After hybridisation there is a total of four hybridised sp^3 orbitals all of equal energy.

According to Hund's rule, all the four hybridised orbitals are half filled with electrons thus there are four unpaired electrons. Four bonds are now possible.

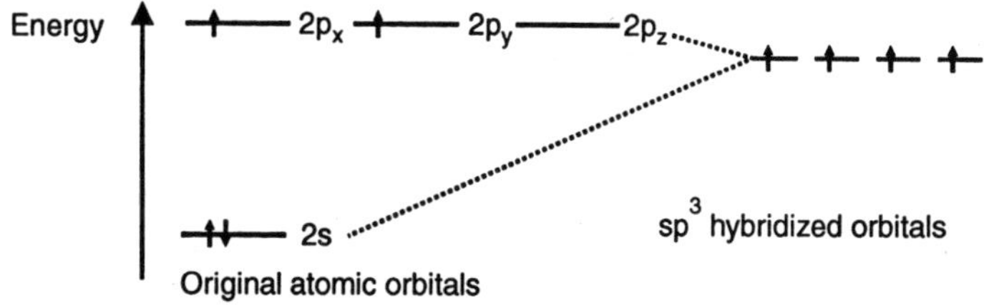

Fig. sp^3 Hybridisation.

Geometry

Each of the sp^3 hybridised orbitals has the same shape. This deformed dumb-bell looks more like a *p* orbital than an *s* orbital since more *p* orbitals were involved in the mixing process, i.e. one lobe is larger as compared to other lobe:

Fig. sp^3 Hybridised orbital.

Each sp^3 orbital occupies a space as far apart from each other as possible by pointing to the concerns of tetrahedron. The angle between each of these lobes is about 109.5°. This is what is meant by the expression *tetrahedral carbon.*

The three-dimensional shape of the tetrahedral carbon can be represented by drawing a normal line for bonds in the plane of the page. Bonds going behind the page are represented by hatched wedge, and bonds coming out the page are represented by a solid wedge:

— Bond in the plane of the page
Bond going behind the page
Bond coming out of the page

Tetrahedral shape

Fig. Tetrahedral shape of an sp^3 hybridised carbon.

Sigma Bonds

A half-filled sp^3 hybridised orbital from one carbon atom can be used to form a bond with a half-filled sp^3 hybridised orbital from another carbon atom. In figure(a) given below the major lobes of the two sp^3 orbitals overlap directly leading to a strong σ bond. Because of their ability to form strong a bonds hybridisation takes place.

The deformed dumb-bell shapes permit a better orbital overlap than would be obtained from a pure *s* orbital or a pure *p* orbital. A σ bond between an sp^3 hybridised carbon atom and a hydrogen atom involves the carbon atom using one of its half-filled sp^3 orbitals and the hydrogen atom using its half-filled 1s orbital.

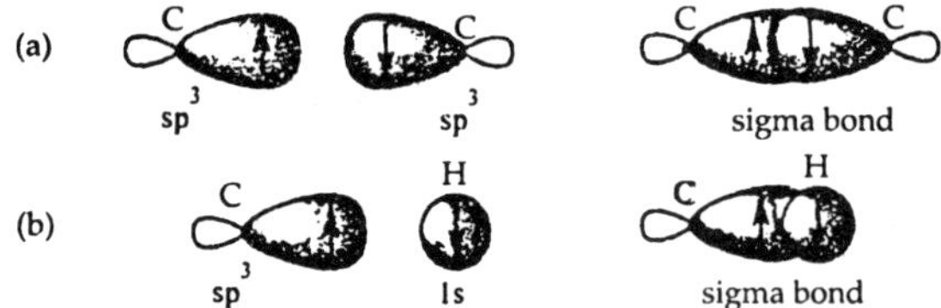

Fig. (a) σ Bond between two sp^3 hybridised carbons; (b) σ bond between an sp^3 hybridised carbon and hydrogen.

Nitrogen, Oxygen and Chloride

The sp^3 type of hybridisation is also observed for nitrogen, oxygen and chlorine atoms in organic structures. Nirtogen has five valence electrons in its second shell (i.e. $2s^22p^3$). After hybridisation, it will have three half-filled sp^3 orbitals and thus can form three bonds. Oxygen has six valence electrons (i.e. $2s^22p^4$). After hybridisation, it will have two half-filled sp^3 orbitals and will form two bonds. Chlorine has seven valence electrons (i.e. $2s^22p^5$). After hybridisation, it will have one half-filled sp^3 orbital and will form one bond.

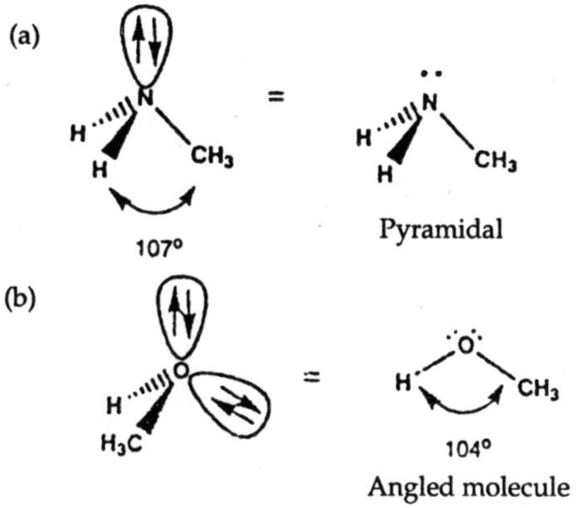

(a)

H CH_3 = H CH_3

107°

Pyramidal

(b)

H H_3C = H CH_3

104°

Angled molecule

Fig. (a) Geometry of sp^3 hybridised nitrogen; (b) geometry of sp^3 hybridised oxygen.

The four sp^3 orbitals for these three atoms (i.e. N, O and Cl) form a tetrahedral arrangement having one or more of the hybridised orbitals occupied by a lone pair of electrons. For an isolated atom, nitrogen forms a pyramidal shape where the bond angles are slightly less than 109.5° (c. 107°) fig. (a). This compression of the bond angles is because of the orbital containing the lone pair of electrons, which requires a slightly greater amount of space than a bond. Oxygen forms an angled or bent shape where two lone pairs of electrons compress the bond angle from 109.5° to 104° (fig. b)].

Alcohols, amines, alkyl halides, and ethers all contain sigma bonds that involve nitrogen, oxygen, or chlorine. Bonds between these atoms and carbon are formed by the overlap of half-filled sp^3 hybridised orbitals from each atom. Bonds involving hydrogen atoms (e.g. O–H and N–H) are formed by the overlap of the half-filled 1s orbital from hydrogen and a half-filled sp^3 orbital from oxygen or nitrogen.

sp^2 *Hybridisation*

Definition: In sp^2 hybridisation, the *2s* orbital is mixed with two of the *2p* orbitals (e.g. $2p_x$ and $2p_z$) to form three sp^2 hybridised orbitals of equivalent energy. The remaining $2p_y$ orbital remains unaffected. The energy of each hybridised orbital is greater than that of the original *s* orbital but less than that of the original *p* orbitals. The remaining *2p* orbital (in this case the $2p_y$ orbital) remains at its original energy level:

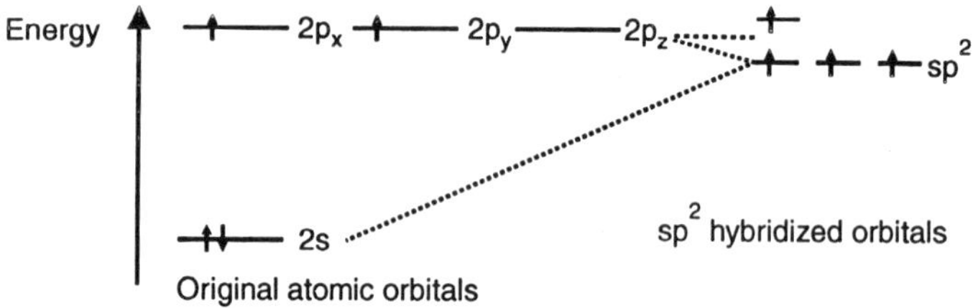

Fig. sp^2 Hybridisation.

Electronic Configuration

In case of carbon, there are four valence electrons (i.e. $2s^2 2p_x^1 2p_y^1 p_z^v$) to fit into the three hybridised sp^2 orbitals and the remaining *2p* orbital. The first three electrons are fitted into each of the hybridised orbitals according to Hund's rule and they are all half-filled this leaves one electron still to place. There is a choice between pairing it up in a half-filled sp^2 orbital or placing it into the vacant $2p_y$ orbital. Generally, we fill up orbitals of equal energy before moving to an orbital of higher energy. However, if the energy difference between orbitals is small (as here) it is easier for the electron to go into the higher energy $2p_y$ orbital resulting in three half-filled sp^2 orbitals and one half-filled orbital. Four bonds are now possible.

Geometry

The unhybridised $2p_y$ orbital has the usual dumb-bell shape. Each of the sp^2 hybridised orbitals has a deformed dumb-bell shape similar to an sp^3 hybridised orbital. The difference between the sizes of the major and minor lobes is larger for the sp^2 hybridised orbital than that in case of sp^3 hybridised orbitals.

The hybridised orbitals and the $2p_y$ orbital occupy spaces as far apart from each other as possible. The lobes of the $2p_y$ orbital occupy the space above and below the plane of the *x* and z axes. The three hybridised sp^2 orbitals (major lobes shown only) will then occupy the remaining space such that they are as far apart from the $2p_y$ orbital and from each other as possible. Thus, they are all placed in the *x-z* plane pointing towards the corner of a triangle (trigonal planar shape). The angle between each of these lobes is 120°. We are now ready to look at the bonding of an sp^2 hybridised carbon.

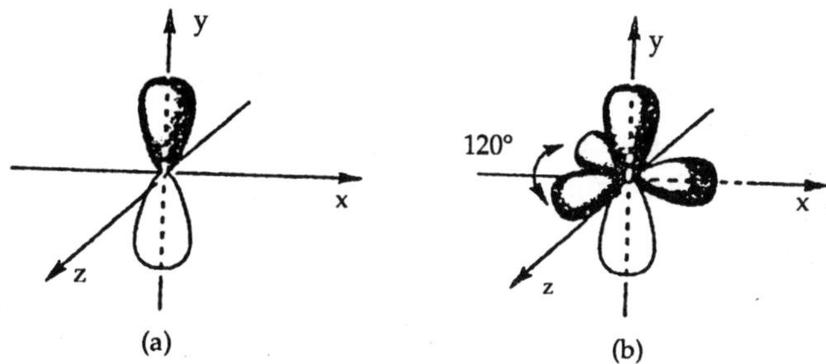

Fig. (a) Geometry of the $2p_y$ orbital; (b) geometry of the $2p_y$ orbital and the sp^2 hybridised orbitals.

Alkenes

In alkenes all the four bonds formed by carbon are not σ-bonds. In this case of sp^2 Hybridisation the three half-filled sp^2 hybridised orbitals form a trigonal planar shape. The use of these three orbitals in bonding explains the shape of an alkene, for example ethene ($H_2C{=}CH_2$). As far as the C–H bonds are concerned, the hydrogen atoms uses a half-filled 1*s* orbital to form a strong σ bond with a half filled sp^2 orbital from carbon. A strong σ bond is also possible between the two carbon atoms of ethene due to the overlap *of* sp^2 hybridised orbitals from each carbon.

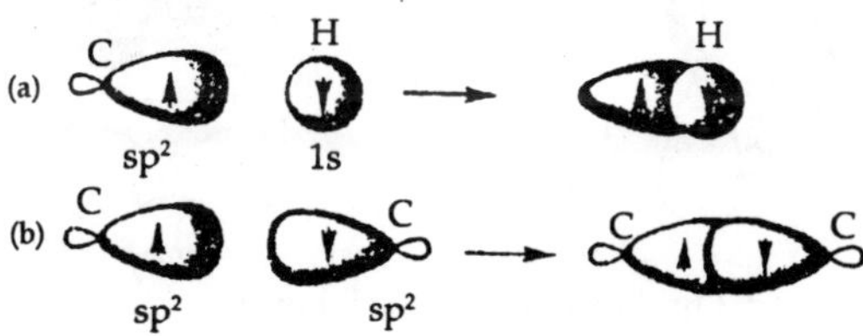

Fig. (a) Formation of a C–H σ bond; (b) formation of a C–C σ bond.

The full σ bonding diagram for ethene is shown in fig. (a) given below and it can be simplified as shown in fig. (b) given below. *Ethene is a flat, rigid molecule where each carbon is trigonal planar.* sp^2 hybridisation explains the trigonal planar carbons but can not explained why the molecule is rigid and planar. If the σ bonds were the only bonds present in ethene, the molecule would not remain planar because rotation can occur round the C–C σ bond.

Therefore, there must be further bonding which 'locks' the alkenes into this planar shape. This bond involves the unhybridised half-filled $2p_y$ orbitals on each carbon that overlap side-on to produce a *pi (π) bond,* with one lobe above and one lobe below the plane of the molecule. This π bond prevents rotation, round the C – C bond since the π bond would have to be broken to allow rotation. A π bond is weaker than a σ bond since the $2p_y$ orbitals overlap side-on, resulting in a weaker overlap. The presence of a π bond can also explains the greater reactivity of alkenes than alkanes, since a π bond is more easily broken and is more likely to take part in reactions.

Fig. (α) σ Bonding diagram for ethene; (b) simple representation of σ bonds for ethene.

Bond rotation

Fig. Bond rotation around a s bond.

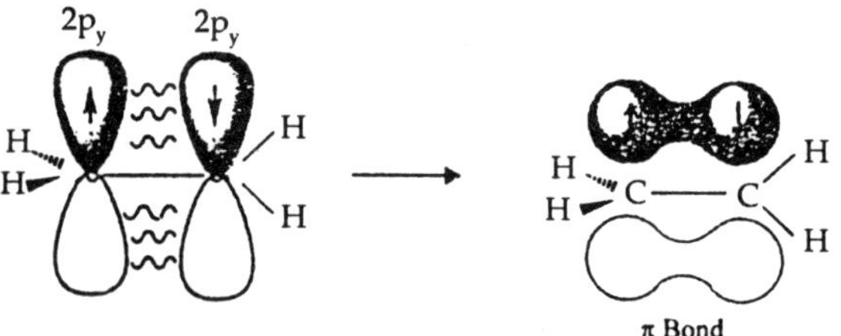

Fig. Formation of a π bond.

Carbonyl Groups

In a carbonyl group (C=O) we find that both the carbon and oxygen atoms are sp^2 hybridised. The energy level diagram given below shows the arrangement of valence electrons of oxygen after sp^2 hybridisation. Two of the sp^2 hybridised orbitals are filled with lone pairs of electrons, that leaves two half-filled orbitals available far bonding.

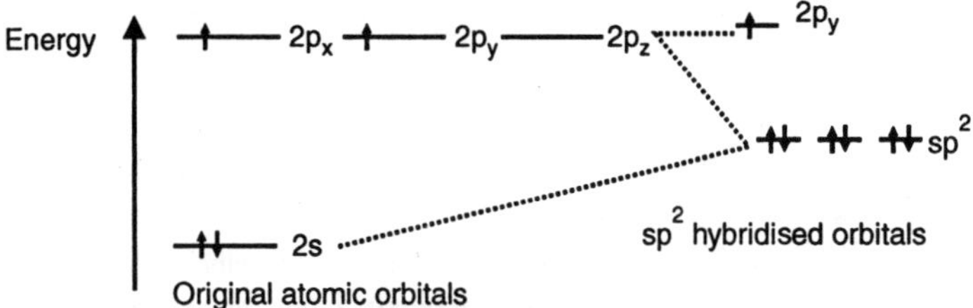

Fig. Energy level diagram for sp^2 hybridised oxygen.

The sp^2 orbital can be used to form a strong σ bond, while the $2p_y$ orbital can be used for the weaker π bond. Figure given below shows how the σ and π bonds are formed in the carbonyl group and it also explains why carbonyl groups are planar with the carbon atom having a trigonal planar space.

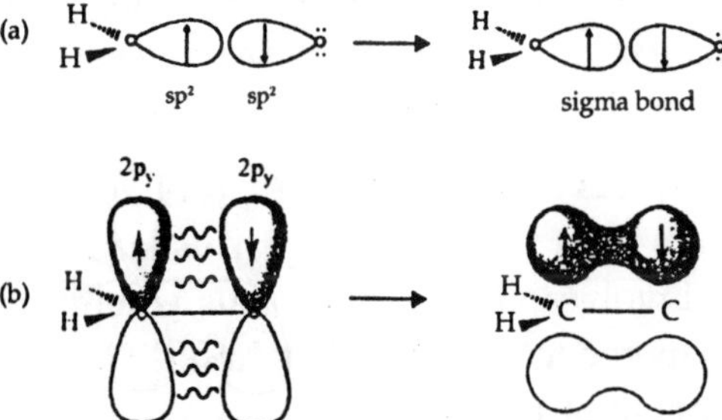

Fig. (a) Formation of the carbonyl σ bond; (b) formation of the carbonyl π bond.

It also explains the reactivity of carbonyl groups since the π bond is weaker than the σ bond and is more likely to be involved in reactions.

Aromatic Rings

All the carbon atoms in an aromatic ring are sp^2 hybridised that means that each carbon can form three σ bonds and one

π bond. All the single bonds are σ while a double bond consists of one σ bond and one π bond.

For example, double bonds are shorter than single bonds and if benzene had this exact structure, the ring would be deformed with longer single bonds than double bonds.

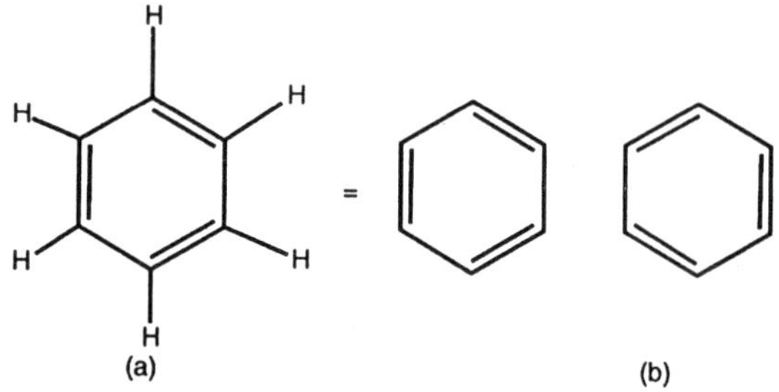

H H H H H H

(a) (b)

Fig. (a) Representation of the aromatic ring: (b) 'deformed' structure resulting from fixed bonds.

Actually all the C–C bonds in benzene are of the same length. To understand this, we must look more closely at the bonding which occurs. Figure given below shows benzene with all its σ bonds and is drawn such that we are looking into the plane of the benzene ring. Since all the carbons are sp^2 hybridised, there is a $2p_y$ orbital left over on each carbon which can overlap with a $2p_y$ orbital on either side of it.

From this, it is clear that each $2p_y$ orbital can overlap with its neighbours right round the ring. This leads to a molecular orbital that involves all the $2p_y$ orbitals where the upper and lower lobes merge to give two doughnut-like lobes above and below the plane of the ring.

The molecular orbital is symmetrical and the six π electrons are said to be delocalised around the aromatic ring since they are not localised between any two particular carbon atoms. The aromatic ring is generally represented as shown in figure given below to represent this delocalisation of the π electrons. Delocalisation increases the stability of aromatic rings and so they are less reactive than alkenes (i.e. it requires more energy to disrupt the delocalised π system of an aromatic ring than it does to break the isolated π bond of an alkene).

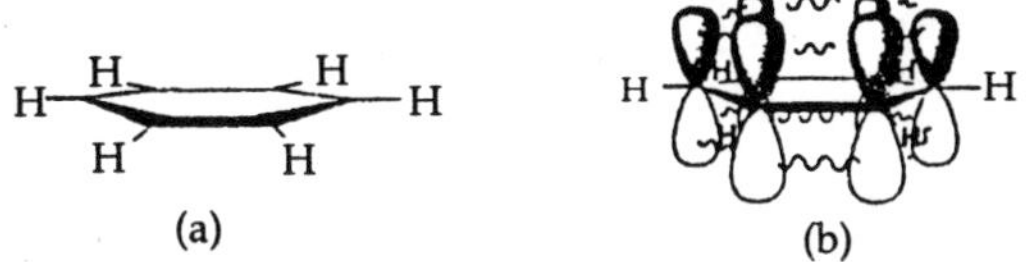

Fig. (a) σ Bonding diagram for benzene, (b) π Bonding diagram for benzene.

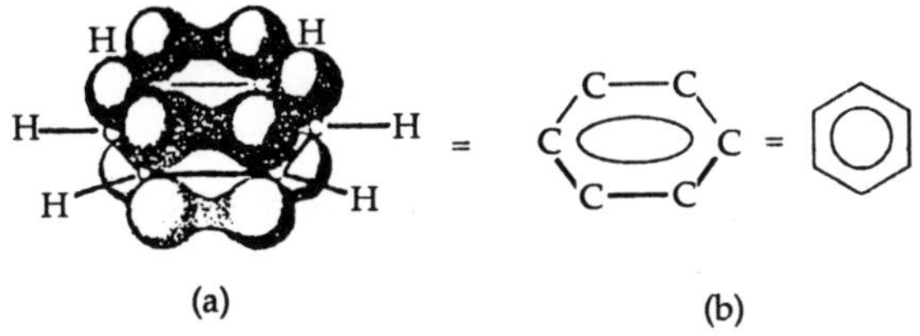

Fig. Bonding molecular orbital for benzene; (b) representation of benzene to illustrate delocalisation.

Conjugated Systems

Aromatic rings are not the only structures where delocalisation of π electrons can occur. Delocalisation can also occur in conjugated systems where there are alternating single and double bonds (e.g. 1,3-butadiene). All four carbons in 1,3-butadiene and sp^2 hybridised and so each of these carbons has a half-filled *p* orbital that can interact to give two π bonds. However, a certain amount of overlap is also possible between the *p* orbitals of the middle two carbon atoms and so the bond connecting the two alkenes has some double bond character.

The delocalisation also increases stability. However, the conjugation in a conjugated alkene is not as great as in the aromatic system. In the latter system, the π electrons are completely delocalised round the ring and all the bonds are equal in length. In 1,3-butadiene, the π electrons are not fully delocalised and are more likely to be found in the terminal C–C bonds. Although there is a certain amount of π character in the middle bond, the latter is more like a single bond than a double bond.

Other example of conjugate systems include α, β-unsaturated ketones and α, β-unsaturated esters. These too have increased stability because of conjugation.

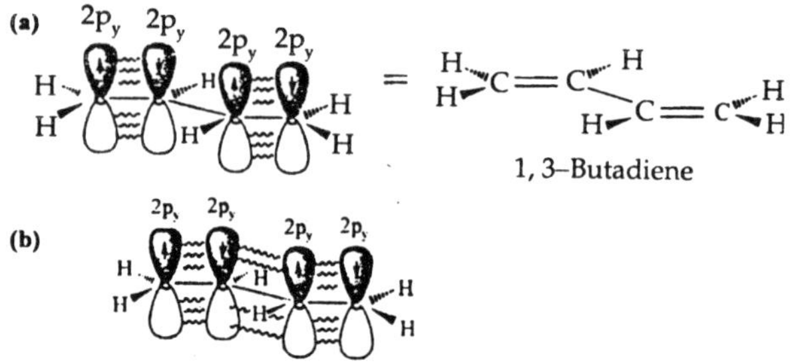

1, 3–Butadiene

Fig. (a) π Bonding in 1,3-butadiene; (b) delocalisation in 1,3-butadiene.

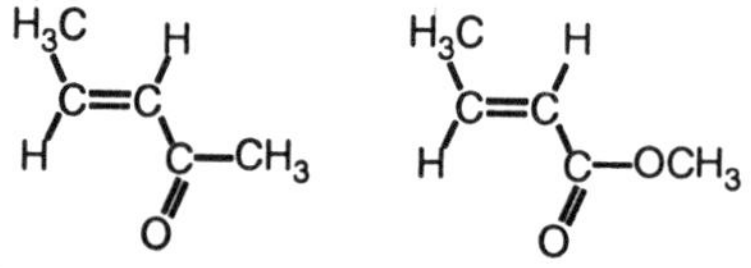

Fig. (a) α, β-unsaturated ketone; (b) α,β-unsaturated ester.

sp-Hybridisation

Definition: In *sp* hybridisation of carbon the *2s* orbital is mixed with one of the *2p* orbitals (e.g. $2p_x$) to form two *sp* hybrid orbitals or equal energy. This leaves two *2p* orbitals unaffected (i.e. unhybridised) ($2p_y$ and $2p_z$) with slightly higher energy than the hybridised orbitals.

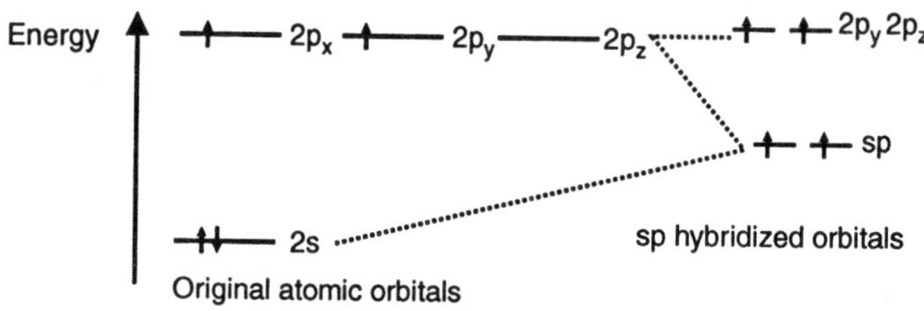

Fig. *sp*-Hybridisation of carbon.

Electronic Configuration

In a carbon atom the first two electrons fit into each *sp* orbital according to Hund's rule such that each orbital has a single unpaired electron. This leaves two electrons that can be paired up in the half-filled *sp* orbitals or placed in the vacant $2p_y$ and $2p_z$ orbitals. The energy difference between the orbitals is small and so it is easier for the electrons to fit into the higher energy orbitals than to pair up. This results in two half-filled *sp* orbitals and two half-filled *2p* orbitals, and so four bonds are possible.

Geometry

The *sp* hybridised orbitals are deformed dumb-bells with one lobe much larger than the other. The $2p_y$ and $2p_z$ orbitals are at right angles to each other. The *sp* hybridised orbitals occupy the space left over and are in the *x* axis pointing in opposite directions (only the major lobe of the *sp* orbitals are shown in black in following figure b).

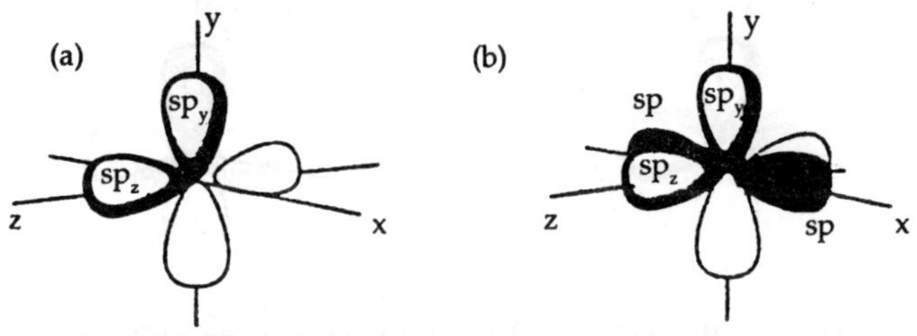

Fig. (a) $2p_y$ and $2p_z$ orbitals of an *sp* hybridised carbon; (b) $2p_y$, $2p_z$ and *sp* hybridised orbitals of an *sp* hybridised carbon.

A molecule making use of the two *sp* orbitals for bonding will be linear in shape. There are two common functional groups where such bonding occurs; alkynes and nitriles.

Alkynes

In ethyne each carbon is *sp* hybridised. The C–H bonds are strong a bonds where each hydrogen atom uses its half-filled 1*s* orbital to bond with a half-filled *sp* orbital on carbon. The remaining *sp* orbital on each carbon is used to form a strong σ carbon-carbon bond. The full σ bonding diagram for ethyne is linear and can be simplified as shown below.

H—C≡C—H

Fig. Ethyne.

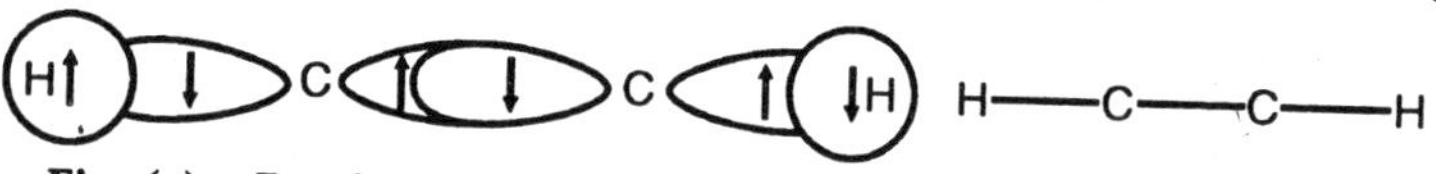

Fig. (a) s Bonding for ethyne; (b) representation of σ bonding.

Further bonding is possible because each carbon has half-filled *p* orbitals. Hence, the $2p_y$ and $2p_z$ orbitals of each atom can overlap side-on to form two π bonds.

The π bond formed by the overlap of the $2p_y$ orbitals is represented in dark gray. The π bond resulting from the overlap of the $2p_z$ orbitals is represented in light gray. Alkynes are linear molecules and are reactive due to the relatively weak π bonds.

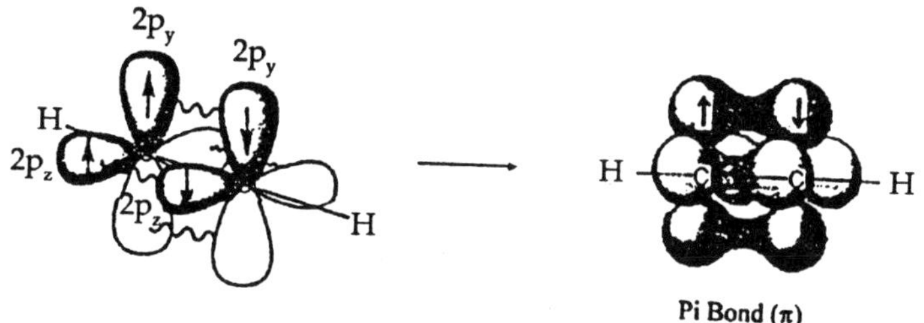

Fig. π-Bonding in ethyne.

Nitrile Groups

To explain the bonding within a nitrile group (C≡N) where both the carbon and the nitrogen are *sp* hybridised. The energy level diagram in figure given below shows how the valence electrons of nitrogen are arranged after *sp* hybridisation. A lone pair of electrons occupies one of the *sp* orbitals, but the other *sp* orbital can be used for a strong σ bond.

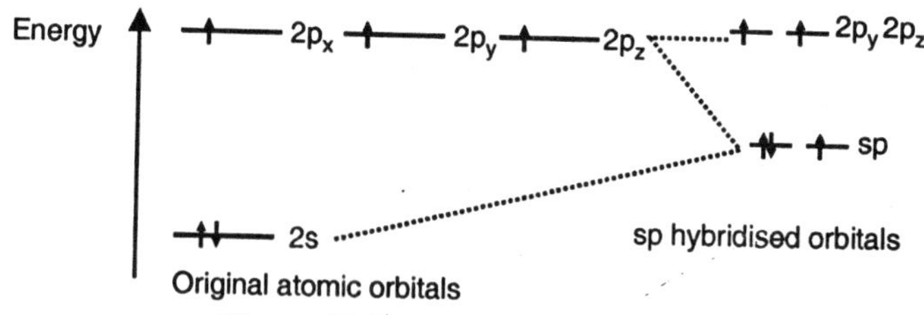

Fig. *sp* Hybridisation of nitrogen.

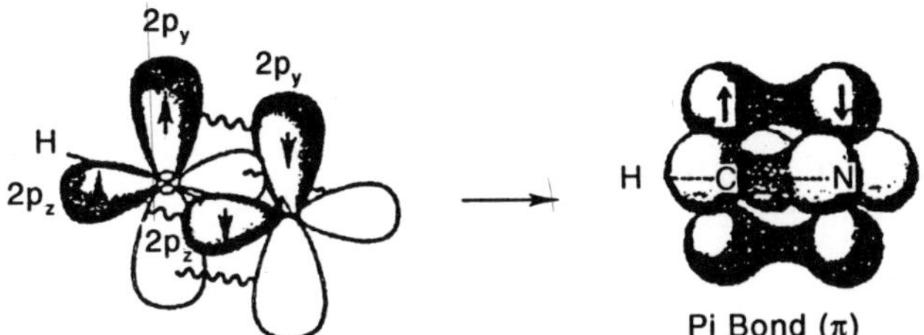

Fig. π-Bonding in HCN.

The $2p_y$ and $2p_z$ orbitals can be used for two σ bonds. Figure given below represents the σ bonds of HCN as lines and how the remaining $2p$ orbitals are used to form two π bonds.

Reactions of Organic Names

Reaction of Barton

The photolysis of nitrites which do not contain γ-hydrogen atoms usually results in the elimination of nitric oxide and the formation of hydroxy and carbonyl compounds. When γ-hydrogen atoms are present than the product is an oxime or the corresponding nitroso dimer. This reaction is known as *Barton S reaction.*

H, ONO, CHR, RCH, CH_2, H_2C $\xrightarrow{h\upsilon}$ NO, H, O, CHR, RCH, CH_2, CH_2 $\longrightarrow$ NO, OH, •CHR, RCH, CH_2, CH_2

NO, OH, CR, RCH, CH_2, H_2C $\longleftarrow$ NO, OH, CHR, RCH, CH_2, H_2C $\longrightarrow$ dimer

The other example of Barton reaction is:

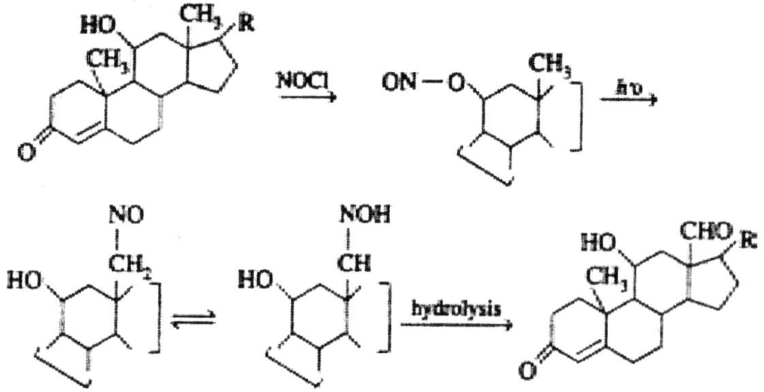

By Barton reaction a methyl group in the δ position to an OH group can be oxidised to a CHO group. The alcohol is first converted to the nitrite ester. Photolysis of the nitrite results in conversion of the nitrite group to the OH group and nitrosation of the methyl group. Hydrolysis of the oxime tautomer gives the aldehyde.

Hofmann-Loffler-Freytag Reaction: It is a photochemically initiated decomposition of N-haloamines in acidic solution. First δ-halo-amines are formed, these are then converted into pyrrolidines by intramolecular nucleophilic substitution, which involves a hydrogen abstraction.

The first step of the reaction is a rearrangement, with the halogen migrating from the nitrogen to the 4 or 5 position of the alkyl group and the second step, the ring closure takes place.

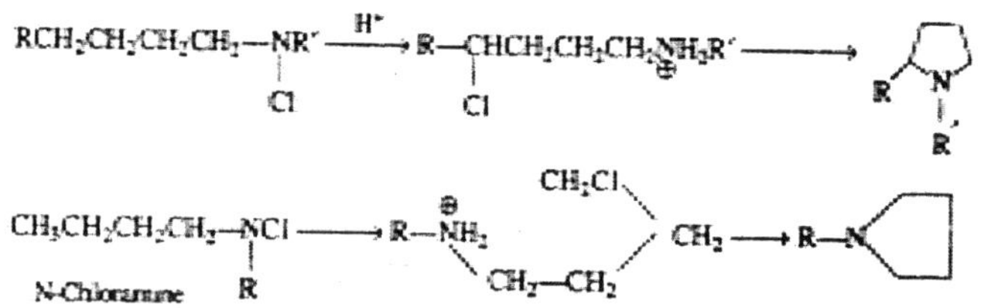

$$RCH_2CH_2CH_2CH_2{-}N(Cl)R' \xrightarrow{H^+} R{-}CH(Cl)CH_2CH_2CH_2\overset{\oplus}{N}H_2R' \longrightarrow \text{(pyrrolidine, R, N–R)}$$

$$CH_3CH_2CH_2CH_2{-}N(R)Cl \longrightarrow R{-}\overset{\oplus}{N}H_2 \; (CH_2Cl,\ CH_2{-}CH_2,\ CH_2) \longrightarrow R{-}N\text{(ring)}$$

N-Chloramine

A similar reaction has been carried out on N-haloamides, which give γ-lactone.

$$RCH_2CH_2CH_2CONHI \xrightarrow{h\nu} RCHCH_2CH_2 \; (O{-}C{=}O)$$

A similar reaction has been carried out on N-haloamides, which give γ-lactone.

$$C_6H_5-CH_2\underset{\underset{NNHTs}{\|}}{C}CH_3 \xrightarrow[\text{reaction}]{\text{Shapiro}} A$$

The Shapiro Reaction with its Mechanism :

Shapiro Reaction : Treatment of the tosylhydrazone of an aldehyde or a ketone with a strong base leads to the formation of an olefin, the reaction being formally an elimination accompanied by a hydrogen shift. This reaction is called *Shapiro reaction.*

$$-\underset{H}{\overset{|}{C}}-\underset{\underset{N}{\|}}{C}- \xrightarrow[2.\ H_2O]{1.\ 2RLi} -\overset{|}{C}=\underset{H}{\overset{|}{C}}-$$

The most useful method synthetically involves treatment of the substrate with at least two equivalent of an organolithium compound (MeLi) in ether, hexane, or tetramethylenediamine. Tosylhydrazones of α, β-unsaturated ketones give conjugated dienes.

The mechanism of the reaction has been formulated as

$$-\underset{H}{\overset{|}{C}}-\underset{\underset{N-NH-Ts}{\|}}{C} \xrightarrow{2RLi} -\overset{|}{\overset{-}{C}}-\underset{\underset{N^{-}-\overset{-}{N}-Ts}{|}}{C} \longrightarrow -\overset{|}{C}=\underset{\underset{N=N\ \ Li^{-}}{|}}{C} \longrightarrow -\overset{|}{C}=\underset{Li}{C}- \xrightarrow{H_2O} -\overset{|}{C}=\underset{H}{C}-$$

$$C_6H_5-CH_2\underset{\underset{NNHTs}{\|}}{C}CH_3 \xrightarrow[(ii)\ H_2O]{(i)\ 2RLi} C_6H_5-\overset{H}{\overset{|}{C}}=\underset{H}{\underset{|}{C}}-CH_3$$

Baeyer-Villiger Oxidation

Both aldehydes and ketones are oxidised by peroxy acids. This reaction, called the *Baeyer-Villiger oxidation, is* especially useful with ketones, because it converts them to carboxylic esters. For example, treating acetophenone with a peroxy acid converts it to the ester, phenyl acetate.

$$C_6H_5-\overset{\overset{O}{\|}}{C}-CH_3 \xrightarrow{RCOOH} C_6H_5-O-\overset{\overset{O}{\|}}{C}-CH_3$$

Acetophenone Phenyl acetate

The mechanism proposed for this reaction involves the following steps—

The carbonyl reactant removes a proton from an acid.

The peroxy acid attacks the protonated carbonyl reactant.

A proton is removed from the oxonium ion

The peroxy acid carbonyl group in this intermediate is protonated, preparing the RCO_2H portion to be a leaving group

The phenyl group migrates with an electron pair to the adjacent oxygen simultaneous with departure of RCO_2H as a leaving group

A proton is removed resulting in the ester product

The oxidation may be carried out by Caro's acid (per monosulphuric acid, H_2SO_5) or with perbenzoic, peracetic or monoperphthalic acid.

The products of this reaction show that a phenyl group has a greater tendency to migrate than a methyl group. Had this not been the case, the product would have been $C_6H_5COOCH_3$ and not $CH_3COOC_6H_5$.

This tendency of a group to migrate is called its *migratory aptitude*. Studies of the Baeyer-Villiger oxidation and other reactions have shown that the migratory aptitude of groups is H > phenyl > 3° alkyl > 2° alkyl > 1° alkyl > methyl. In all cases, this order is for groups migrating with their electron pairs, that is, as anions.

Completed Reactions:

(i) 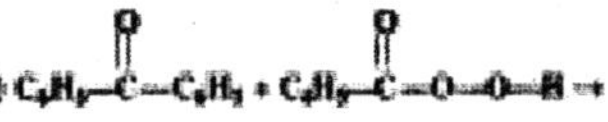 $C_6H_5-\overset{O}{\overset{\|}{C}}-C_6H_5 + C_6H_5-\overset{O}{\overset{\|}{C}}-O-O-H \rightarrow$

(ii) 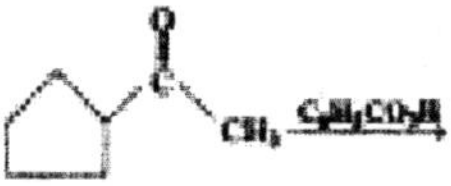 Cyclopentyl–C(=O)–CH_3 $\xrightarrow{C_6H_5CO_3H}$

(iii) 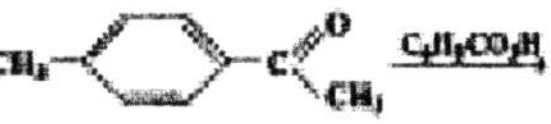 CH_3–C_6H_4–C(=O)CH_3 $\xrightarrow{C_6H_5CO_3H}$

(iv) 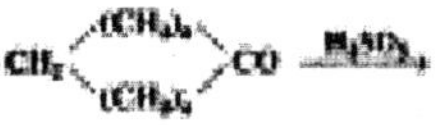 $CH_2\langle^{(CH_2)_6}_{(CH_2)_6}\rangle CO$ $\xrightarrow{H_2SO_4}$

(v) 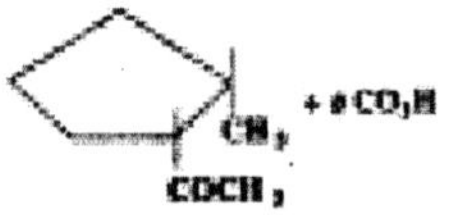 Cyclopentane (CH_3, $COCH_3$) $+ \phi CO_3H$

(vi)  Cyclohexanone $\xrightarrow{C_6H_5CO_3H}$

(vii) 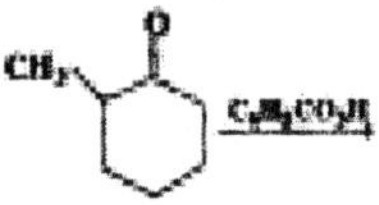 2-Methylcyclohexanone (CH_3) $\xrightarrow{C_6H_5CO_3H}$

(viii) C_6H_5–C(=O)–C_6H_4–OCH_3 $\xrightarrow{\text{Perbenzoic acid}}$

(ix) 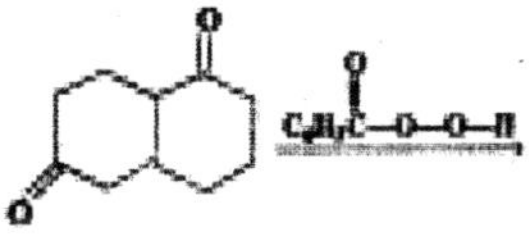 Bicyclic diketone $\xrightarrow{C_6H_5\overset{O}{\overset{\|}{C}}-O-O-H}$

(i) 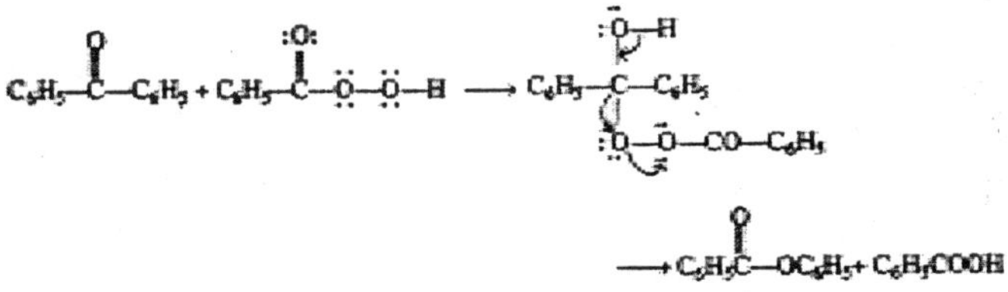 $C_6H_5-\overset{O}{\overset{\|}{C}}-C_6H_5 + C_6H_5-\overset{:O:}{\overset{\|}{C}}-\ddot{O}-\ddot{O}-H \longrightarrow C_6H_5-\overset{:\bar{O}-H}{C}-C_6H_5$ with $:\ddot{O}-\bar{O}-CO-C_6H_5$

$\longrightarrow C_6H_5\overset{O}{\overset{\|}{C}}-OC_6H_5 + C_6H_5COOH$

(ii)  Cyclopentyl–O–C(=O)–CH_3

(iii) 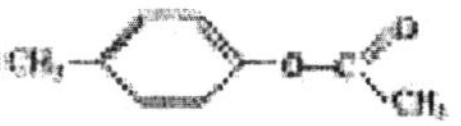 CH_3–C_6H_4–O–C(=O)CH_3

(iv)  $CH_2\langle^{(CH_2)_6-CO}_{(CH_2)_6-O}$ $\xleftarrow[\text{[O]}]{H_2SO_4}$ $CO\langle^{(CH_2)_6}_{(CH_2)_6}\rangle CH_2$

Lactone — Cyclic ketone

(v) 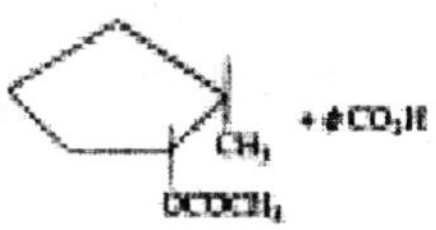 Cyclopentane (CH_3, $OCOCH_3$) $+ \phi CO_3H$

(vi) 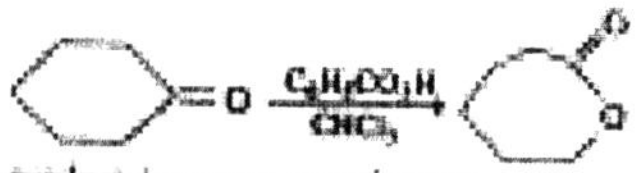 $\xrightarrow[CHCl_3]{C_6H_5CO_3H}$

The product is omega-caprolactone, a seven membered lactone that cannot be made by an intramolecular ring closure.

(vii) 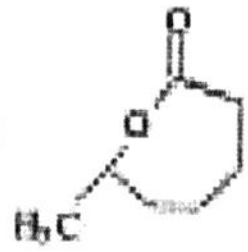 O, O, H_3C

(viii) C—O—OCH_3

(ix) 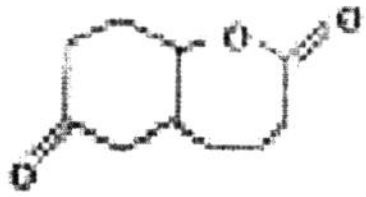 O, O, O

(b) (i) PhCOOMe + MeCOOPh

(ii) p-$O_2NC_6H_4C(=O)$—OPh

(iii) $PhC(=O)OC_6H_4OCH_3$-p

Reaction of Chichibabin

Pyridine and other heterocyclic nitrogen compounds an be aminated with the alkali-metal amides by chichibabin reaction. The attack is always in the 2 position unless both such positions are filled, in which case the 4 position is attacked. Substituted alkali-metal amides have also been used.

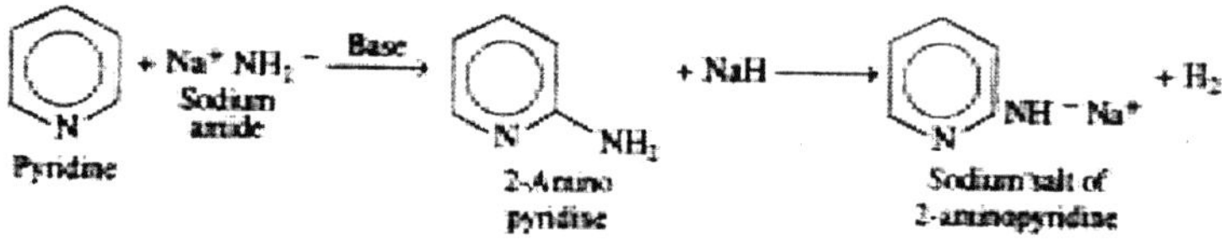

Pyridine + $Na^+ NH_2^-$ (Sodium amide) $\xrightarrow{Base}$ 2-Amino pyridine (NH_2) + NaH ⟶ Sodium salt of 2-aminopyridine ($NH^- Na^+$) + H_2

A similar reaction of this type as

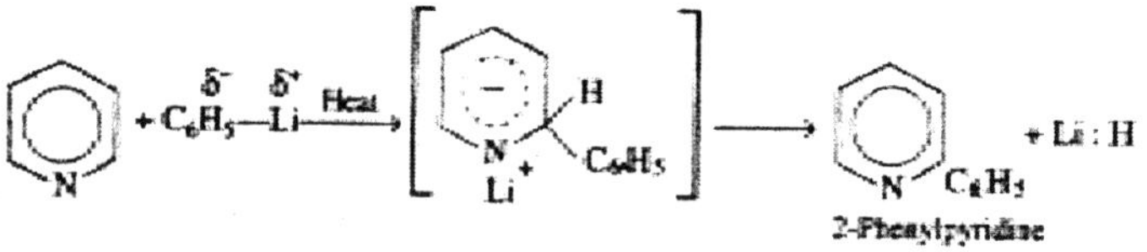

Pyridine + $\overset{\delta^-}{C_6H_5}$—$\overset{\delta^+}{Li}$ $\xrightarrow{Heat}$ [N^+, Li^+, H, C_6H_5] ⟶ 2-Phenylpyridine (C_6H_5) + Li : H

Completed Reactions:

(i) $\xrightarrow{CF_3C(=O)—O—O—H}$

(ii) CO—CH_3 $\xrightarrow{R—COOOH}$

H

Optically active

(iii) A + $H_2C(COOEt)_2$ $\xrightarrow{Base}$ O, H_5C_6, C_6H_5, EtOOC, COOEt

(iv) N $\xrightarrow{NaNH_2}$

(v) H_3C N CH_3 $\xrightarrow{NaNH_2}$

(vi) N + Br—CH_2—C(=O)OC_2H_5 $\xrightarrow{Heat}$ A $\xrightarrow{H_2O}$ B.

(vii) O + CH_2O + $(CH_3)_2$ NH ⟶

(viii) $C_6H_5COCH_3 + CH_2O +$ pyrrolidine $\longrightarrow$

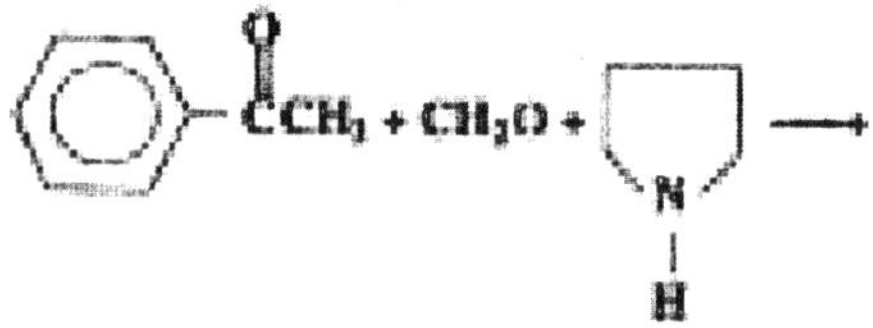

(ix) 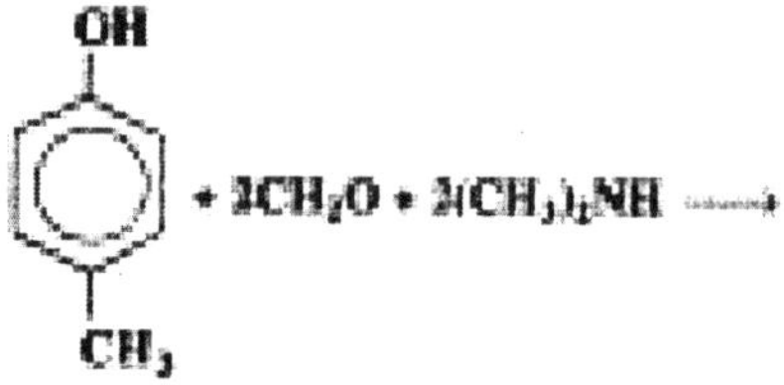 p-cresol $+ 2CH_2O + 2(CH_3)_2NH \longrightarrow$

(i) 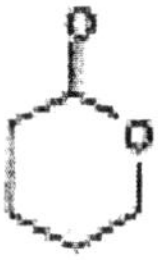 δ-valerolactone

(ii) 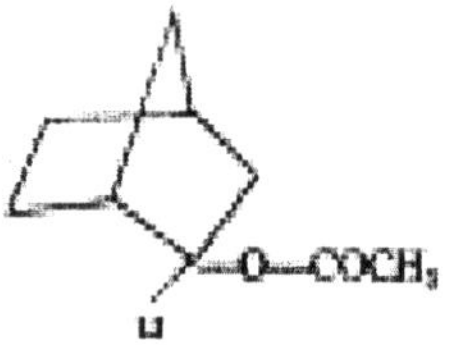 bicyclic acetate (—O—COCH$_3$, H)

Reaction (i) and (ii) are Baeyer-Villiger reaction.

(iii) 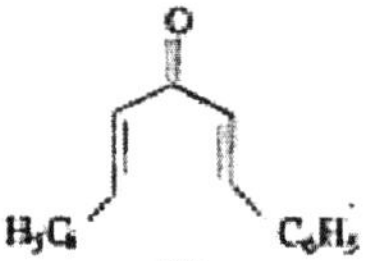 H_5C_6–CH=CH–CO–CH=CH–C_6H_5 (A)

(iv) 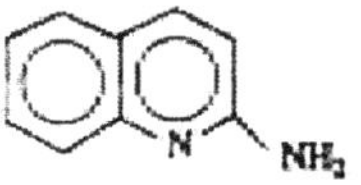 2-aminoquinoline (NH_2)

(v) 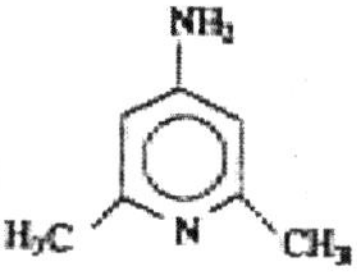 4-amino-2,6-dimethylpyridine (NH_2, H_3C, CH_3)

(vi) 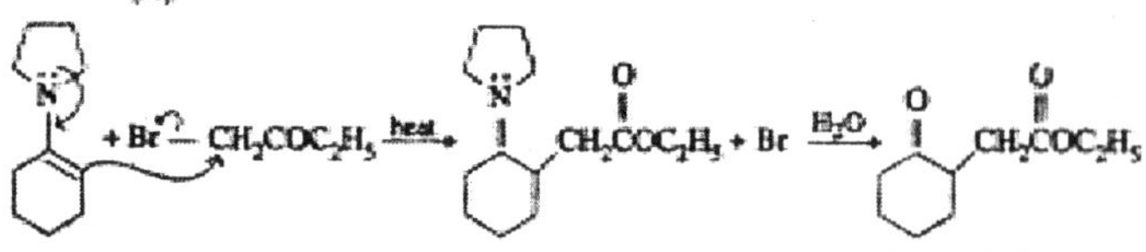 1-pyrrolidinocyclohexene $+ Br—CH_2COC_2H_5 \xrightarrow{heat}$ iminium salt with $CH_2COOC_2H_5$ $+ Br^- \xrightarrow{H_2O}$ 2-($CH_2COOC_2H_5$)cyclohexanone

A γ-keto ester (75%)

(Stork enamine reaction)

(vii) 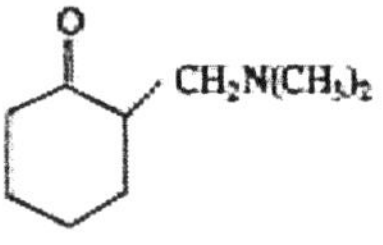 2-($CH_2N(CH_3)_2$)cyclohexanone

(viii) $C_6H_5CCH_2CH_2—N$ (pyrrolidine), with C=O

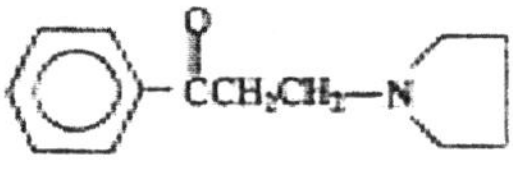

(ix) 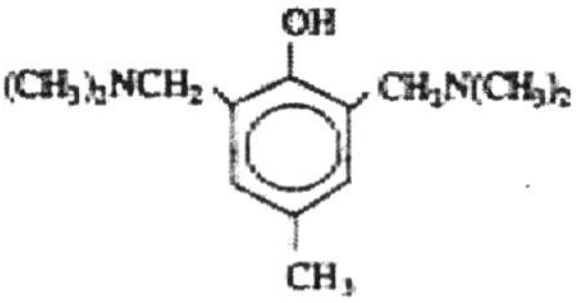 OH, $(CH_3)_2NCH_2$, $CH_2N(CH_3)_2$, CH_3 substituted benzene

(The reaction (vii)—(ix) are examples of Mannich reaction)

The product and name the reaction :

(i) 2-chlorocyclohexanone + OR^- ⟶

(ii) 1-(cyclohex-1-enyl)pyrrolidine + $CH_2=CH—CO_2Et$ → A $\xrightarrow{H_3O^+}$ B

(iii) HO, H, H, Me $\xrightarrow[\text{(R, R)DET} \\ \text{Bu}^t\text{O—OH}]{\text{Ti(OPr}^i)_4}$

(E)-hex-2-en-1-ol

(iv) Ph, OH, Me $\xrightarrow[\text{Ti (O—}i\text{ Pr)}_4]{t\text{-Bu—OOH}}$ A + B
Chiral diethyl tartrate

(v) CHO—CH_2—CH_2—CHO + $MeNH_2$ + $(CH_3)_2C=O$ ⟶

(vi) $C_4H_9—N(Cl)—(CH_2)_3—CH_3 \xrightarrow[\text{(ii) Base}]{\text{(i) } H_2SO_4}$

(i) cyclopentyl—COOR

Favorskii Rearrangement

(ii) (A) 1-(cyclohex-1-enyl)pyrrolidine bearing $CH_2CH_2CO_2Et$ $\xrightarrow{H_3O^+}$ (B) 2-oxocyclohexyl with CH_2CH_2COOEt, H

Stork Enamine reaction

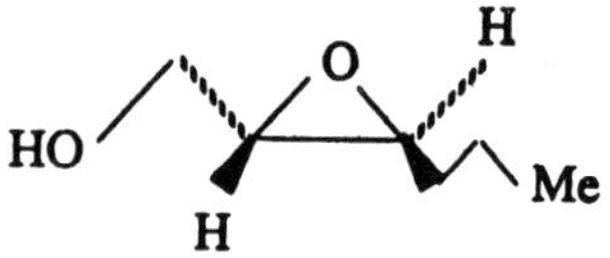

H

HO Me

H

(iii) Propyloniranmethanol

Sharpless asymmetric epoxidation

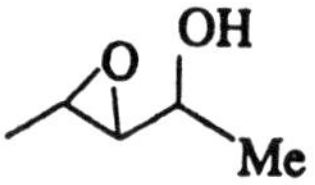

OH O Me OH Me O

(iv) A B

Sharpless asymmetric synthesis

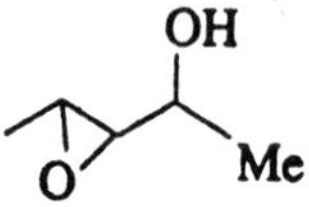

N—Me

O

(v)

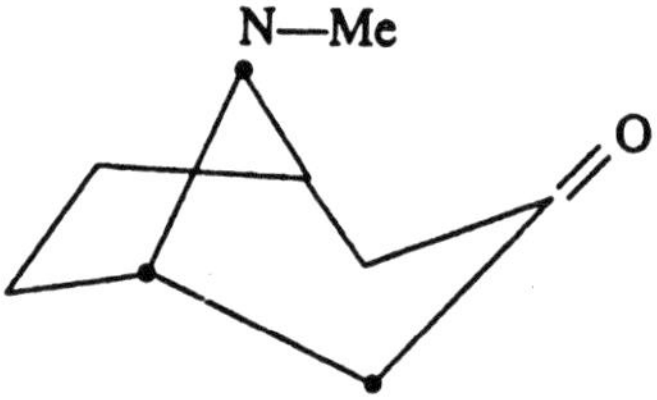

Tropinone

Robinson synthesis of tropinone (example of Mannich reaction in synthesis)

N

(vi) C_4H_{10}

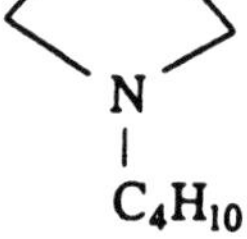

Hofmann-Loffler-Freytag Reaction

Completed sequence of reaction :

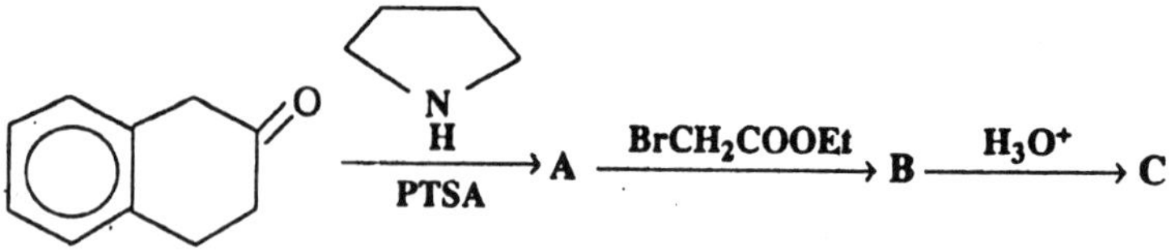

$$\xrightarrow[\text{PTSA}]{\text{N-H (pyrrolidine)}} A \xrightarrow{BrCH_2COOEt} B \xrightarrow{H_3O^+} C$$

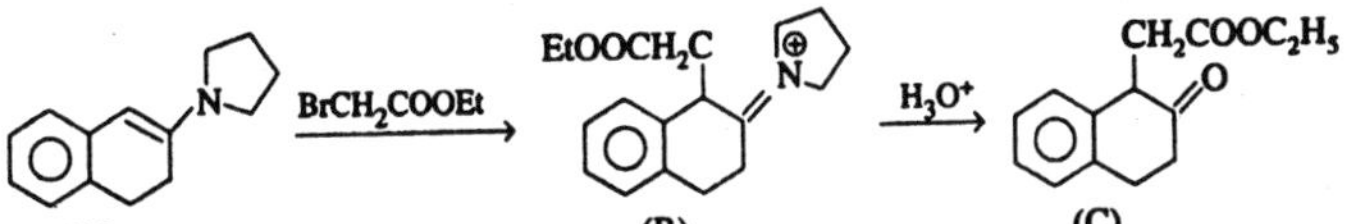

BrCH$_2$COOEt → (B) EtOOCH$_2$C ... $\xrightarrow{H_3O^+}$ (C) CH$_2$COOC$_2$H$_5$

(A) (B) (C)

(i) A reasonable mechanism for the following reaction :

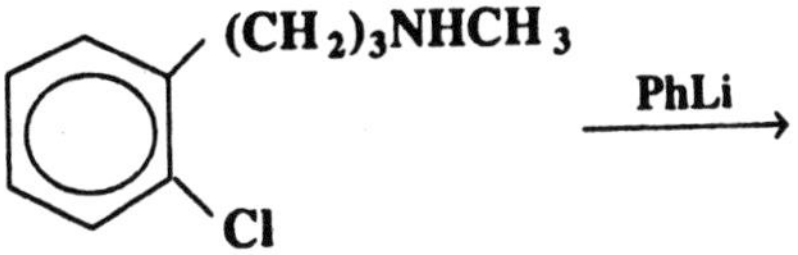

(CH$_2$)$_3$NHCH$_3$ / Cl $\xrightarrow{PhLi}$

(ii) The way one may link together, phenol, formaldehyde and a secondary amine.

(iii) The structure of the product

Indole + CH_2O + $(C_2H_5)_2NH$

(iv) Starting from (I and II), one can get III (4,4-dimethyl cyclohexanone)

Me_2CHCHO
I

$H_2C = CHCOCH_3$
II

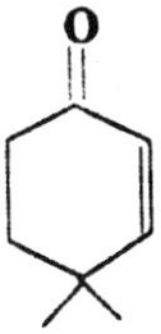

III

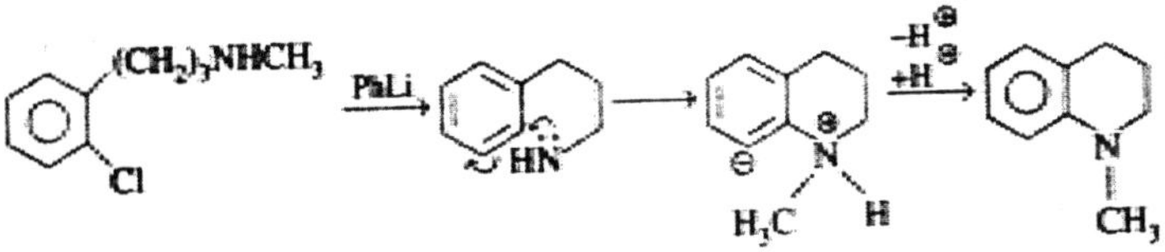

(CH$_2$)$_3$NHCH$_3$ / Cl $\xrightarrow{PhLi}$ ... HN → ... N, H$_3$C, H $\xrightarrow{-H^{\oplus}, +H^{\oplus}}$ ... N–CH$_3$

(ii) This is Mannich reaction, the active hydrogen compound being phenol (the hydrogens in the o- and p-positions are active) and the result is amino alkylation in the ortho position which is preferred.

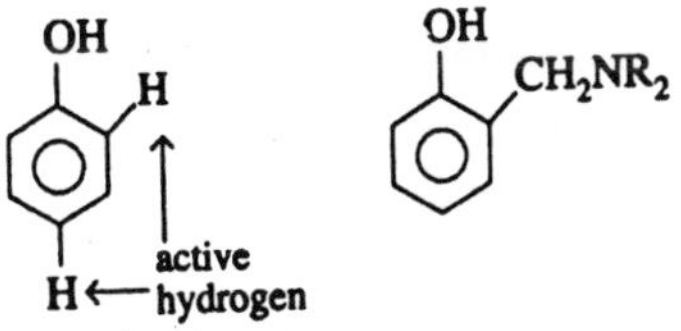

OH, H, H ← active hydrogen

OH, CH$_2$NR$_2$

(iii) Indole undergoes Mannich reaction with CH_2O and $(C_2H_5)_2NH$ to give 3-N, N′-diethylamino-ethylindole.

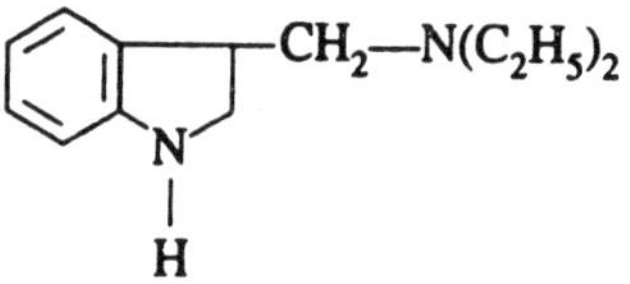

$-CH_2-N(C_2H_5)_2$

N

H

(iv) The first reaction is the Michael addition of the enolate anions to the methyl vinyl ketone followed by intramolecular aldol condensation.

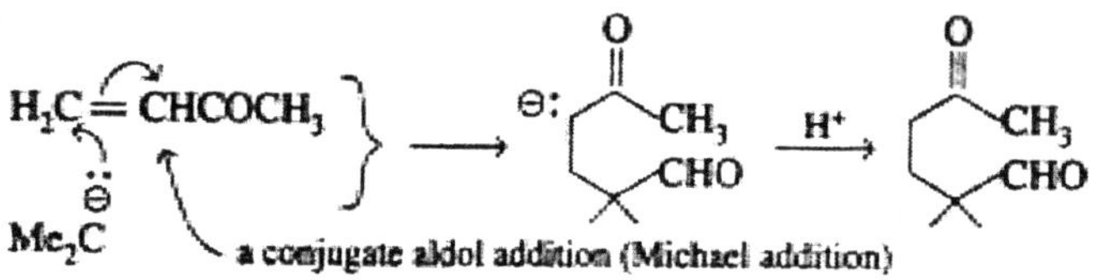

$H_2C=CHCOCH_3$ $Me_2C^{\ominus}$ → CH_3 CHO $\xrightarrow{H^+}$ CH_3 CHO

a conjugate aldol addition (Michael addition)

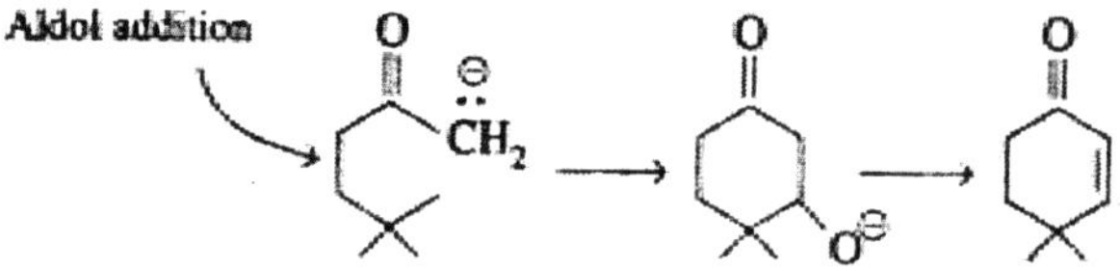

Aldol addition → $\ddot{C}H_2$ → →

The following canverted:

(a) NO–O–C–C–C–CH_3 → OH–C–C–C–CH_2–NO

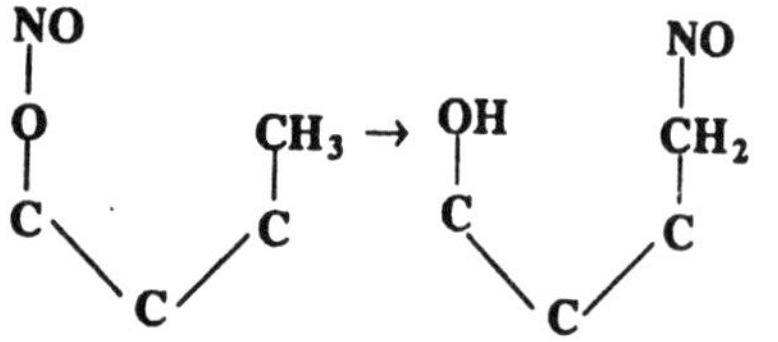

(b) $CH_3C(=NNHTs)(CH_2)_5CH_3 \longrightarrow CH_2=CH(CH_2)_5CH_3$

The structures of the major product in the following reactions:

(a) $\xrightarrow[CH_3COOH]{CH_3CO_3H}$ **(b)** $\xrightarrow{I_2/AgOAC}$

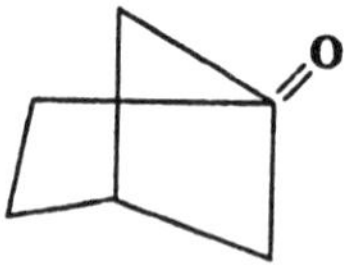

(c) Me, Me, Me, Me, Br Br $\xrightarrow{n\text{-BuLi}}$

(a) NO–O–C–C–C–CH_3 $\xrightarrow{h\nu}$ NO• + •O–C–C–C–CH_3 $\xrightarrow{\text{Abstraction}}$ OH–C–C–C–$\dot{C}H_2$ + NO• → OH–C–C–C–CH_2–NO

This is an example of Barton reaction.

(b) TsNNH=C(CH_3)$(CH_2)_5CH_3$ (*p*-toluenesulfonyl-hydrazone from 2-Octanone) ⟶ Li···$TsN^{\ominus}$–N=C(CH_2)$(CH_2)_5CH_3$ → $CH_2=CH(CH_2)_5CH_3$ (1-Octene)

This is an example of Shapiro reaction.

(a) + CH_3COOH (b) HO, OH (c) $H_3C-C(CH_3)_2-C(CH_3)=CBr_2$

The transformations carried out:

(a) ⟶ (b) COOEt ⟶ CH_2OH

(c) OCH_3, CH_3 $\xrightarrow{?}$ O, CH_3 (d) O $\xrightarrow{?}$ O

(e) CHO $\xrightarrow{?}$ CH_2CH_2COOEt

(a)

$\xrightarrow{\text{ozonolysis}}$

$\xrightarrow{\text{base}}$ (Carbanion)

$\xrightarrow{\text{Protonation}}$ OH $\xrightarrow[\text{(Elimination of water)}]{\text{Base}}$ or

(b)

COOEt $\xrightarrow[\text{dry HCl}]{HOCH_2CH_2OH}$ (Ketal) COOEt $\xrightarrow{LiAlH_4}$ $\xrightarrow[H_2O]{H^{\oplus}}$ CH_2OH

(c) Birch reduction

OCH_3, CH_3 $\xrightarrow[\text{EtOH}]{NaBH_4}$ OCH_3, CH_3 $\xrightarrow{\text{acid hydrolysis}}$ O, CH_3

(d) $\xrightarrow{H_2O_2/OH^{\ominus}}$

(e) C_6H_5–CHO $\xrightarrow[\text{(Perkin reaction)}]{(CH_3CO)_2O/CH_3COONa}$ C_6H_5–CH=CH–COOH

$\downarrow$ EtOH/H^+

C_6H_5–CH_2–CH_2–COOEt $\xleftarrow[\Delta]{H_2/Ni}$ C_6H_5–CH=CH–COOEt

Rearrangement of Favorskii

The reaction of a-haloketones (chloro, bromo or iodo) with alkoxide ions to give rearranged esters is called *Favorskii rearrangement*. For example, 2-chlorocyclohexanone is converted to the methyl ester of cyclopentane carboxylic acid by treatment with sodium methoxide in ether.

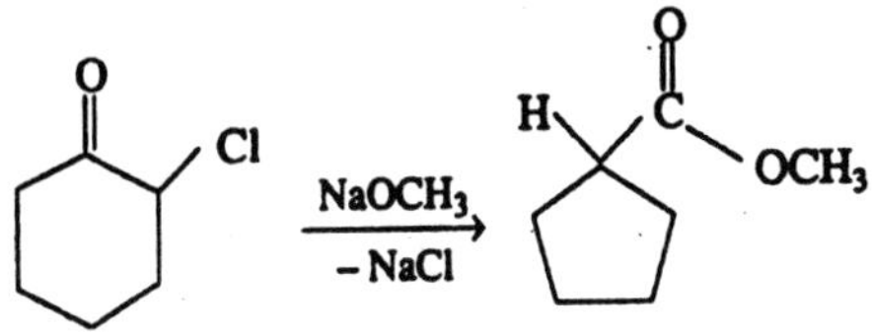

$\xrightarrow[-NaCl]{NaOCH_3}$

The Favorskii rearrangement is an example of a migration to an electron rich carbon atom. Its mechanism with 2-chlorocyclohexanone can be represented as follows—

(i) 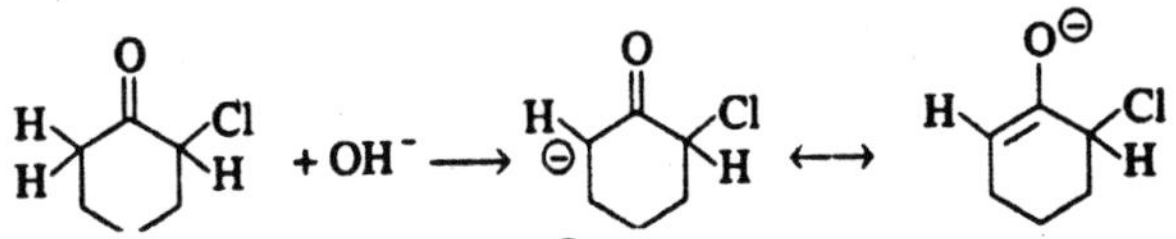 $+ OH^- \longrightarrow$ $\longleftrightarrow$

(ii) 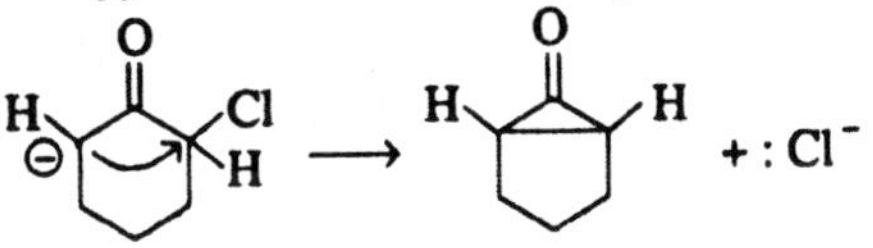 $\longrightarrow$ $+ : Cl^-$

(iii) 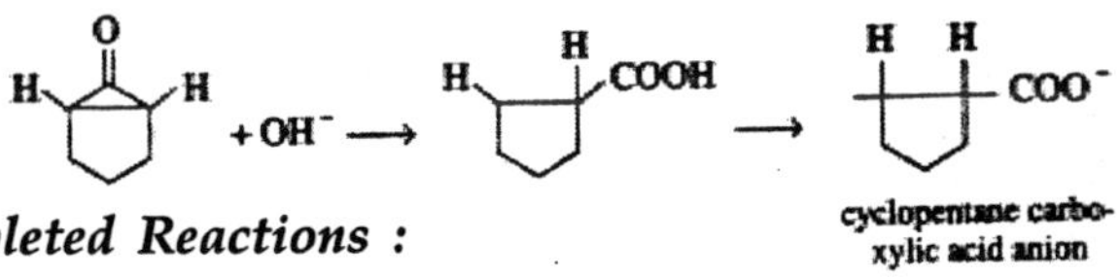 $+ OH^- \longrightarrow$ $\longrightarrow$

cyclopentane carboxylic acid anion

Completed Reactions :

(i) 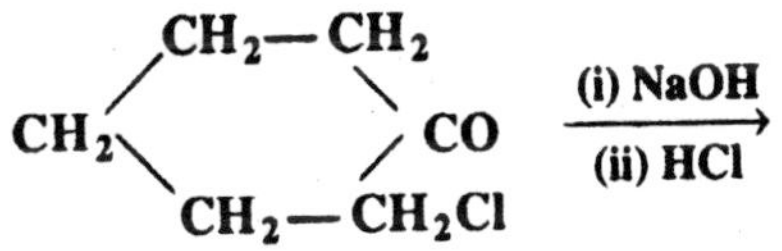 $CH_2(CH_2-CH_2)(CH_2-CH_2Cl)CO$ $\xrightarrow[(ii)\ HCl]{(i)\ NaOH}$

(ii) $\xrightarrow[(ii)\ Et_2O]{(i)\ NaOEt}$

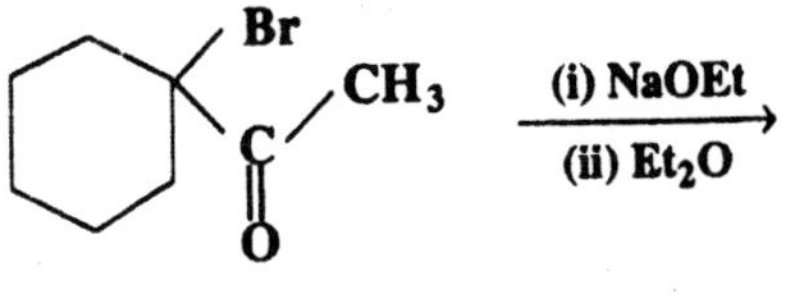

(i) 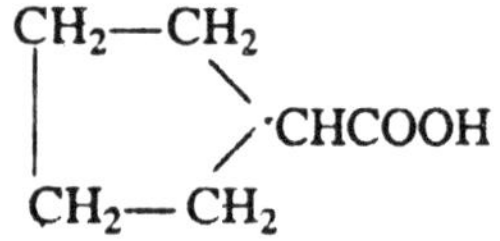 $CH_2—CH_2$ / $CH_2—CH_2$ >·CHCOOH

(ii) 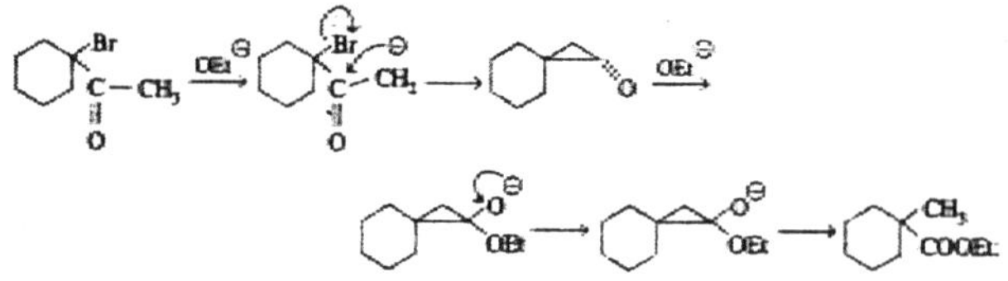 [reaction scheme: 2-bromo-2-acetylcyclohexane with OEt^- → cyclopropanone intermediate → with OEt^- → ring opening → cyclohexane bearing CH_3 and COOEt]

Explanation:

Both $PhCH_2COCH_2Cl$ and $PhCHClCOCH_3$ form $PhCH_2CH_2COOH$ when reacted with OH^-, followed by acidification.

A common intermediate is required in this Favorskii rearrangement to set the same product. Removal of an αH by OH^- is followed by an S_N2 displacement of Cl^- to give a cyclopropanone ring. Ring opening occurs to give the more stable benzylic carbanion.

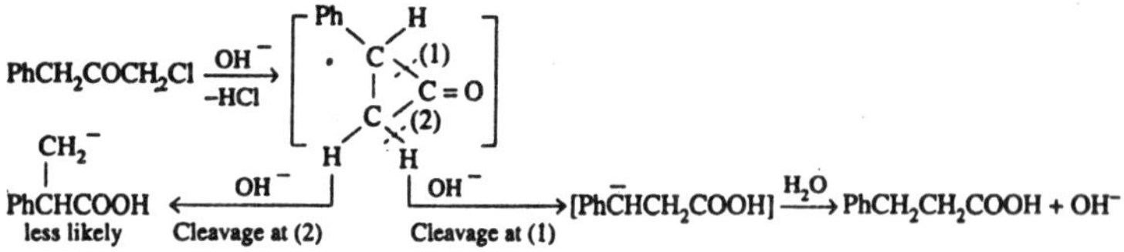

$$PhCH_2COCH_2Cl \xrightarrow[-HCl]{OH^-} \left[\text{Ph(H)C}^{(1)}\text{–C(=O)–}^{(2)}\text{CH}_2 \text{ cyclopropanone}\right]$$

$$\xrightarrow[\text{Cleavage at (1)}]{OH^-} [Ph\bar{C}HCH_2COOH] \xrightarrow{H_2O} PhCH_2CH_2COOH + OH^-$$

$$\xrightarrow[\text{Cleavage at (2)}]{OH^-} PhCH(CH_2^-)COOH \text{ (less likely)}$$

Favorskii Reaction: The reaction of aldehydes with terminal acetylenes is known as *Favorskii reaction.* Sodium acetylides are the most common reagents, but lithium, magnesium and other metallic acetylides have also been used. A particularly convenient reagent is lithium acetylide-ethylene diamine complex. Alternatively, the substrate may be treated with the alkyne itself in the presence of a base, so that the acetylide is generated in situ. 1,4-Diols can be prepared by treatment of aldehyde with dimetalloacetylenes.

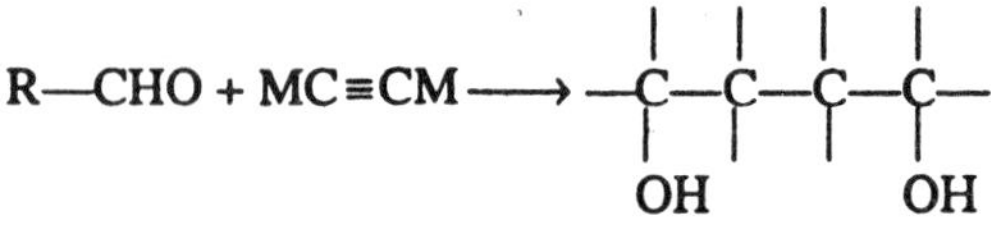

$$R—CHO + MC{\equiv}CM \longrightarrow —\underset{OH}{C}—C—C—\underset{OH}{C}—$$

The Stork Enamine Reaction with its Mechanism:

Stork Enamine Reaction : Aldehydes and ketones react with secondary amines to form compounds called *enamines.* The general reaction for enamine formation can be written as

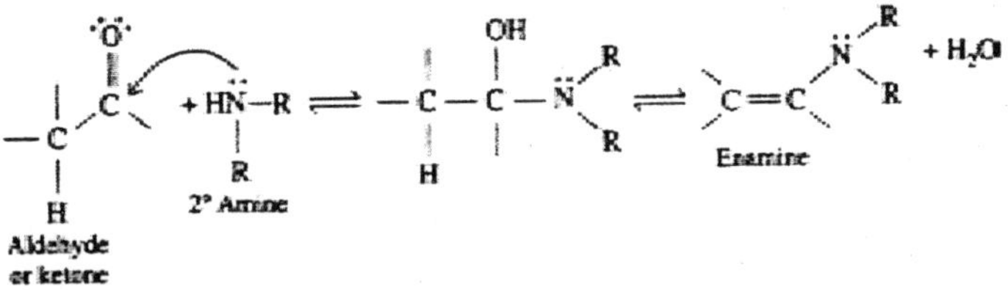

Aldehyde or ketone + 2° Amine ⇌ ⇌ Enamine + H_2O

Water is removed as an azeotrope or by a drying agent. Enamine formation is also catalysed by the presence of traces of an acid. The secondary amines commonly used are pyrrolidine, piperidine and morphine.

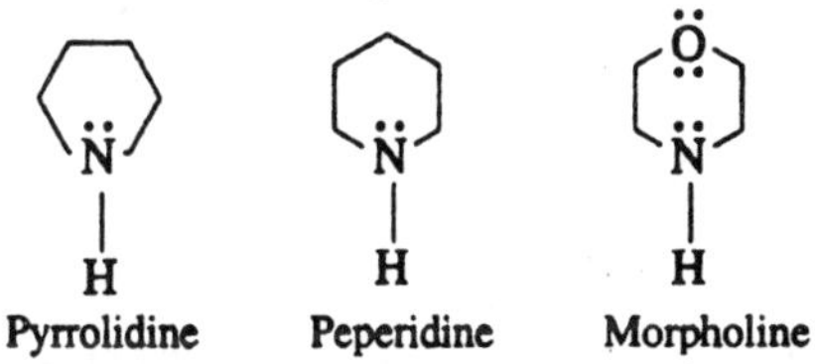

Pyrrolidine   Peperidine   Morpholine

Cyclohexanone, for example, reacts with pyrrolidine in the following way

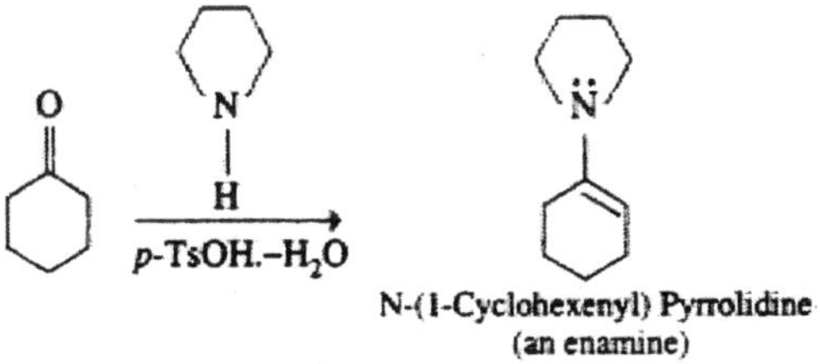

p-TsOH.–H_2O

N-(1-Cyclohexenyl) Pyrrolidine
(an enamine)

Enamines are good nucleophiles and have both a nucleophilic nitrogen and a nucleophilic carbon.

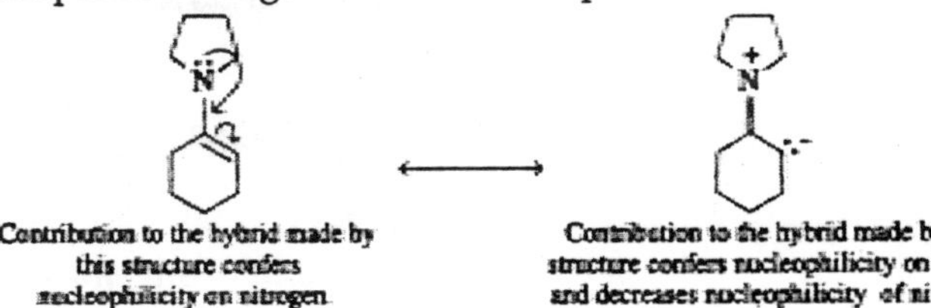

Contribution to the hybrid made by this structure confers nucleophilicity on nitrogen

Contribution to the hybrid made by this structure confers nucleophilicity on carbon and decreases nucleophilicity of nitrogen.

The nucleophilicity of the carbon of enamines makes them particularly useful reagents in organic synthesis because they can be acylated, alkylated, and used in Michael addition.

When an enarnine reacts with an acyl halide or acid anyhydrides the product is the C-acylated compound. The iminium ion that forms hydrolyses when water is added, and overall reaction provides a synthesis of β-diketones.

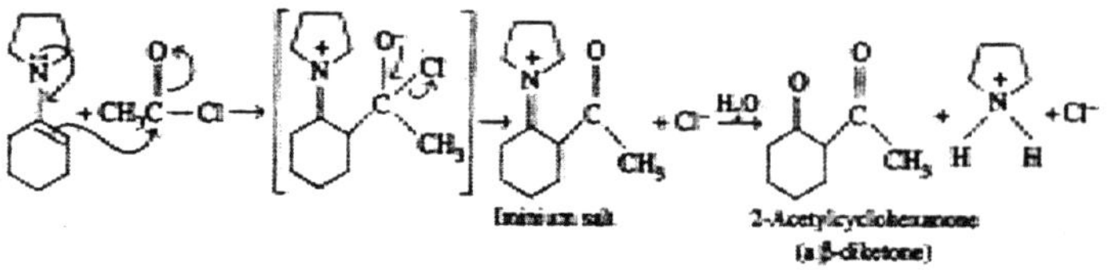

Iminium salt

2-Acetylcyclohexanone (a β-diketone)

Although N-acylatinn may occur in this synthesis, the N-acyl product is unstable and can act as an acylating agent itself.

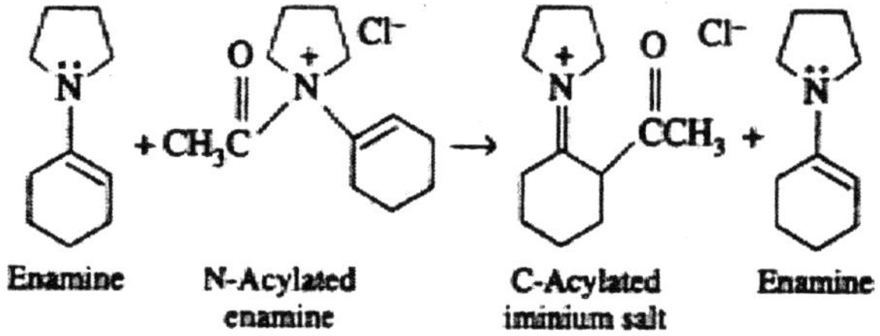

Enamine | N-Acylated enamine | C-Acylated iminium salt | Enamine

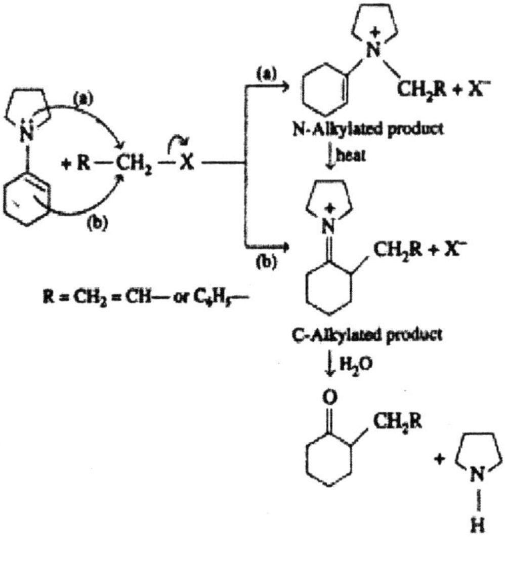

N-Alkylated product

heat

C-Alkylated product

H_2O

$R = CH_2{=}CH—$ or $C_6H_5—$

As a consequence, the yields of C-acylated products are generally high.

Alkylation of enamines may lead to the formation of N-alkylated product, which on heating is converted to C-alkyl compound (This rearrangement is common with allylic halide, alkyl halide or a-haloacetic ester.

Addition Reaction of Michael

When a new carbon-carbon bond is produced by nucleophilic addition to conjugated systems, the process is called *Michael addition*. The generalised process involves an α, β-unsaturated compound and a compound containing an active hydrogen attached to a carbon atom (*e.g.*, malonic ester, acetoacetic ester, nitrocompounds, aldehydes, ketones, etc.) These are condensed in the presence of a base.

The overall reaction and its mechanism can be represented as follows :

Mechanism:

Overall reaction

$$CH_3\underset{\underset{CH_3}{|}}{C}=CH\overset{\overset{O}{\|}}{C}OC_2H_5 + CH_2(\overset{\overset{O}{\|}}{C}OC_2H_5)_2 \xrightarrow[\substack{C_2H_5OH \\ 25°C}]{C_2H_5O^-Na^+} CH_3\underset{\underset{CH(CO_2C_2H_5)_2}{|}}{\overset{\overset{CH_3}{|}}{C}}—CH_2\overset{\overset{O}{\|}}{C}OC_2H_5$$

(70%)

Mechanism:

Step 1

$$C_2H_5\bar{O} + H—CH(\overset{\overset{O}{\|}}{C}OC_2H_5)_2 \rightleftharpoons C_2H_5OH + \bar{:}CH(\overset{\overset{O}{\|}}{C}OC_2H_5)_2$$

An alkoxide ion removes a proton from the anion of the active methylene compound

Step 2

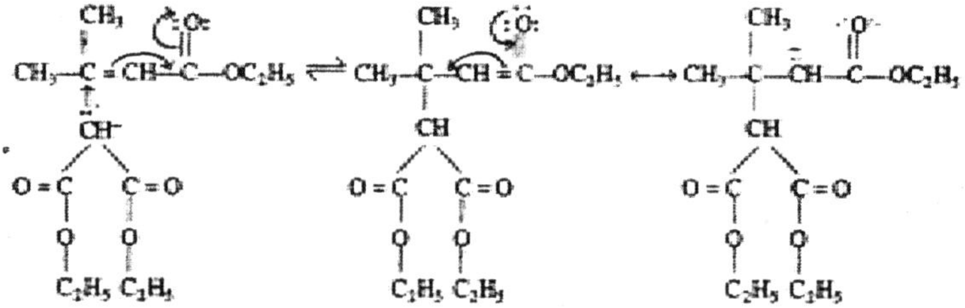

$$CH_3-\overset{\overset{CH_3}{|}}{\underset{\underset{\overset{\cdot\cdot}{C}H^-}{\uparrow}}{C}}=CH-\overset{\overset{O}{\|}}{C}-OC_2H_5 \rightleftharpoons CH_3-\overset{\overset{CH_3}{|}}{\underset{\underset{CH}{|}}{C}}-CH=C-OC_2H_5 \longleftrightarrow CH_3-\overset{\overset{CH_3}{|}}{\underset{\underset{CH}{|}}{C}}-\overset{-}{C}H-\overset{\overset{O}{\|}}{C}-OC_2H_5$$

(each CH bearing two $C(=O)OC_2H_5$ groups)

Conjugate addition of the anion to the α,β-unsaturated Ester leads to a new enolate anion

Step 2

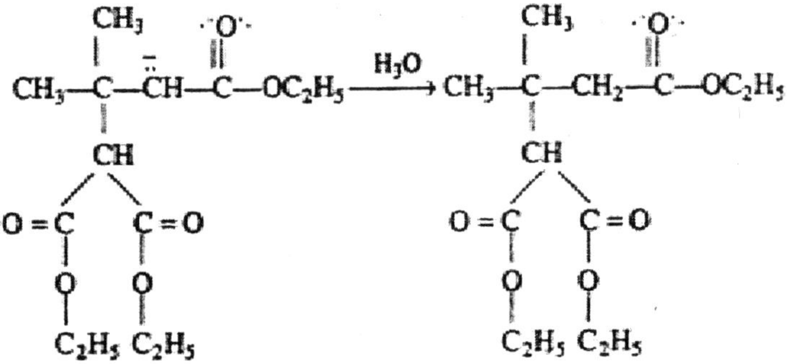

$$CH_3-\overset{\overset{CH_3}{|}}{\underset{\underset{CH(COOC_2H_5)_2}{|}}{C}}-\overset{\cdot\cdot}{\overset{-}{C}}H-\overset{\overset{O}{\|}}{C}-OC_2H_5 \xrightarrow{H_3O} CH_3-\overset{\overset{CH_3}{|}}{\underset{\underset{CH(COOC_2H_5)_2}{|}}{C}}-CH_2-\overset{\overset{O}{\|}}{C}-OC_2H_5$$

Other examples are as follows :

(i) 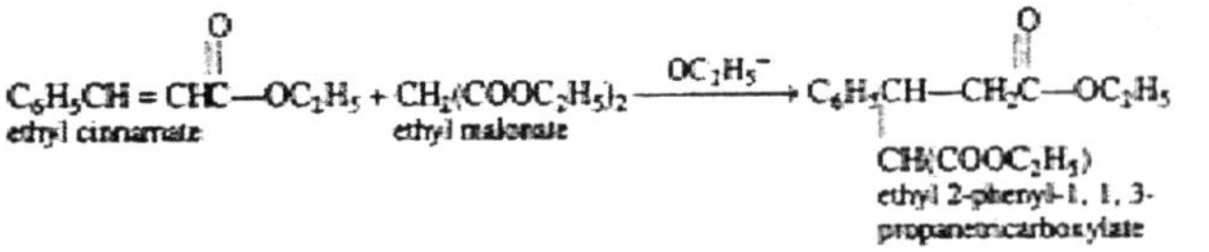 $C_6H_5CH=CH\overset{\overset{O}{\|}}{C}-OC_2H_5$ (ethyl cinnamate) $+ CH_2(COOC_2H_5)_2$ (ethyl malonate) $\xrightarrow{OC_2H_5^-} C_6H_5\underset{\underset{CH(COOC_2H_5)}{|}}{C}H-CH_2\overset{\overset{O}{\|}}{C}-OC_2H_5$

ethyl 2-phenyl-1, 1, 3-propanetricarboxylate

(ii) 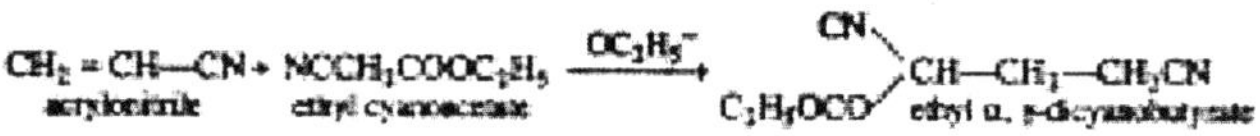 $CH_2=CH-CN$ (acrylonitrile) $+ NCCH_2COOC_2H_5$ (ethyl cyanoacetate) $\xrightarrow{OC_2H_5^-}$ $(CN)(C_2H_5OCO)CH-CH_2-CH_2CN$ ethyl α, γ-dicyanobutyrate

(iii) 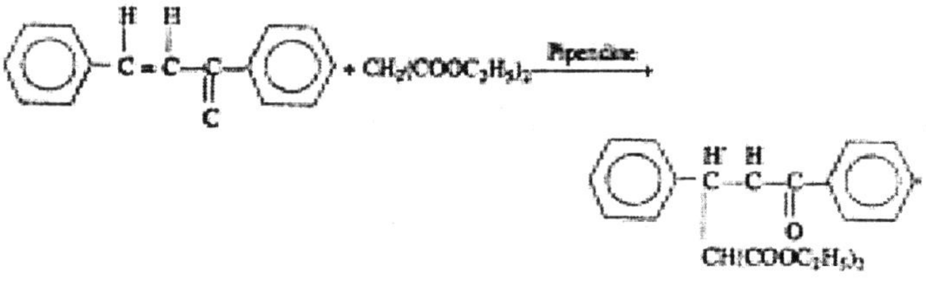 $C_6H_5-\overset{H}{C}=\overset{H}{C}-\underset{\underset{C}{\|}}{C}-C_6H_5 + CH_2(COOC_2H_5)_2 \xrightarrow{\text{Piperidine}}$

$$C_6H_5-\underset{\underset{CH(COOC_2H_5)_2}{|}}{\overset{H}{C}}-\overset{H}{C}-\underset{\underset{O}{\|}}{C}-C_6H_5$$

(iv) 2-Methylcyclohexanone (CH_3, H) $\xrightleftharpoons{KOH}$ enolate carbanion (CH_3, $\ominus$) $\xrightarrow{CH_2=CHCOC_6H_5}$ 2-methyl-2-($CH_2-\overset{\ominus}{C}HCOC_6H_5$)cyclohexanone $\xrightarrow[-C_2H_5O^-]{C_2H_5OH}$ 2-methyl-2-($CH_2CH_2COC_6H_5$)cyclohexanone

The products of the following Michael addition, predicted:

(i) Ethyl crotonate + Malonic ester $\longrightarrow \xrightarrow{OH^-} \xrightarrow{H^+} \xrightarrow{heat}$ **B.**

(ii) Ethyl acrylate + ethylacetoacetate $\longrightarrow$ **A** $\xrightarrow{H_2O, H^+}$ **B.**

(iii) Methyl vinyl ketone + malonic ester X.

(iv) Benzalacetophenone + acetophenone $\longrightarrow$ *Y.*

(v) Acrylonitrile + allyl cyanide $\longrightarrow$ **A** $\xrightarrow{H_2O, H^+}$ **B + $2NH_4^+$.**

(i) $CH_3CH=CHCOOEt + H_2C(COOEt)COOEt \longrightarrow CH_3CH(CH_2COOEt)-HC(COOEt)COOEt$ (A) $\longrightarrow CH_3CH(CH_2COOH)-CH_2-COOH$ (B)

(ii) $CH_2=CHCOOEt + H_2C(COCH_3)COOEt \longrightarrow H-C(CH_2CH_2COOEt)(COCH_3)COOEt$ (A) $\xrightarrow[\text{cleavage}]{\text{Ketonic}} CH_2CH_2COOH-CH_2COCH_3$ (B)

(iii) $CH_2=CHCOCH_3 + H_2C(COOEt)COOEt \longrightarrow H-C(CH_2CH_2COCH_3)(COOEt)COOEt$ (X)

(iv) $PhCH=CHCOPh + HCH_2COPh \longrightarrow PhCH(CH_2COPh)-CH_2COPh$ (Y)

(v) $CH_2=CHCN + H_2C(CN)CH=CH_2 \longrightarrow H-C(CH_2CH_2CN)(CN)CH=CH_2$ (A) $\xrightarrow[-H^+]{H_2O} H-C(CH_2CH_2COOH)(COOH)CH=CH_2$ (B)

The acrolein, $H_2C{=}CHCHO$ is epoxidised much more readily with a basic solution of H_2O_2 than with a peroxyacid:

(b) What is Robinson annelation reaction ?

(c) Give the steps for this transformation—

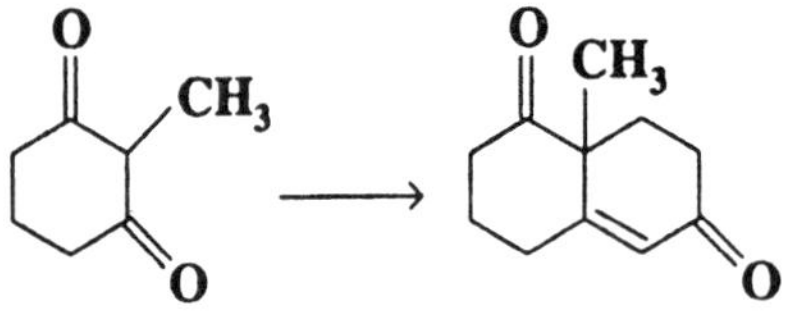

(a) Base converts H_2O_2 to conjugate base, HOO^-, that undergoes a Michael addition with acrolein to give an α-carbanion that then displaces HO^- from the HOO^- group leaving epoxide.

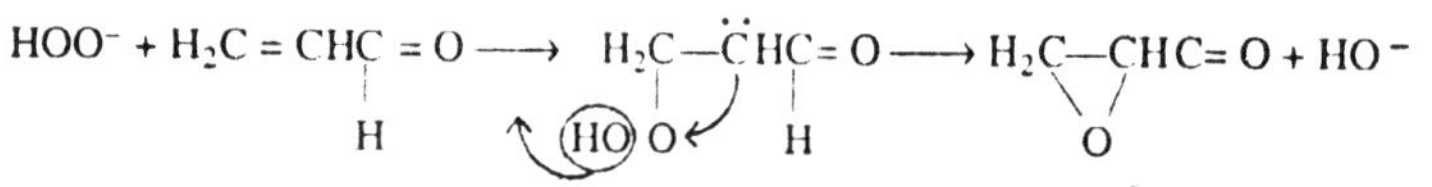

The acid catalysed epoxidation goes by the typical electrophilic attack by HO^+ (from H_2O_2) on C=C.

(b) Robinson annelation reaction involves a Michael addition to an α, β-unsaturated ketone immediately followed by an aldol addition. For example,

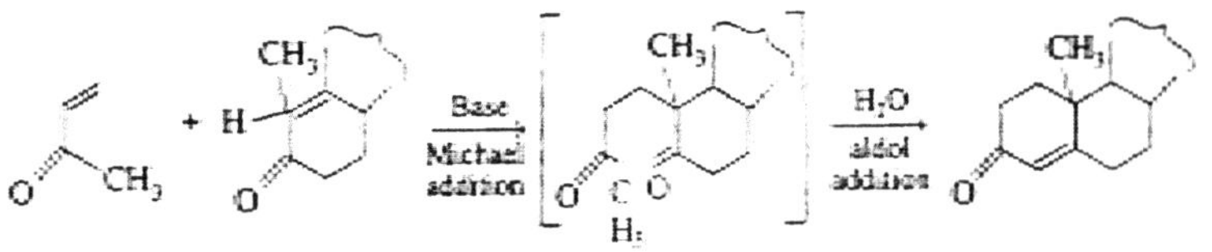

(c) The transformation can be achieved by Michael addition using methyl vinyl ketone (Robinson annelation).

The sequence that follows illustrates how a conjugate aldol addition (Michael addition) followed by a simple aldol condensation may be used to build one ring onto another. This procedure is known as the *Robinson annulation* (ring forming) reaction (after the English chemist Sir Robert Robinson, who won the Nobel Prize in Chemistry in 1947 for his research on naturally occurring compounds).

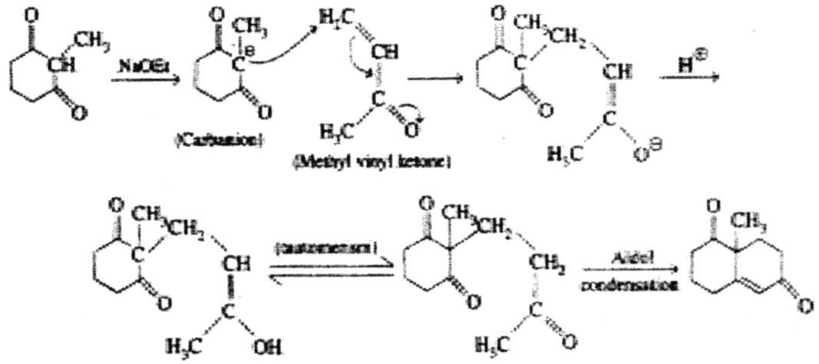

The mechanism of the following reaction :

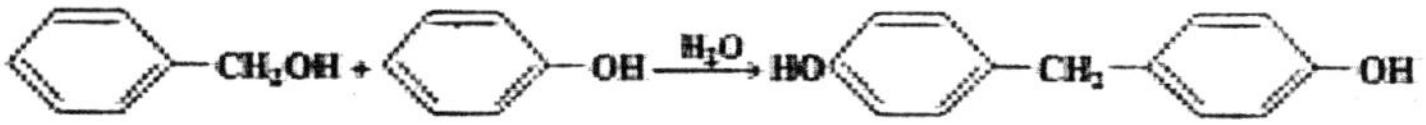

The reaction is a Michael type of addition to a base-induced dehydration product of the (4-hydroxy phenyl) methanol.

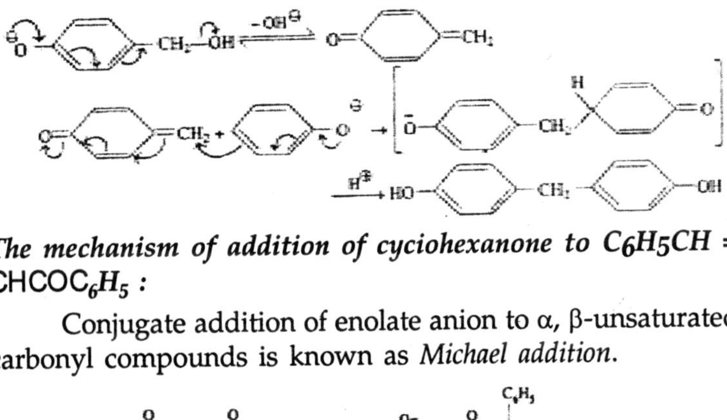

The mechanism of addition of cyciohexanone to C6H5CH = CHCOC6H5 :

Conjugate addition of enolate anion to α, β-unsaturated carbonyl compounds is known as *Michael addition.*

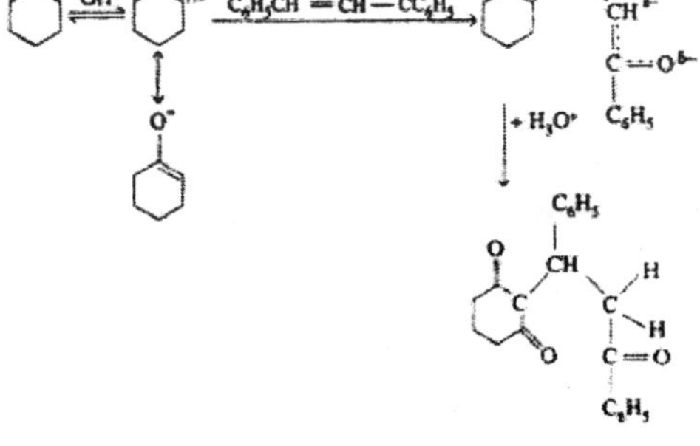

The Mannich Reaction with its Mechanism:

Mannich Reaction : This reaction takes place between an amine, an aldehyde (or ketone) and a highly nucleophilic carbon atom. The reaction is the addition of a nucleophilic carbon to an imonium ion intermediate.

(The reaction is usually carried out in acid solution, but may also be base catalysed. This is the condensation between aldehydes, ammonia or a primary or secondary amine and a compound containing at least one active hydrogen atom, *e.g.*, ketones, β-ketoesters, β-cyanoesters, nitroalkanes, alkynes with C≡H). For example,

$$C_6H_5-CO-CH_3 + CH_2O + HN(CH_3)_2 \xrightarrow{HCl} C_6H_5-CO-CH_2-CH_2-N(CH_3)_2 + H_2O$$

Mechanism : Step 1

$$R_2\ddot{N}H + \underset{H}{\overset{H}{\diagdown}}C=\ddot{O} \rightleftharpoons R_2\ddot{N}-\underset{H}{\overset{H}{\underset{|}{\overset{|}{C}}}}-\ddot{O}-H \overset{HA}{\rightleftharpoons} R_2\ddot{N}-\underset{H}{\overset{H}{\underset{|}{\overset{|}{C}}}}-\overset{H}{\overset{|}{\overset{+}{O}}}-H \overset{-H_2O}{\rightleftharpoons} R_2\overset{+}{N}=CH_2$$

Iminium cation

Reaction of the secondary amine with the aldehyde forms a hemiaminal.

The hemiaminal loses a molecule of water to form an iminium cation.

Step 2

$$CH_3-\overset{O}{\overset{\|}{C}}-CH_3 \overset{HA}{\rightleftharpoons} CH_3-\overset{O-H}{\overset{|}{C}}=CH_2 \longrightarrow CH_3-\overset{O}{\overset{\|}{C}}-CH_2-CH_2-\ddot{N}R_2 + HA$$

Enol   Mannich base

$$CH_2=\overset{+}{N}R_2$$

Iminium cation

The enol form of the active hydrogen compound reacts with the iminium cation to form a β-aminocarbonyl compound (a Mannich base)

The products of the Mannich reaction (Mannich bases) are useful intermediate in organic synthesis e.g.

(i) $PhCOCH_2CH_2NHR_2^+\,Cl^- \xrightarrow{Heat} PhCOCH=CH_2 \xrightarrow[cat]{H_2} PhCOCH_2CH_3$

(ii) $PhCOCH_2CH_2MR_2 \xrightarrow{CN^-} PhCOCH_2CH_2CN$

Completed reactions :

(i) $C_6H_5COCH_3 \xrightarrow[\text{(ii) NaOH}]{\text{(i) HCHO} + (CH_3)_2NH_2^+\,Cl^-}$

(ii) Indole + A + B ⟶ 3-(CH_2NMe_2)indole (N–H)

(iii) $NMe_2CHNO_2 + HCHO + EtNH_2 \xrightarrow{HCl}$

Give its mechanism also

(iv) 2,6-dimethylphenol (OH, H_3C, CH_3) $+ CH_2O +$ piperidine (N–H) $\xrightarrow{H^+}$

(v) $C_6H_5-C{\equiv}CH + HCHO + Me_2NH \xrightarrow{H^+}$

(vi) Phenol (OH) $+ HCHO + R_2NH \longrightarrow$

(i) $C_6H_5COCH_2CH_2\overset{H}{N}(CH_3)_2^+Cl^- \xrightarrow{NaOH} C_6H_5COCH_2CH_2-N(CH_3)_2$

(ii) $\underset{A}{CH_2O} + \underset{B}{Me_2NH}$

(iii) $EtNH_2 + CH_2O \underset{}{\overset{H^+}{\rightleftharpoons}} H_2O + Et\overset{+}{N}H{=}CH_2 \longleftrightarrow EtNH-\overset{+}{C}H_2$

$Me_2CH-\overset{+}{N}(=O)O^- \rightleftharpoons Me_2C{=}\overset{+}{N}(OH)O^-$

$Me_2C{=}\overset{+}{N}(O-H)O^- \xrightarrow{-H^+} Me_2C(NO_2)CH_2NHEt \xrightarrow{HCl}$

$\overset{+}{C}H_2-NHEt$

$Me_2C(NO_2)CH_2\overset{+}{N}H_2EtCl^-$

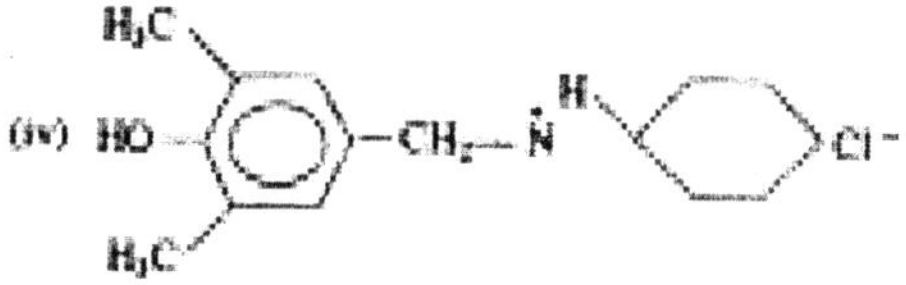

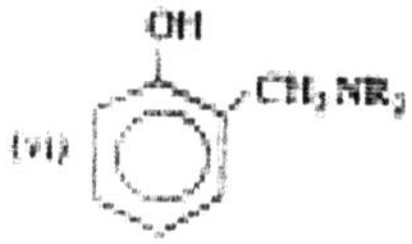

(iv) HO–$C_6H_2(CH_3)_2$–CH_2–NH–(cyclohexyl) Cl^-

(v) C_6H_5–C≡C–CH_2–NMe_2

(vi) OH, CH_2NR_2

The Sharpless Asymmetric Epoxidation with an Example:

Sharpless Asymmetric Epoxidation: This is a method of converting allylic alcohols to chiral epoxy alcohols with very high enantioselectivity (*i.e.*, with preference for one enantiomer rather than formation of racemic mixture). It involves treating the allylic alcohol with tert-butyl hydroperoxide, titanium(IV) tetra isopropoxide [Ti(O—*i*Pr)4] and a specific stereoisomer of tartaric ester. For example,

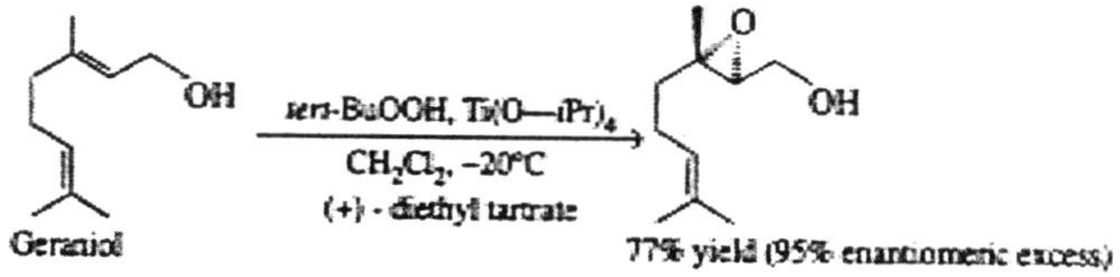

Geraniol —— tert-BuOOH, Ti(O—iPr)$_4$, CH_2Cl_2, –20°C, (+) - diethyl tartrate ——→ 77% yield (95% enantiomeric excess)

The oxygen that is transferred to the allylic alcohol to form epoxide is derived from tert-butyl hydroperoxide. The enantioselectivity of the reaction results from a titanium complex among the reagents that includes the enantiomerically pure tartrate ester as one of the ligands. The choice whether to use (+) or (–) tartrate ester for stereochemical control depends on which enantiomer of epoxide is desired.

The stereochemical preferences are such that it is possible to prepare either enantiomer of a chiral epoxide in high enantiomeric excess, simply by choosing the appropriate (+) or (–) tartrate stereoisomer as the chiral ligand.

OH

(S)-Methylglycidol

(+)-dialkyl tartrate

OH Sharpless asymmetric epoxidation

(–)-dialkyl tartrate

OH

(R)-Methylglycidol

The Ene Reaction with Example

How an Ene Reaction is Related to Diels-Alder Reaction?: Ene reaction (Hydro-allyl addition)-A reaction of an allylic compound with an olefin which resembles both a cycloaddition and a [1, 5]-sigmatropic shift of hydrogen. In this reaction an alkene having an allylic hydrogen atom reacts thermally with a dienophile (enophile) C = C, C = O, N = N, N = O, C = S *etc.* with the formation of a new σ bond to the terminal carbon of the allyl group. The interaction of a hydrogen atom with the HOMO of the allyl radical and the LUMO of the enophile is a symmetry allowed process.

H H

As related to Diels-Alder reaction this also represents a 6π electron electrocyclic reaction. Like Diels-Alder reaction the ene reaction is reversible. For example, the produce) 1-pentene of the ene reaction between ethene and propene, gives back ethene and propene on decomposition at 400° C.

$$\xrightarrow{400°C} H_2C = CH_2$$

H

CH_2

A reaction with structure between .malefic anhydride and propene under thermal conditions:

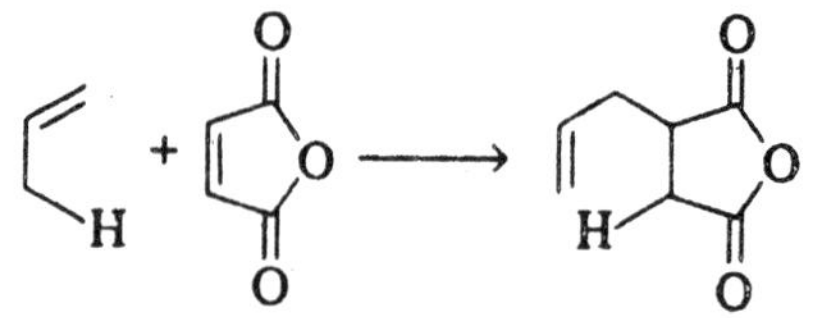

Propene Maleic anhydride

It is an ene reaction.

Different Elements

Thiols Chemistry

Thiols can be prepared by the action of alkyl halides with an excess of KOH and hydrogen sulphide. It is an S_N2 reaction and involves the generation of a hydrogen sulphide anion (HS^-) as nucleophile. In this reaction, there is the possibility of the product being ionised and reacting with a second molecule of alkyl halide to produce a thioether (RSR) as a by-product. An excess of hydrogen sulphide is normally used to avoid this problem.

a) $$R{-}X \xrightarrow{KOH\,/\,H_2S} \underset{\text{Thiol}}{R{-}SH}$$

b) $$R{-}X + S{=}C(NH_2)_2 \longrightarrow \underset{\text{S-Alkylisothiouronium salt}}{\left[R{-}\overset{\oplus}{S}{=}C(NH_2)_2 \right] X^{\ominus}} \xrightarrow[H_2O]{NaOH} \underset{\text{Thiol}}{R{-}SH} + S{=}C(NH_2)_2$$

c) $$R{-}S{-}S{-}R \xrightarrow{Zn\,/\,H^{\oplus}} \underset{\text{Thiol}}{R{-}SH}$$

Fig. Synthesis of thiols.

The formation of thioether can also be avoided by using an alternative procedure that involves thiourea. The thiourea acts as the nucleophile in an S_N2 reaction to produce an *S*-alkylisothiouronium salt that is then hydrolysed with aqueous base to give the thiol.

Thiols can also be obtained by reducing disulphides with zinc in the presence of acid.

Properties

Thiols form extremely weak hydrogen bonds—much weaker than alcohols — and so thiols have boiling points that are similar to comparable thioethers and which are lower than comparable alcohols, e.g. ethanethiol boils at 37°C whereas ethanol boils at 78°C.

Low molecular weight thiols are process disagreeable odours.

Reactivity

Thiols are the sulphur equivalent of alcohols (RSH). The sulphur atom is larger and more polarisable than oxygen which means that sulphur compounds as a whole are more powerful nucleophiles than the corresponding oxygen compounds. Thiolate ions (e.g. $CH_3CH_2S^-$) are stronger nucleophiles and weaker bases than corresponding alkoxides ($CH_3CH_2O^-$). Conversely, thiols are stronger acids than corresponding alcohols.

The relative size difference between sulphur and oxygen also shows that sulphur's bonding orbitals are more diffuse than oxygen's bonding orbitals. Due to this, there is a poorer bonding interaction between sulphur and hydrogen, than between oxygen and hydrogen. Because, the S–H bond of thiols is weaker than the O–H bond of alcohols (80 kcal mol^{-1} vs 100 kcal mol^{-1}). This means that the S–H bond of thiols is more prone to oxidation than the O–H bond of alcohols.

Reactions

Thiols can be easily oxidised by mild oxidising agents like bromine or iodine to give disulphides:

$$\underset{\text{Thiol}}{R{-}SH} \xrightarrow{Br_2 \text{ or } I_2} \underset{\text{Disulphide}}{R{-}S{-}S{-}R}$$

Fig. Oxidation of thiols.

Thiois react with base to form thilate ions which can act as powerful nucleophiles:

$$\underset{\text{Thiol}}{R{-}SH} \xrightarrow{\text{Base}} \underset{\text{Thiolate}}{R{-}S^{\ominus}} \xrightarrow{R'X} R{-}S{-}R'$$

Fig. Formation of thiolate ions.

Preparation of Ethers, Epoxides and Thioethers

Ethers

For the synthesis of ether, the Williamson ether synthesis is considered as the best method. It involves the S_N2 reaction between a metal alkoxide and a primary alkyl halide or tosylate. The alkoxide needed for the reaction is obtained by treating an alcohol with a strong base like sodium hydride. An alternative procedure is to treat the alcohol directly with the alkyl halide in the presence of silver oxide, thus avoiding the need to prepare the alkoxide beforehand.

a) $R{-}X \xrightarrow{NaoR''} \underset{\text{Ether}}{R{-}OR''}$ b) $R{-}X \xrightarrow[Ag_2O]{HOR''} \underset{\text{Ether}}{R{-}OR''}$

Fig. Synthesis of ethers.

For synthesis of an unsymmetrical ether, the most hindered alkoxide should be reacted with the simplest alkyl halide rather than the other way round. As this is an S_N2 reaction, primary alkyl halides react better then secondary or tertiary alkyl halides.

$$(CH_3)_3C{-}O^{\ominus} \longrightarrow H_3C{-}O{-}C(CH_3)_3 \not\longleftarrow H_3C{-}O^{\ominus} + (CH_3)_3C{-}O$$

Fig. Choice of synthetic routes to an unsymmetrical ether.

Alkenes can be converted to ethers by the electrophilic addition of mercuric trifluoroacetate, followed by addition of an alcohol. An organomercuric intermediate is obtained that can be reduced with sodium borohydride to yield the ether:

$$(H_3C)_2C{=}CH_2 \xrightarrow[\text{ROH/THF}]{Hg(O_2CCF_3)_2} RO{-}C(H_3C)_2{-}C(HgO_2CCF_3)H_2 \xrightarrow[HO^{\ominus}]{NaBH_4} RO{-}C(H_3C)_2{-}CH_3$$

Fig. Synthesis of an ether from an alkene and an alcohol.

Epoxides

Epoxides can be synthesised by the action of aldehydes or ketones with sulphur-ylides. They can also be prepared from alkenes by reaction with m-chloroperoxybenzoic acid.

$$R_2C{=}CR_2 \xrightarrow{X_2/H_2O} X{-}CR_2{-}CR_2{-}OH \xrightarrow{\text{NaoH}} \text{epoxide } R_2C{-}O{-}CR_2$$

Fig. Synthesis of an epoxide via a halohydrin.

$$R_2C^{\delta+}(X^{\delta-}){-}CR_2{-}OH + {}^{\ominus}OH \longrightarrow \left[R_2C(X){-}CR_2{-}O^{\ominus}\right] \xrightarrow{S_N2} \text{epoxide } R_2C{-}O{-}CR_2 + X^{\ominus}$$

Fig. Mechanism of epoxide formation from a halohydrin.

They can also be obtained from alkenes in a two-step process. The first step involves electrophilic addition of a halogen in aqueous solution to form a halohydrin. Treatment of the halohydrin with base then ionises the alcohol group, that can then act as a nucleophile. The oxygen uses a lone pair of electrons to form a bond to the neighbouring electrophilic carbon, thus displacing the halogen by an intramolecular S_N2 reaction.

Thioethers

Thioethers (or sulphides) can be prepared by the S_N2 reaction of primary or secondary alkyl halides with a thiolate anion (RS^-). The reaction is similar to the Williamson ether synthesis.

$$R\text{–}X \xrightarrow{NaSR''} R-SR''$$

Fig. Synthesis of a disulphide from an alkyl halide.

Symmetrical thioethers can be prepared by treating an alkyl halide with KOH and an equivalent of hydrogen sulphide. The reaction produces a thiol which is ionised again by KOH and reacts with another molecule of alkyl halide.

Ether, Epoxides and Thioethers: Properties

Ethers

Ethers are made up an oxygen linked to two carbon atoms by σ bonds. In aliphatic ethers (ROR), the three atoms involved are sp^3 hybridised and have a bond angle of 112°. In Aryl ethers the oxygen is linked to one or two aromatic rings (ArOR or ArOAr) and in such a case the attached carbon(s) is sp^1 hybridised.

The C—O bonds are polarised in such a way that the oxygen is slightly negative and the carbons are slightly positive. Because of the slightly polar C—O bonds, ethers have a small dipole moment. However, ethers have no X—H groups (X=heteroatom) and cannot interact by hydrogen bonding. Therefore, they have lower boiling points than comparable alcohols and similar boiling points to comparable alkanes. However, hydrogen bonding is possible to protic solvents and their solubilities are similar to alcohols of comparable molecular weight.

The oxygen of an ether is a nucleophilic centre and the neighbouring carbons are *electrophilic centres,* but in both cases the nucleophilicity or electrophilicity is weak. Therefore, ethers are relatively unreactive.

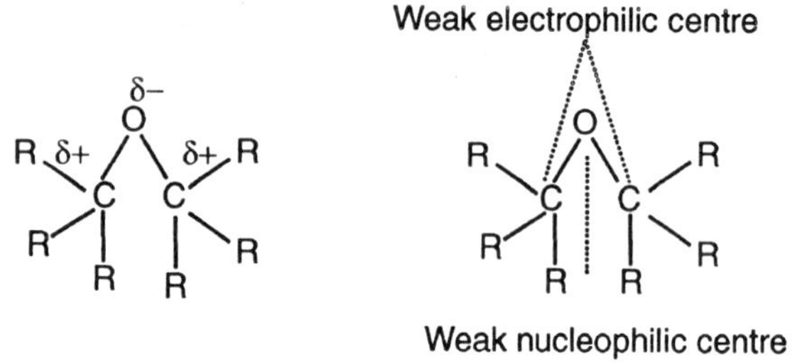

Weak electrophilic centre

Weak nucleophilic centre

Fig. Properties of ethers.

Epoxides

Epoxides (or oxiranes) are three-membered cyclic ethers and differ from other cyclic and acyclic ethers in that they are reactive with different reagents. The reason for this difference in reactivity is the strained three-membered ring.

Reactions with nucleophiles can result in ring opening and relief of strain. Nucleophiles will attack either of the electrophilic carbons present in an epoxide by an S_N2 reaction:

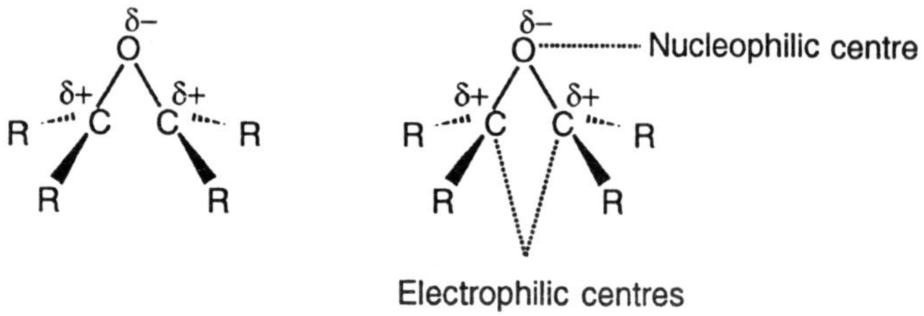

Nucleophilic centre

Electrophilic centres

Fig. Properties of an epoxide.

Thioethers

Thioethers (or sulphides; RSR) are the sulphur equivalents of ethers (ROR). Because the sulphur atoms are polarisable, they can stabilise a negative charge on an adjacent carbon atom. Thus, hydrogens on this carbon are more acidic than those on comparable ethers.

Study of Amines and Nitriles

Preparation of Amines

Reduction: Nitriles and amides can be easily reduced to alkylamines using lithium aluminium hydride ($LiAlH_4$). In the case of a nitrile, a primary amine is the only possible product. Primary, secondary, and tertiary amines can be prepared from primary, secondary and tertiary amides, respectively.

Substitution with NH_2

Primary alkyl halides and some secondary alkyl halides can undergo S_N2 nucleophilic substitution with an azide ion (N_3^-) to yield an alkyl azide. The azide can then be reduced with $LiAlH_4$ to give a primary amine:

$$CH_3CH_2{-}Br \xrightarrow[\text{Ethanol}]{NaN_3} \underset{\text{Alkyl azide}}{CH_3CH_2{-}N{=}\overset{\oplus}{N}{=}\overset{\ominus}{N}} \xrightarrow[\text{b) } H_2O]{\text{a) } LiAlH_4} CH_3CH_2{-}NH_2$$

Fig. Synthesis of a primary amine from an alkyl halide via an alkyl azide.

The overall reaction involves replacing the halogen atom of the alkyl halide with an NH, unit. Another method is the Gabriel synthesis of amines. This involves treating phthalimide with KOH to abstract the N–H proton. The N–H proton of phthalimide is more acidic (pK_a9) than the N–H proton of an amide since the anion formed can be stabilised by resonance with both neighbouring carbonyl groups. The phthalimide ion can then be alkylated by treating it with an alkyl halide in nucleophilic substitution.

Phthalimide $\xrightarrow[\text{Ethanol}]{KOH}$ [resonance structures of phthalimide anion]

Fig. Ionisation of phthalimide.

Subsequent hydrolysis releases a primary amine. Still other possible method is to react an alkyl halide with ammonia, but this is less satisfactory because overalkylation is possible. The reaction of an aldehyde with ammonia by reductive amination is another method of obtaining primary amines.

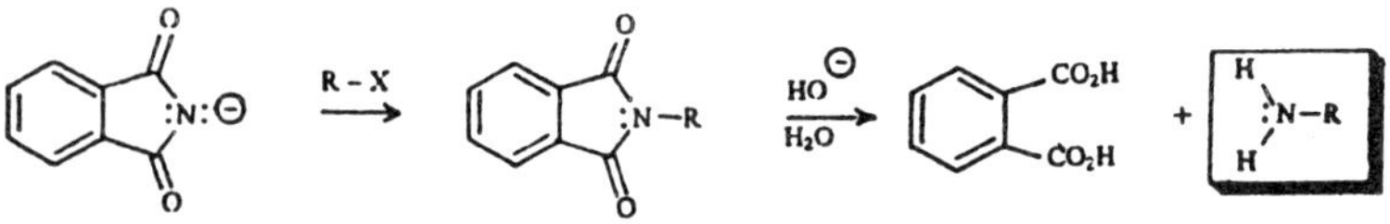

Fig. Gabriel synthesis of primary amines.

Alkylation of Alkylamines

We can convert primary and secondary amines to secondary and tertiary amines, respectively, by alkylation with alkyl halides by the S_n2 reaction. However, overalkylation may occur and so better methods of amine synthesis which are available are used.

Reductive Amination: It is a more controlled method of adding an extra alkyl group to an alkylamine. Primary and secondary alkylamines can be treated with a ketone or an aldehyde in the presence of a reducing agent known as *sodium cyanoborohydride*. The alkylamine reacts with the carbonyl compound by nucleophilic addition followed by elimination to give an imine or an iminium ion which is immediately reduced by sodium cyanoborohydride to yield the final amine. This is the equivalent of adding one extra alkyl group to the amine.

Therefore, primary amines get converted to secondary amines and secondary amines are converted to tertiary amine. The reaction is suitable for the synthesis of primary amines if ammonia is used instead of an alkylamine. The reaction goes through an imine intermediate if ammonia or a primary amine is used. When a secondary amine is used, an iminium ion intermediate is involved.

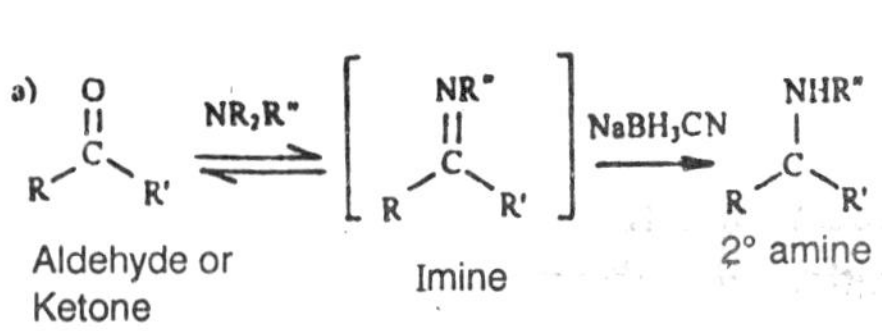

b) Aldehyde or Ketone $\xrightarrow{NHR''_2}$ [Iminium ion] $\xrightarrow{NaBH_3CN}$ 3° amine

Another method of alkylating an amine is to acylate the amine to yield an amide and then carry out a reduction with $LiAlH_4$. Although two steps are involved, there is no risk of overalkylation since acylation can only occur once.

$R{-}NH_2 \xrightarrow[\text{Pyridine}]{CH_3COCl} R{-}NH{-}C(=O){-}CH_3 \xrightarrow[\text{b) } H_2O]{\text{a) } LiAlH_4} R{-}NH{-}CH_2{-}CH_3$

Fig. Alkylation of an amide via an amine.

Rearrangements

The following two rearrangement reactions can be used to convert carboxylic acid derivatives into primary amines in which the carbon chain in the product has been shortened by one carbon unit. These are called the *Hofmann and the Curtius rearrangements*. The Hofmann rearrangement involves the treatment of a primary amide with bromine under basic conditions, while the Curtius rearrangement involves heating an acyl azide. In both cases we get a primary amine with loss of the original carbonyl group.

Primary amide $\xrightarrow{4\ NaOH,\ Br_2,\ H_2O}$ $R{-}NH_2$ (+ Na_2CO_3, 2 NaBr, 2 H_2O) $\xleftarrow{H_2O,\ \text{heat}}$ Acyl azide (+ CO_2 (g), N_2 (g))

Fig. Hofmann rearrangement (left) and Curtius rearrangement (right).

In both reactions, the alkyl group (R) gets transferred from the carbonyl group to the nitrogen to form an intermediate isocyanate (O=C=N–R). This is then hydrolysed by water to form carbon dioxide and the primary amine. The Curtius rearrangement has the advantage that nitrogen is lost as a gas that helps to take the reaction to completion.

Arylamines

The direct introduction of an amino group to an aromatic ring is not possible. But nitro groups can be added directly by electrophilic substitution and then reduced to the amine. The reduction is done under acidic conditions yielding an arylaminium ion as product. The free base can be isolated by basifying the solution with sodium hydroxide to precipitate the arylamine.

NO_2 NH_2

HNO_3 / H_2SO_4 → a) $SnCl_2$ / $H_3O^{\oplus}$ b) $NaOH$, H_2O →

Fig. Introduction of an amine to an aromatic ring.

Once an amino group has been introduced to an aromatic ring, it can be alkylated with an alkyl halide, acylated with an acid chloride or converted to a higher amine by reductive animation as already described for an alkylamine.

Amines' Properties

Structure

Amines are made up of an *sp*^{3} hybridised nitrogen linked to three substituents by three σ bonds. The substituents can be hydrogen, alkyl or aryl groups, but at least one of the substituents must be an alkyl or aryl group. If only one such group is present, the amine is known as *primary*. If two groups are present, the amine is secondary.

If three groups are present, the amine is tertiary. If the substituents are all alkyl groups, the amine is referred as being an alkylamine. If there is at least one aryl group directly attached to the nitrogen, then the amine is known as an *arylamine*.

The nitrogen atom has four *sp*^{3} Hybridised orbitals pointing to the corners of a tetrahedron in the same way as an *sp*^{3}

hybridised carbon atom. However, one of the sp^3 orbitals is occupied by the nitrogen's lone pair of electrons.

Therefore, the atoms in an amine functional group are pyramidal in shape. The C–N–C bond angles are approximately 109° which is consistent with a tetrahedral nitrogen. However, the bond angle is slightly less than 109° since the lone pair of electrons demands a slightly greater amount of space than a σ bond.

Pyramidal Inversion

Because amines are tetrahedral so they are chiral if they have three different substituents. However, it is not possible to separate the enantiomers of a chiral amine because amines can easily undergo pyramidal inversion. This process interconverts the enantiomers. The inversion involves a change of hybridisation where the nitrogen becomes sp^2 hybridised rather than sp^3 hybridised. Because of this, the molecule becomes planar and the lone pair of electrons occupy a *p* orbital. Once the hybridisation reverts back to sp^3, the molecule can either revert back to its original shape or invert.

Although the enantiomers of chiral amines cannot be separated, such amines can be alkylated to form quaternary ammonium salts where the enantiomers can be separated. Once the lone pair of electrons is locked up in a σ bond, pyramidal inversion becomes impossible and the enantiomers can no longer interconvert.

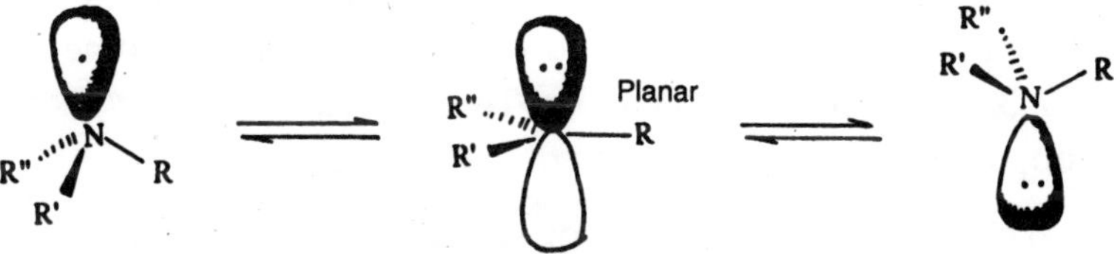

Fig. Pyramidal inversion.

Physical Properties

Amines are polar compounds and intermolecular hydrogen bonding is possible for primary and secondary amines.

Therefore, primary and secondary amines have higher boiling points than alkanes of similar molecular weight. Tertiary amines also have higher boiling points than comparable alkanes, but have slightly lower boiling points than comparable primary or secondary amines as they cannot participate in intermolecular hydrogen bonding.

However, all amines can participate in hydrogen bonding with protic solvents, so amines have similar water solubilities to comparable alcohols. Low molecular weight amines are freely miscible with water. Low molecular weight amines have an offensive fishlike smell.

Basicity

Amines are weak bases but they are more basic than alcohols, ethers, or water. Due to this, amines act as bases when they are dissolved in water and an equilibrium is set up between the ionised form (the ammonium ion) and the unionised form (the free base).

$$R_3N: + H_2O \rightleftharpoons R_3N^{\oplus}\text{—}H + HO^{\ominus}$$

Fig. Acid-base reaction between an amine and water.

The basic strength of an amine can be measured by its pK_b value (typically 3-4). The lower the value of pK_b, the stronger the base. The pK_b for ammonia is 4.74, which compares with pK_b values for methylamine, ethylamine, and propylamine of 3.36, 3.25 and 3.33, respectively. This shows that larger alkyl groups increase base strength.

This is an inductive effect by which the ion is stabilised by dispersing some of the positive charge over the alkyl group. This shifts the equilibrium of the acid base reaction towards the ion, which means that the amine is more basic. The larger the alkyl group, the more significant this effect.

$$R{-}NH_2 + H_2O \rightleftharpoons R{\rightarrow}\ddot{N}H_3^{\oplus} + HO^{\ominus}$$

Fig. Inductive effect of an alkyl group on an alkylammonium ion.

Moreover alkyl substituents should have an even greater inductive effect and we can expect secondary and tertiary amines to be stronger bases than primary amines. This is not necessarily the case and there is no direct relationship between basicity and the number of alkyl groups attached to nitrogen. The inductive effect of more alkyl groups is counterbalanced by a salvation effect.

Once the ammonium ion is formed, it is solvated by water molecules — a stabilising factor that involves hydrogen bonding between the oxygen atom of water and any N–H group present in the ammonium ion. The more hydrogen bonds that are possible, the greater the stabilisation. Due to this, solvation and solvent stabilisation is stronger for alkylaminium ions formed from primary amines than for those formed from tertiary amines. The solvent effect tends to be more important than the inductive effect as far as tertiary amines are concerned and so tertiary amines are generally weaker bases than primary or secondary amines.

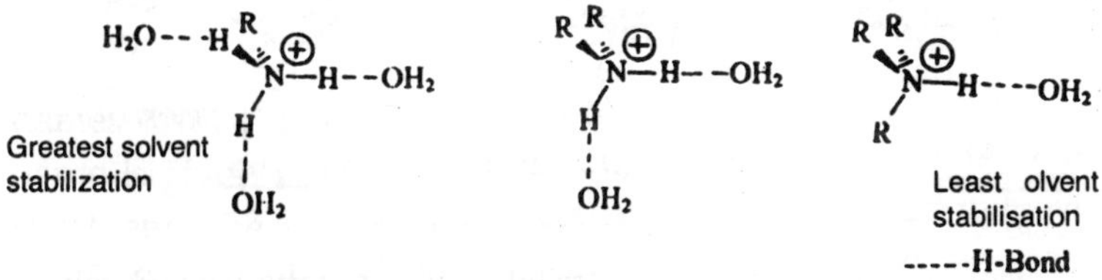

Greatest solvent stabilization

Least olvent stabilisation

-----H-Bond

Fig. Solvent effect on the stabilisation of alkylammonium ions.

Aromatic amines (anylamines) are weaker bases than alkylamines as the orbital containing nitrogen's lone pair of electrons overlaps with the π system of the aromatic ring. In terms of resonance, the lone pair of electrons can be used to form a double bond to the aromatic ring, resulting in the possibility of three zwitterionic resonance structures. (A zwitterion is a molecule containing a positive and a negative charge). Since nitrogen's lone pair of electrons is involved in this interaction. It is less available to form a bond to a proton and so the amine is less basic.

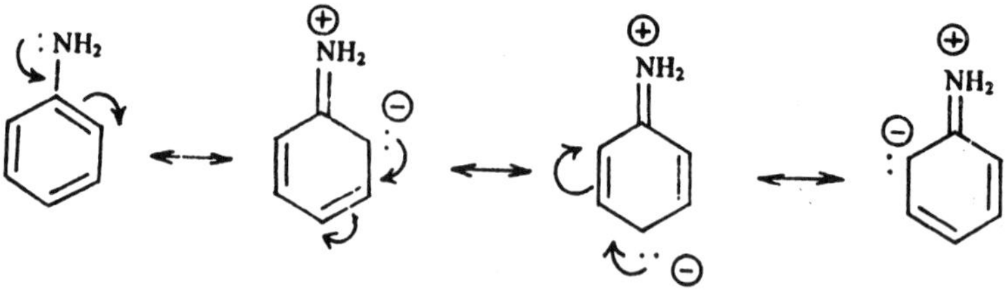

Fig. Resonance interaction between nitrogen's lone pair and the aromatic ring.

The nature of aromatic substituent also affects the basicity of aromatic amines. Substituents that deactivate aromatic rings (e.g. NO_2, Cl or CN) lower electron density in the ring, which means that the ring will have an electron-withdrawing effect on the neighbouring ammonium ion. Because of this the charge will be destabilised and the amine will be a weaker base. Substituents that activate the aromatic ring enhance electron density in the ring and the ring will have an electron-donating effect on the neighbouring charge. This has a stabilising effect and so the amine will be a stronger base. The relative position of aromatic substituents can be important if resonance is possible between the aromatic ring and the substituent.

In such cases, the substituent will have a greater effect if it is at the *ortho* or *para* position, e.g., *para*-nitroaniline is a weaker base than *meta*-nitroaniline. This is because one of the possible resonance structures for the *para* isomer is highly disfavoured since it places a positive charge immediately next to the ammonium ion. Therefore, the number of feasible resonance structures for the *para* isomer is limited to three, compared to four for the meta isomer. Due to this the *para* isomer experience less stabilisation and so the amine will be less basic.

If an activating substituent is present that is capable of interacting with the ring by resonance, the opposite holds true and the *para* isomer will be a stronger base than the meta isomer. This is because the crucial resonance structure mentioned above would have a negative charge immediately next to the ammonium ion and this would have a stabilising effect.

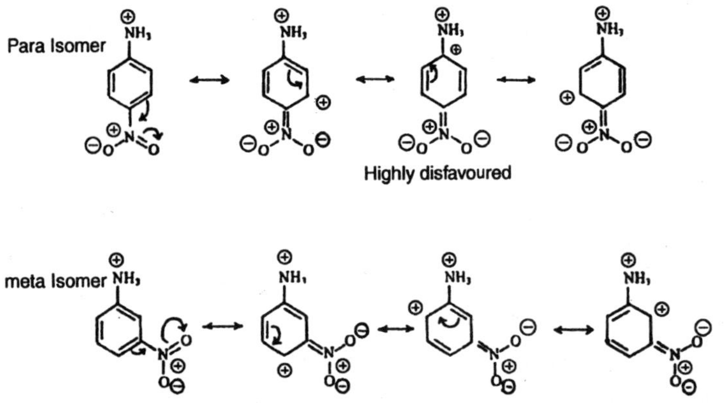

Fig. Resonance structures for *para*-nitroaniline and *meta*-nitroaniline.

Reactivity

Amines react as nucleophiles or bases, since the nitrogen atom has a readily available lone pair of electrons that can participate in bonding. Because of this the amines react with acids to form water soluble salts. This permits the easy separation of amines from other compounds. A crude reaction mixture can be extracted with dilute hydrochloric acid such that any amines present are protonated and dissolve into the aqueous phase as water-soluble salts. The free amine can be recovered by adding sodium hydroxide to the aqueous solution such that the free amine precipitates out as a solid are as an oil.

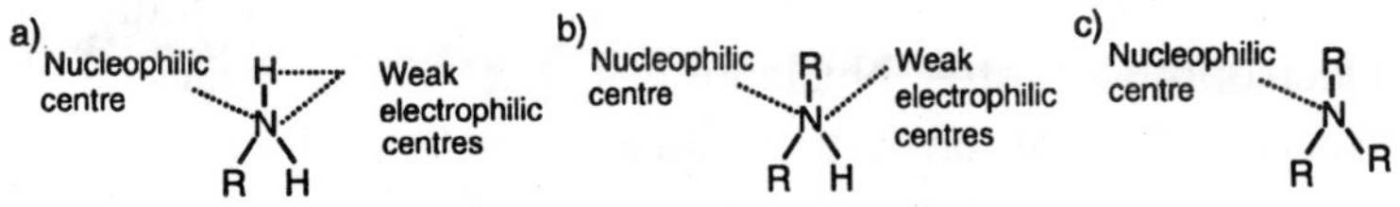

Fig. Nucleophilic and electrophilic centres in (a) primary, (b) secondary, and (c) tertiary amines.

Amines will also react as nucleophiles with a wide range of electrophiles including alkyl halides, aldehydes, ketones, and acid chlorides.

The N–H protons of primary and secondary amines are weakly electrophilic or acidic and will react with a strong base to form amide anions. For example, diisopropylamine (pK_a~40) reacts with butyllithium to give lithium diisopropylamide (LDA) and butane.

Nitrilis Chemistry

Preparation

Nitriles can be prepared by the S_N2 reaction of a cyanide ion with a primary alkyl halide. However, this limits the nitriles that can be synthesised to those having the following general formula RCH_2CN. A more general synthesis of nitriles involves the dehydration of primary-amides with reagents such as thionyl chloride or phosphorus pentoxide:

$$H_3C\text{–}CH_2\text{–}CH(CH_3)\text{–}C(=O)NH_2 \xrightarrow{SOCl_2} H_3C\text{–}CH_2\text{–}CH(CH_3)\text{–}C{\equiv}N + SO_2\,(g) + 2HCl$$

Fig. Dehydration of primary amides with thionyl chloride.

Properties

The nitrile group (CN) is linear in shape with both the carbon and the nitrogen atoms being *sp* hybridised. The triple bond linking the two atoms consists of one σ bond and two π bonds. Nitriles are strongly polarised.

The nitrogen is a nucleophilic centre and the carbon is an electrophilic centre. Nucleophiles react with nitriles at the electrophilic carbon. Generally, the nucleophile will form a bond to the electrophilic carbon resulting in simultaneous breaking of one of the π bonds. The π electrons end up on the nitrogen to form an sp^2 hybridised imine anion which then react further to give different products depending on the reaction conditions used.

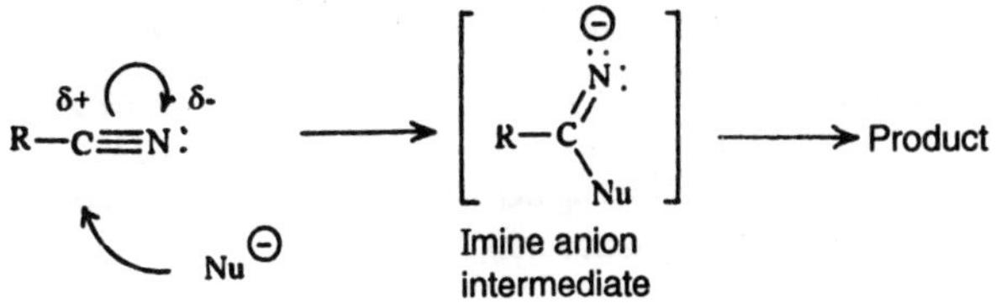

Fig. Reaction between nucleophile and nitriles.

Reactions

Nitriles (RCN) get hydrolysed to carboxylic acids (RCO_2H) in acidic or basic aqueous solutions. The mechanism of the acid-catalysed hydrolysis involves initial protonation of the nitrile's nitrogen atom. This activates the nitrile group towards nucleophilic attack by water at the electrophilic carbon. One of the nitrile π bonds breaks simultaneously and both the π electrons move onto the nitrogen yielding a hydroxyl imine. This rapidly isomerises to a primary amide which is hydrolysed under the reaction conditions to form the carboxylic acid and ammonia.

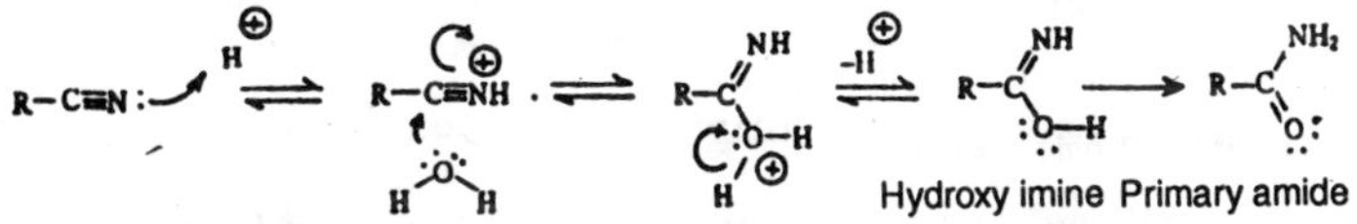

Fig. Acid-catalysed hydrolysis of nitrile to carboxylic acid.

Nitriles (RCN) can be reduced to primary amines (RCH_2HN_2) with lithium aluminium hydride that provides the equivalent of a nucleophilic hydride ion. The reaction can be explained by the nucleophilic attack of two hydride ions:

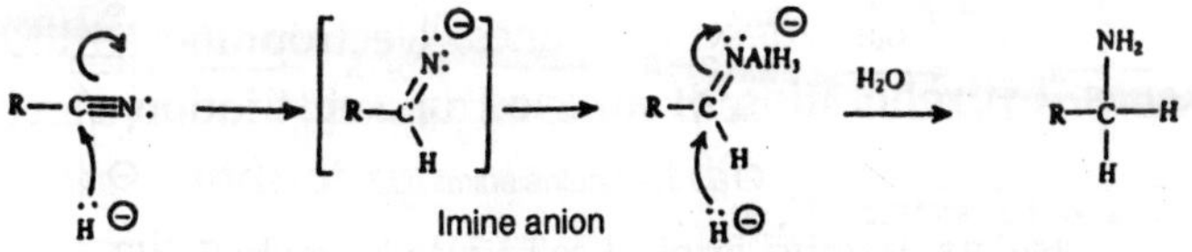

Fig. Reduction of nitriles to form primary amines.

With a milder reducing agent like DIBAH (diisobutylaluminium hydride), the reaction stops after the addition of one hydride ion, and an aldehyde is obtained instead (RCHO).

Grignard Reaction

Nitriles can be treated with Grignard reagents or organolithium reagents to give ketones:

$$R{-}C{\equiv}N \xrightarrow[\text{b) } H_3O^{\oplus}]{\text{a) } CH_3MgI} R{-}\overset{O}{\overset{\|}{C}}{-}CH_3 \xleftarrow[\text{b) } H_3O^{\oplus}]{\text{a) } CH_3Li} R{-}C{\equiv}N$$

Fig. Nitriles react with Grignard reagent or organolithium reagents to produce ketones.

Grignard reagents provide the equivalent of a nucleophilic carbanion which can attack the electrophilic carbon of a nitrile group. One of the π bonds is broken simultaneously forming an intermediate imine anion that is converted to a ketone on treatment with aqueous acid.

$$R{-}C{\equiv}N: \;(\ddot{C}H_3^{\ominus}) \xrightarrow{CH_3MgI} \left[R{-}C(CH_3){=}\ddot{N}:^{\ominus} \right] \xrightarrow{H_3O^{\oplus}} R{-}\overset{O}{\overset{\|}{C}}{-}CH_3$$

Imine anion

Fig. Mechanism of the Grignard reaction on a nitrile group.

Formation of Alcohols, Phenols and Thiols

Preparation of Alcohols

Functional Group Transformation: Alcohols can be prepared by nucleophilic substitution of alkyl halides, hydrolysis of esters, reduction of carboxylic acids or esters, reduction of aldehydes or ketones, electrophilic addition of alkenes, hydroboration of alkenes, or substitution of ethers.

C–C Bond Formation

Alcohols can also be obtained from epoxides, aldehydes, ketones, esters, and acid chloride as a consequence of C–C bond formation. These reactions involve the addition of carbanion equivalents through the use of Grignard or organolithium reagents.

Making of Phenols

Incorporation

Phenol groups can be introduced into an aromatic ring by sulphonation of the aromatic ring followed by reacting the product with sodium hydroxide to convert the sulphonic acid group to a phenol. The reaction conditions are drastic and only alkyl-substituted phenols can be prepared by this method.

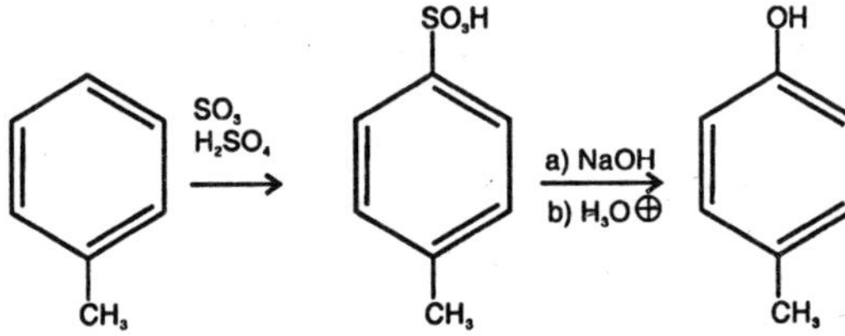

Fig. Synthesis of a phenol via sulphonation.

Another general method of preparing phenols is to hydrolyse a diazonium salt, prepared from an aniline group (NH_2):

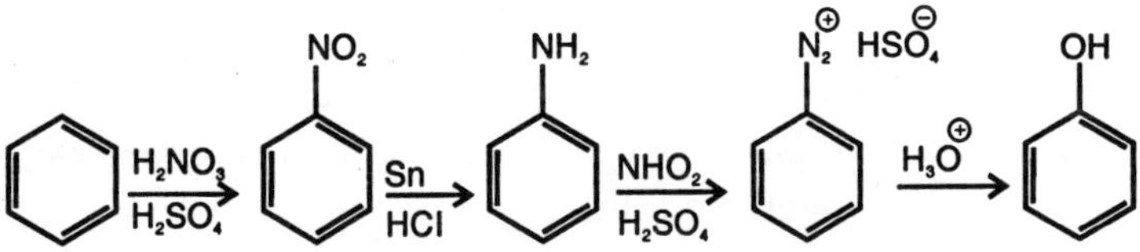

Fig. Synthesis of a phenol via diazonium salt.

Functional Group Transformation

A number of functional group can be converted to phenols, e.g. Sulphonic acids and amino groups which have already been mentioned. Phenyl esters can be hydrolysed. Aryl ethers can be cleaved. The bond between the alkyl group and oxygen is specifically cleaved because the Ar–OH bond is too strong to be cleaved.

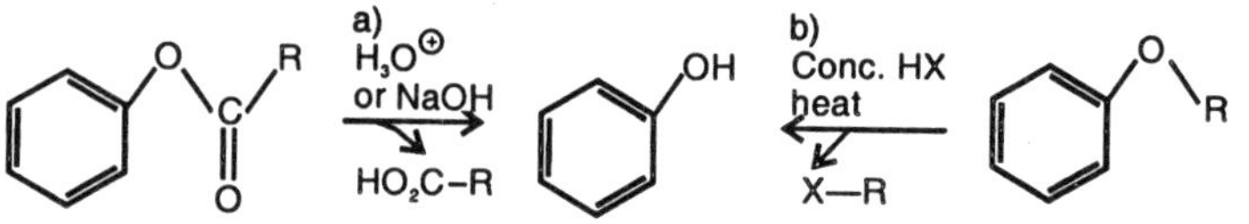

Fig. Functional group transformations to a phenol.

Alcohols and Phenols: Properties

Alcohols

The alcohol functional group (R_3C–OH) has the same geometry as water, with a C–O–H bond angle of approximately 109°. Both the carbon and the oxygen are sp^3 hybridised. Due to the presence of the O–H group intermodular hydrogen bonding is possible that accounts for the higher boiling points of alcohols compared with alkanes of similar molecular weight. Due to hydrogen bonding, alcohols are more soluble in protic solvents than alkenes of similar molecular weight. Actually, the smaller alcohols (methanol, ethanol, propanol, and *tert*-butanol) are completely miscible in water. With larger alcohols, the hydrophobic character of the bigger alkyl chain takes precedence over the polar alcohol group and so larger alcohols are insoluble in water.

The O–H and C–O bonds are both polarised because of the electronegative oxygen, in such a way that oxygen is slightly negative and the carbon and hydrogen atoms are slightly positive. Due to this, the oxygen serves as a nucleophilic centre while the hydrogen and the carbon atoms serve as weak electrophilic centres:

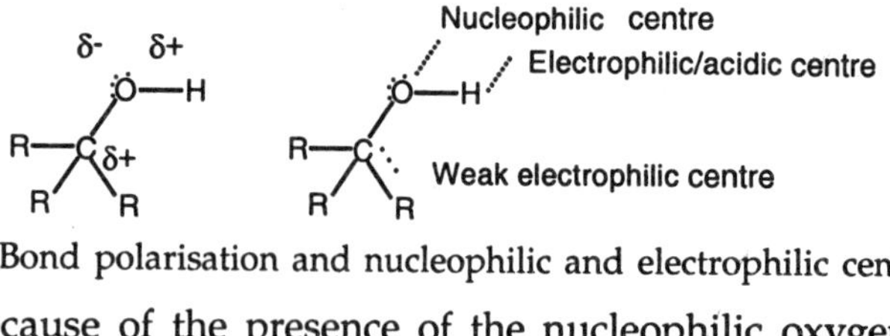

Fig. Bond polarisation and nucleophilic and electrophilic centres.

Because of the presence of the nucleophilic oxygen and electrophilic proton, alcohols can act both as weak acids and as weak bases when dissolved in water. However, the equilibrium in both cases is virtually completely weighted to the unionised form.

Fig. Acid-base properties of alcohols.

Alcohols generally react with stronger electrophiles than water. However, they are less likely to react with nucleophiles unless the latter are also strong bases, in that case the acidic proton is abstracted to form an alkoxide ion (RO^-) alkoxide ions are quite the oxygen atom acting as the nucleophilic centre. The intermediate formed can then react more readily as an electrophile at the carbon centre.

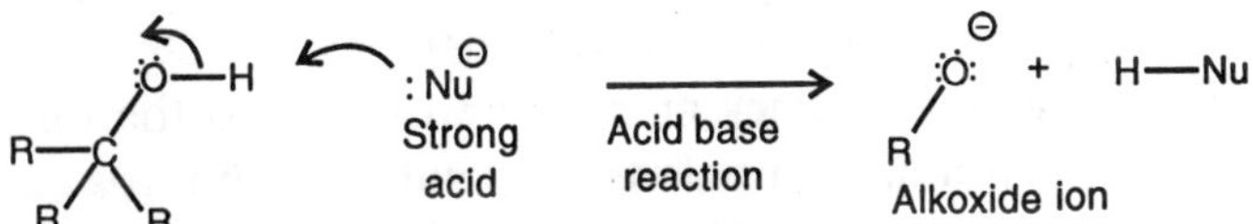

Fig. Formation of an alkoxide ion.

Phenols

Phenols are compounds that have an OH group directly attached to an aromatic ring. Therefore, the oxygen is *sp^3* hybridised and the aryl carbon is *sp^2* hybridised. Although phenols share some characteristics with alcohols, they have distinct properties and reactions that set them apart from that functional group.

Phenols can participate in intermolecular hydrogen bonding that means that they have a moderate water solubility and have higher boiling points than aromatic compounds lacking the phenolic group. Phenols are weakly acidic, and in aqueous solution an equilibrium exists between the phenol and the phenoxide ion [Following fig(a)]. When treated with a base, the phenol gets converted to the phenoxide ion [Following fig(b)].

Phenol + H_2O ⇌ Phenoxide + H_3O^+ Phenol →(Base) Phenoxide ion

Fig. Acidic reactions of phenol.

The phenoxide ion is stabilised by resonance and delocalisation of the negative charge into the ring, therefore phenoxide ions are weaker bases than alkoxide ions. This means that phenols are more acidic than alcohols, but less acidic than carboxylic acids. The pK_a useful reagents in organic synthesis. However, they cannot be used if water is the solvent since the alkoxide ion would act as a base and abstract a proton from water to regenerate the alcohol. Therefore, an alcohol would have to be used as solvent instead of water.

Nucleophiles that are also strong bases react with the electrophilic hydrogen of an alcohol rather than the electrophilic carbon. Nucleophilic attack at carbon would need the loss of a hydroxide ion in a nucleophilic substitution reaction. However, this is not favoured as the hydroxide ion is a strong base and a poor leaving group. However, reactions which involve the cleavage of an alcohol's C–O bond are possible if the alcohol is first 'activated' such that the hydroxyl group is converted into a better leaving group.

One method is to react the alcohol under acidic conditions such that the hydroxyl group is protonated before the nucleophile makes its attack. Cleavage of the C–O bond would then be more likely because the leaving group would be a neutral water molecule that is a much better leaving group. Alternatively, the alcohol can be treated with an electrophilic reagent to convert the OH group into a different group (OY) that can then act as a better leaving group.

In both cases, the alcohol must first act as a nucleophile with values of most phenols is in the order of 11, compared to 18 for alcohols and 4.74 for acetic acid. This means the phenols can be ionised with weaker bases than those needed to ionise alcohols, but need stronger bases than those needed to ionise carboxylic acids. For example, phenols are ionised by sodium hydroxide but not by the weaker base sodium hydrogen carbonate.

Alcohols beings less acidic are not ionised by either base but carboxylic acids are ionised by both sodium hydroxide and sodium hydrogen carbonate solutions.

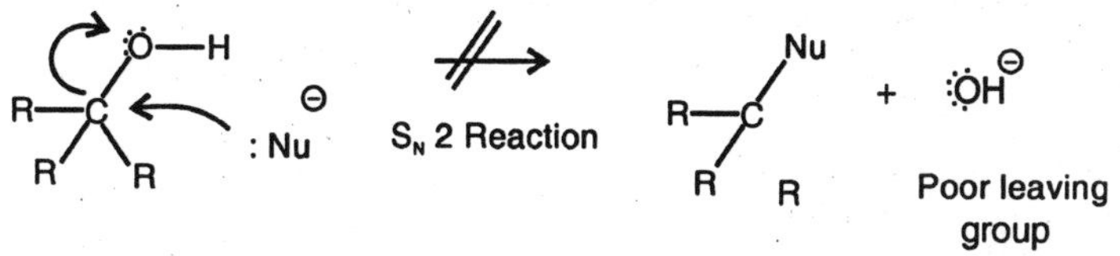

S_N 2 Reaction

Poor leaving group

Fig. Nucleophilic substitution of alcohols is not favoured.

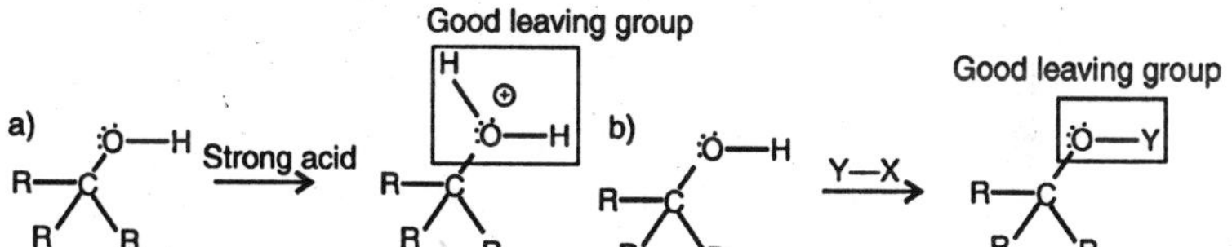

a) Strong acid — Good leaving group

b) Y—X — Good leaving group

Fig. Activation of an alcohol.

These acid-base reactions allow a simple way distinguishing between most carboxylic acids, phenols, and alcohols. Since the salts formed from the acid-base reaction are water soluble, compounds containing these functional groups can be distinguished by testing their solubilities in sodium hydrogen carbonate and sodium hydroxide solutions. This solubility test is not valid for low molecular weight structures like methanol or ethanol since these are water soluble and dissolve in basic solution because of their water solubility rather than their ability to form salts.

Alcohol's Reaction

Acid-base Reactions

Alcohols are slightly weaker acids than water and thus the conjugate base generated from an alcohol (like alkoxide ion) is a stronger base than the conjugate base of water (the hydroxide ion). Due to this, it is not possible to generate an alkoxide ion using sodium hydroxide as base. Alcohols do not react with sodium bicarbonate or amines, and a stronger base like sodium hydride or sodium amide is needed to generate the alkoxide ion.

Alcohols can also be converted to alkoxide ions on treatment with potassium, sodium lithium metal. Some organic reagents can also act as strong bases, e.g. Grignard reagents and organolithium reagents.

$$R\text{-}\ddot{O}\text{-}H + {}^{\ominus}\text{:Base} \longrightarrow R\text{-}\ddot{O}\text{:}^{\ominus} \quad H\text{—Base}$$

Fig. Generation of alkoxide ion.

Alkoxide ions are neutralised in water and so reactions involving these reagents should be accomplished in the alcohol from which they were derived, that is reactions involving sodium ethoxide are best carried out in ethanol. Alcohols have a typical pK_a of 15.5-18.0 compared to pK_a values of 25 for ehtyne, 38 for ammonia and 50 for ethane.

Elimination

Alcohols, like alkyl halides, can undergo elimination reactions to form alkenes. Since water is eliminated, the reaction is also called a *dehydration.*

$$HO\text{—}C(CH_3)_2\text{—}CH_3 \xrightarrow{20\%\ H_2SO_4} (H_3C)_2C{=}CH_2 + H_2O$$

Fig. Elimination of an alcohol.

Like alkyl halides, the elimination reaction of an alcohol needs the presence of a susceptible proton at the β-position:

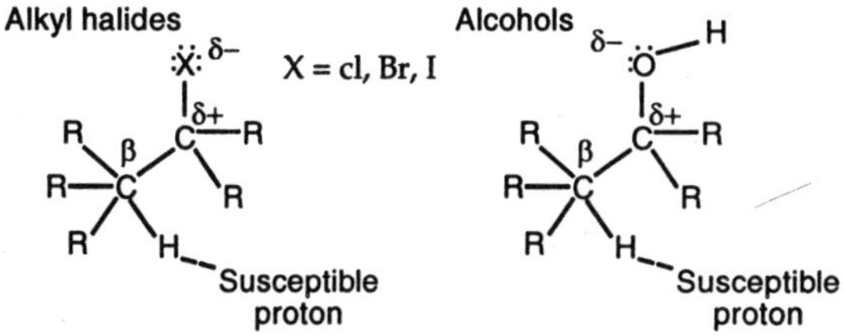

Alkyl halides

X = cl, Br, I

Alcohols

Susceptible proton

Susceptible proton

Fig. Susceptible β-protons in an alkyl halide and an alcohol.

The elimination of alkyl halides is done under basic conditions, the elimination of alcohols is done under acid conditions. Under basic conditions, an E2 elimination would require the loss of a hydroxide ion as a leaving group. Since the hydroxide ion is a strong base, it is not a good leaving group and so the elimination of alcohols under basic conditions is difficult to achieve.

Elimination under acidic conditions is more successful because the hydroxyl group is first protonated and then it departs the molecule as a neutral water molecule (dehydration) that is a much better leaving group. If different isomeric alkenes are possible, the most substituted alkene will be favoured. The reaction occurs best with tertiary alcohols as the elimination proceeds by the E1 mechanism.

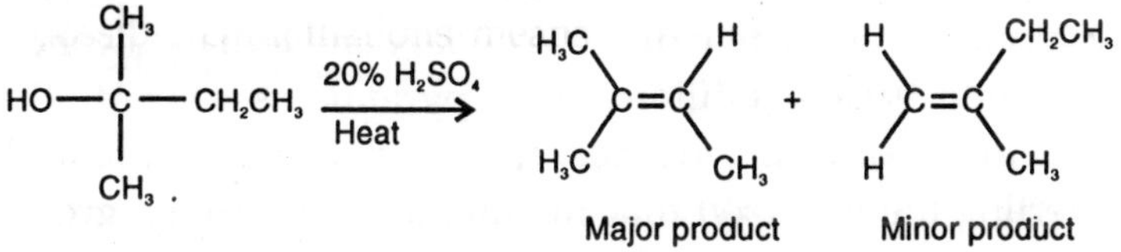

20% H_2SO_4

Heat

Major product

Minor product

Fig. Elimination of alcohols obeys Zaitsev's rule.

The mechanism shown below involves the nucleophilic oxygen of the alcohol making use of one of its lone pairs of electrons to form a bond to a proton to yield a charged intermediate (step 1). When the oxygen gets protonated, the molecule has a much better leaving group because water can be ejected as a neutral molecule.

The E1 mechanism can now proceed as normal. Water is lost and a carbocation is formed (step 2). Water then acts as a base in the second step, making use of one of its lone pairs of electrons to form a bond to the β-proton of the carbocation. The C–H bond is broken and both the electrons in that bond are used to form a π bond between the two carbons. Because this is an E1 reaction, tertiary alcohols react better than primary or secondary alcohols.

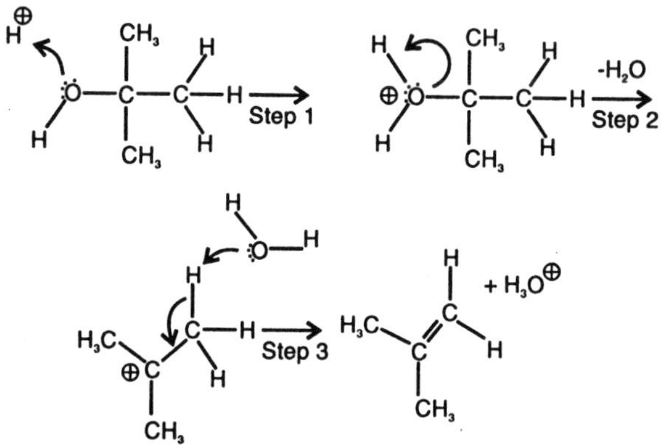

Fig. E1 Elimination mechanism for alcohols.

The E1 reaction is not ideal for the dehydration of primary or secondary alcohols since vigorous heating is needed to force the reaction and this can result in rearrangement reactions. In alternative methods which are useful, reagents like phosphorus oxychloride ($POCl_3$) dehydrate secondary and tertiary alcohols under mild basic conditions using pyridine as solvent. The phosphorus oxychloride serves to activate the alcohol, converting the hydroxyl function into a better leaving group. The mechanism involves the alcohol acting as a nucleophile in the first step. Oxygen uses a lone pair of electrons to form a bond to the electrophilic phosphorus of $POCl_3$ and a chloride ion is lost (step 1). Pyridine then removes a proton from the structure to form a dichlorophosphate intermediate (step 2). The dichlorophosphate group is a much better leaving group than the hydroxide ion and so a normal E2 reaction can occur.

Pyridine acts as a base to remove a β-proton and as this is happening, the electrons from the old C–H bond are used to form a π bond and eject the leaving group (step 3).

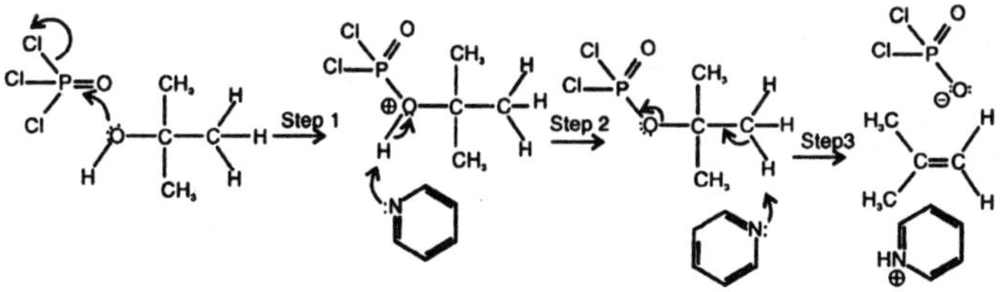

Fig. Mechanism for the $POCl_3$ dehydration of an alcohol.

Synthesis of Alkyl Halides

Tertiary alcohols may undergo the S_N1 reaction to produce tertiary alkyl halides. Since the reaction needs the loss of the hydroxide ion (a poor leaving group), so to convert the hydroxyl moiety into a better leaving group acidic conditions are achieved with the use of HCl or HBr. The acid serves to protonate the hydroxyl moiety as the first step and then a normal S_N1 mechanism occurs where water is lost from the molecule to form an intermediate carbocation. A halide ion then forms a bond to the carbocation centre in the third step.

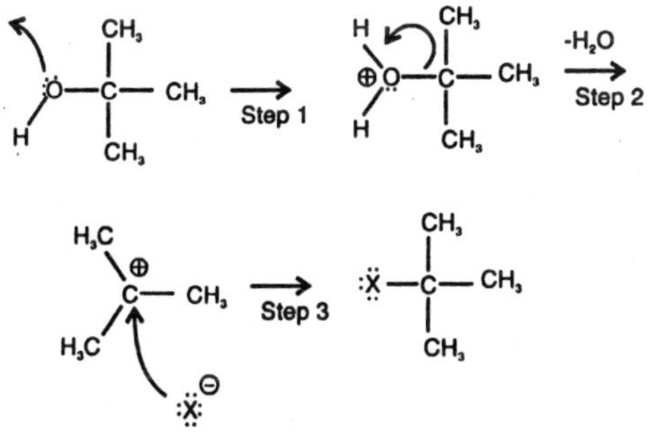

Fig. Conversion of alcohols to alkyl halides.

The first two steps of this mechanism are the same as the elimination reaction. Both reactions are carried out under acidic conditions. The difference is that halide ion serve as good nucleophiles and are present in high concentration. The elimination reaction is carried out using concentrated sulphuric acid and only weak nucleophiles are present (i.e. water) in low concentration. Thus, some elimination may occur and although the reaction of alcohols with HX produces mainly alkyl halide, some alkene by-product is usually present.

Since primary alcohols and some secondary alcohols do not undergo the S_N1 reaction, nucleophilic substitution of these compounds must involve an S_N2 mechanism. Once again, protonation of the OH group is needed as a first step, then the reaction involves simultaneous attack of the halide ion and loss of water. The reaction proceeds with good nucleophiles like the iodide or bromide ion, but fails with the weaker nucleophilic chloride ion. In this case, a Lewis acid has to be added to the reaction mixture. The Lewis acid forms a complex with the oxygen of the alcohol group which results in a much better leaving group for the subsequent S_N2 reaction.

However, the reaction of primary and secondary alcohols with hydrogen halides can generally be a problem since unwanted rearrangement reactions generally occurs.

a) $Cl_2S{=}O + CH_3CH_2OH \longrightarrow$ $Cl_2S(O){-}\overset{\oplus}{O}H{-}CH_2CH_3$ (Et₃N:) $\xrightarrow{-Et_3NH^{\oplus}}$ Alkyl chlorosulphide $ClS(O){-}O{-}CH_2CH_3$ (:Cl:⁻) $\longrightarrow$ $:Cl{-}CH_2CH_3 + SO_2(g) + Cl^{\ominus}$

b) $PBr_3 + CH_3CH_2OH \longrightarrow$ $Br_2P{-}\overset{\oplus}{O}H{-}CH_2CH_3$ (:Br:⁻) $\longrightarrow$ $:Br{-}CH_2CH_3 + HOPBr_2$

Fig. Conversion of an alcohol to an alkyl halide using (a) thionyl chloride; (b) phosphorus tribromide.

To avoid this, the reaction is carried out under milder basic conditions using reagents like thionyl chloride or phosphorus tribromide. These reagents act as electrophiles and react with the alcoholic oxygen to form an intermediate where the OH moiety gets converted into a better leaving group. A halide ion is released from the reagent in this process, and this can act as the nucleophile in the subsequent S_N2 reaction.

In the reaction with thionyl chloride, triethylamine is present to mop up the HCl formed during the reaction. The reaction is also helped by presence of one of the products (SO_2) as a gas which gets expelled thus driving the reaction to completion.

Phosphorus tribromide has three bromine atoms present and each PBr_3 molecule can react with three alcohol molecules to form three molecules of alkyl bromide.

Synthesis of Mesylates and Tosylates

Sometimes it is convenient to synthesise an activated alcohol that can be used in nucleophic substitution reactions like an alkyl halide. Mesylates and tosylates are such sulphonate compounds which serve this purpose. They can be synthesised by action of alcohols with sulphoxyl chlorides in the presence of a base like pyridine or triethylamine. The base serves to 'mop up' the HCl that is formed and avoids acid-catalysed rearrangement reactions.

a) $RCH_2—OH + Cl—SO_2—C_6H_4—CH_3 \xrightarrow{\text{pyridine}} RCH_2—O—SO_2—C_6H_4—CH_3$

p- Toluenesulphfonyl chloride Tosylate

b) $RCH_2—OH + Cl—SO_2—CH_3 \xrightarrow[CH_2Cl_2]{NEt_3} RCH_2—O—SO_2—CH_3$

Methanesulfphonyl chloride Mesylate

Fig. Synthesis of (a) tosylate and (b) mesylate.

The reaction mechanism involves the alcohol oxygen acting as a nucleophilic centre and substituting the chloride ion from the sulphonate. The base then removes a proton from the intermediate to give the sulphonate product. Neither of these steps affects the stereochemistry of the alcohol carbon and so the stereochemistry of chiral alcohols is retained.

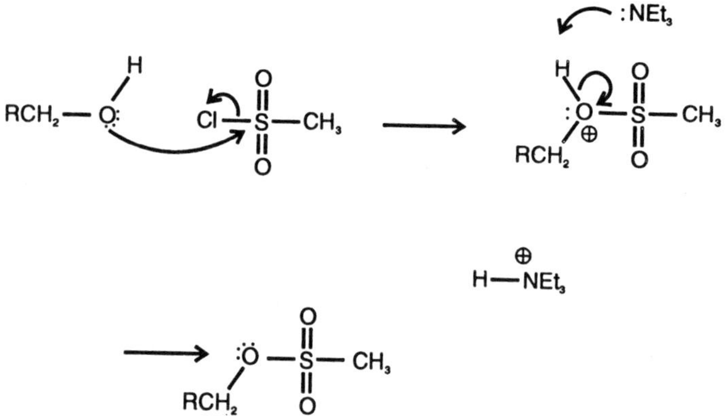

Fig. Mechanism for the formation of a mesylate.

The mesylate and tosylatc groups are excellent leaving groups and can be considered as the equivalent of a halide. Therefore, mesylates and tosylates can undergo the S_N2 reaction in the same way as alkyl halides:

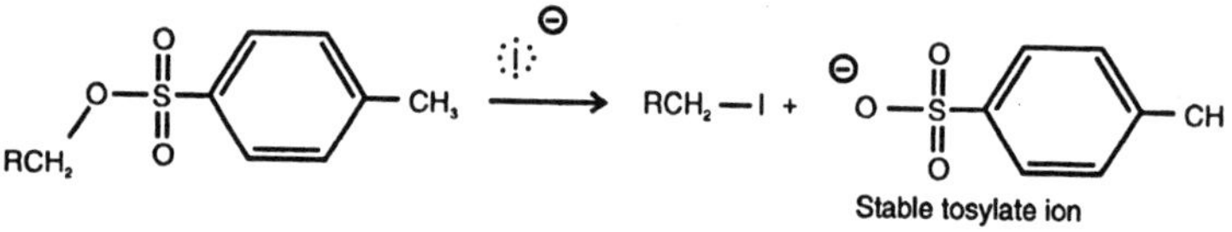

Stable tosylate ion

Fig. Nucleophilic substitution of a tosylate.

Oxidation

The oxidation of alcohols is quite an important reaction in organic synthesis. Primary alcohols can be oxidised to aldehydes, but the reaction is difficult because there is the danger of overoxidation to carboxylic acids. In case of volatile aldehydes, the aldehydes can be distilled from the reaction solution as they are formed. However, this is not possible for less volatile aldehydes. To overcome this problem by a mild oxidising agent like pyridinium chlorochromate (PCC) is used. If a stronger oxidising agent is used in aqueous conditions (e.g. CrO_3 in aqueous sulphuric acid), primary alcohols are oxidised to carboxylic acids, while secondary alcohols are oxidised to ketones.

a) $R{-}CH_2OH \xrightarrow{PCC} R{-}C(=O)H$ (PCC = pyridinium $CrO_3Cl^{\ominus}$)

1° Alcohol → Aldehyde

b) $R{-}CH_2OH \xrightarrow[H_2SO_4]{CrO_3} R{-}C(=O)OH$

1° Alcohol → Carboxylic acid

c) $R{-}CH(R)OH \xrightarrow[H_2SO_4]{CrO_3} R{-}C(=O)R$

2° Alcohol → Ketone

Fig. Oxidation of alcohols.

The success of the PCC oxidation in stopping at the aldehyde stage is solvent dependent. The reaction is done in methylene chloride, whereas oxidation with CrO_3 is done in aqueous acid. Under aqueous conditions, the aldehyde that is formed by oxidation of the alcohol is hydrated and this structure is more sensitive to oxidation than the aldehyde itself. In methylene chloride, hydration cannot take place and the aldehyde is more resistant to oxidation.

$$RCHO \underset{}{\overset{H_2O}{\rightleftharpoons}} RCH(OH)_2$$

Sensitive to oxidation

Fig. Hydration of an aldehyde.

The mechanism of oxidation for a secondary alcohol with CrO_3 involves the nucleophilic oxygen reacting with the oxidising agent to produce a charged chromium intermediate. Elimination then takes place where an α-proton is lost along with the chromium moiety to produce the carbonyl group.

The mechanism can be considered as an E2 mechanism, the difference being that different bonds are being created and broken. As the mechanism needs an α-proton to be removed from the alcoholic carbon, tertiary alcohols cannot be oxidised because they do not contain such a proton.

The mechanism also explains why an aldehyde product is resistant to further oxidation when methylene chloride is the solvent (i.e. no OH present to react with the chromium reagent). When aqueous conditions are used the aldehyde is hydrated and this generates two OH groups that are available to bond to the chromium reagent and result in further oxidation.

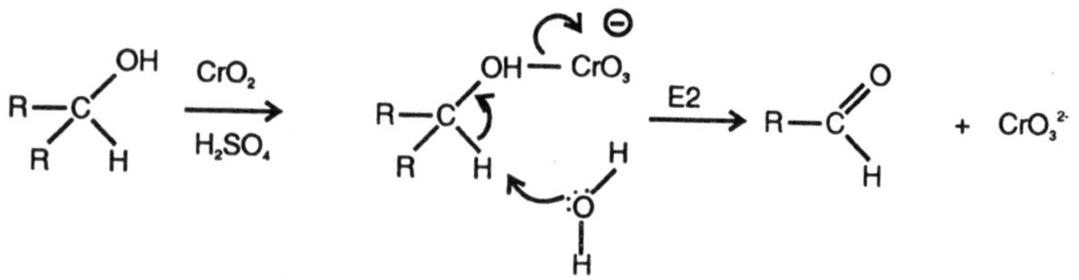

Fig. Mechanism of oxidation of a secondary alcohol with CrO_3.

Phenol's Reaction

Acid-base Reactions

Phenols are stronger acids than alcohols. They react with bases like sodium hydroxide to form phenoxide ions. However, they are weaker acids than carboxylic acids and do not react with sodium hydrogen carbonate.

Phenols are acidic because the oxygen's lone pair of electrons can participate in a resonance mechanism involving the adjacent aromatic ring. Three resonance structures are possible in which the oxygen gains a positive charge and the ring gains a negative charge. The net result is a slightly positive charge on the oxygen that accounts for the acidity of its proton. There are also three aromatic carbons with slightly negative charges.

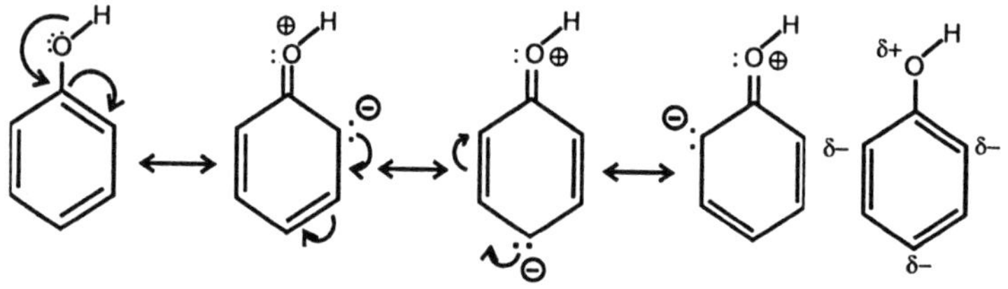

Fig. Resonance structures for phenol.

The type of substitutes present on the aromatic ring can effect the acidity of the phenol. This is because substituents can either stabilise or destabilise the partial negative charge on the ring. The better the partial charge is stabilised, the more effective the resonance will be and the more acidic the phenol will be. Electron-withdrawing groups like a nitro substituent increase the acidity of the phenol since they stabilise the negative charge by an inductive effect. Nitro groups that are *ortho* or *para* to the phenolic group have an even greater effect. This is because fourth resonance structure is possible that delocalises the partial charge even further.

Electron-donating substituents (e.g. alkyl-groups) have the opposite effect and decrease the acidity of phenols.

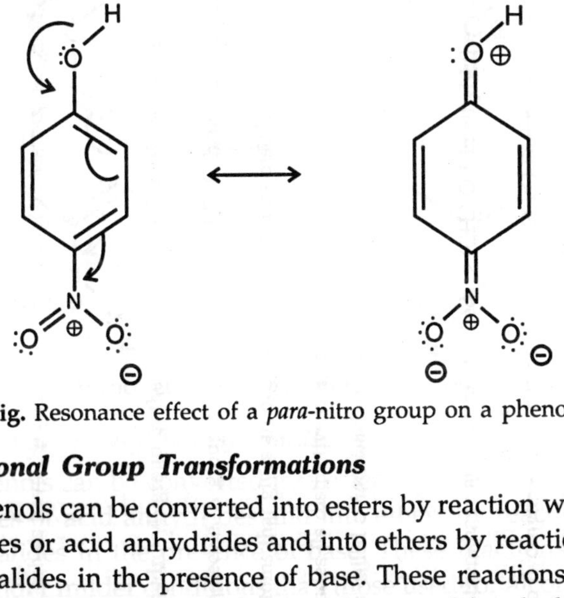

Fig. Resonance effect of a *para*-nitro group on a phenol.

Functional Group Transformations

Phenols can be converted into esters by reaction with acid chlorides or acid anhydrides and into ethers by reaction with alkyl halides in the presence of base. These reactions can be done under milder conditions than those used for alcohols due to the greater acidity of phenols. Thus, phenols can be converted to phenoxide ions with sodium hydroxide rather than metallic sodium.

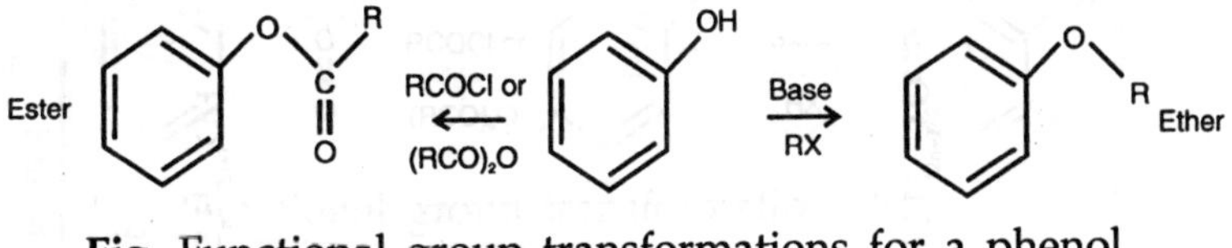

Fig. Functional group transformations for a phenol.

Although the above reactions are common to alcohols and phenols, there are several reactions that can be done on alcohols but not phenols, and vice versa. For example, unlike alcohols, phenols cannot be converted to esters by reaction with a carboxylic acid under acid catalysis. Reactions involving the cleavage of the C–O bond are also not possible for phenols. The aryl C–O bond is stronger than the alkyl C–O bond of an alcohol.

Electrophilic Substitution

Electrophilic substitution is helped by the phenol group that acts as an activating group and directs substitution to the *ortho* and *para* positions. Sulphonation and nitration of phenols are both possible to give *ortho* and *para* substitution products. Sometimes the phenolic groups can be too powerful an activating group and it is difficult to control the reaction to one substitution, e.g., the bromination of phenol leads to 2,4,6-tribromophenol even in the absence of a Lewis acid:

OH $\xrightarrow[H_2O]{Br_2}$ Br, OH, Br, Br

Fig. Bromination of phenol.

The activating power of the phenolic group can be decreased by converting the phenol to an ester that can be removed by hydrolysis once the electrophilic substitution reaction had been carried out. Since the ester is a weaker activating group, substitution takes place only once. Moreover, since the ester is a bulkier group than the phenol, *para* substitution is favoured over *ortho* substitution.

OH $\xrightarrow{CH3COCl}$ OAc $\xrightarrow{Br_2}$ OAc, Br $\xrightarrow[b)\ H_3O^{\oplus}]{a)\ NaOH}$ OH, Br

Fig. Synthesis of *para*-bromophenol.

Oxidation

Phenols are susceptible to oxidation to quinones:

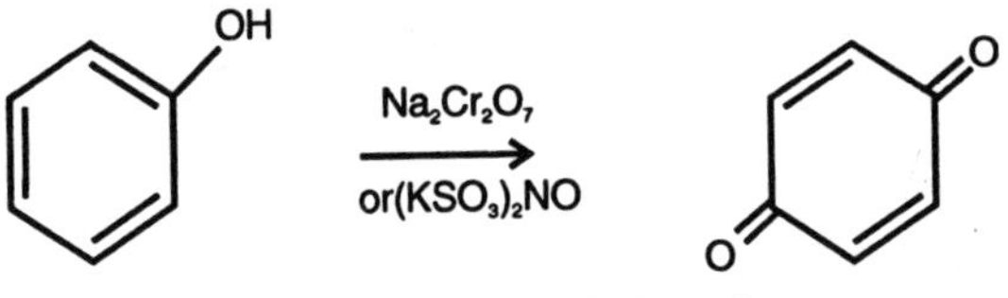

Fig. Oxidation of phenol.

Claisen Rearrangement

It is a useful method of introducing an alkyl substituent to the *ortho* position of a phenol. The phenol gets converted to the phenoxide ion, then treated with 3-bromopropene (an alkyl bromide) to form an ether.

On heating, the allyl group ($-CH_2-CH=CH_2$) is transferred from the phenolic group to the *ortho* position of the aromatic ring. The mechanism involves a concerted process of bond formation and bond breaking known as a *pericyclic reaction.* This yields a ketone structure that immediately tautomerises to the final product. Different allylic reagents can be used in the reaction and the double bond in the final product can be reduced to form alkane substituent without affecting the aromatic ring.

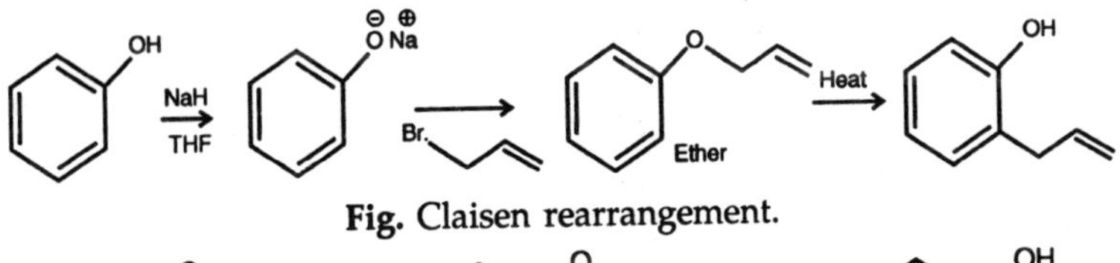

Fig. Claisen rearrangement.

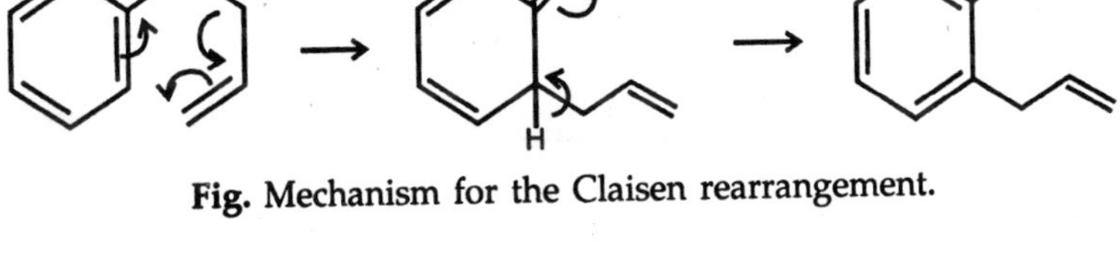

Fig. Mechanism for the Claisen rearrangement.

Process of Reaction

Acid-base Reactions

Because carboxylic acids have an acidic proton (CO_2H), they form water soluble carboxylate salts when treated with a base (e.g. sodium hydroxide or sodium bicarbonate):

$$RCOOH \xrightarrow[NaHCO_3]{NaOH\ or} RCOO^{\ominus}\ Na^{\oplus}$$

Fig. Salt formation

Interconversion of Acid Derivatives

$$H_3C{-}CO{-}Cl \xrightarrow{CH_3CO_2^{\ominus}Na^{\oplus}} H_3C{-}CO{-}O{-}CO{-}CH_3 + HCl$$

$$H_3C{-}CO{-}Cl \xrightarrow[Pyridine]{ROH} H_3C{-}CO{-}O{-}R + HCl$$

$$H_3C{-}CO{-}Cl \xrightarrow[Pyridine\ or\ NaOH]{2\,R_2NH} H_3C{-}CO{-}NR_2 + HCl$$

Fig. Nucleophilic substitutions of an acid chloride.

Reactive acid derivatives can be converted to less reactive acid derivatives by nucleophilic substitution.

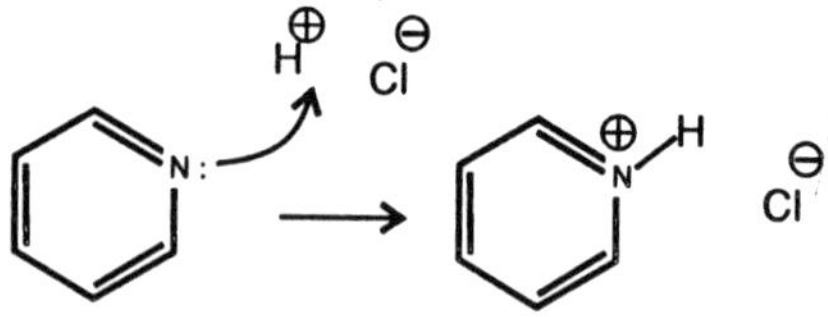

Fig. Role of pyridine in 'mopping up' protons.

Thus, acid chlorides can be converted to acid anhydride, esters, and amides. Hydrochloric acid is released in all these reactions and this may lead to side reactions. Therefore, pyridine or sodium hydroxide may be added so as to mop up the hydrochloric acid.

Acid anhydrides can be converted to esters and amides but not to acid chlorides:

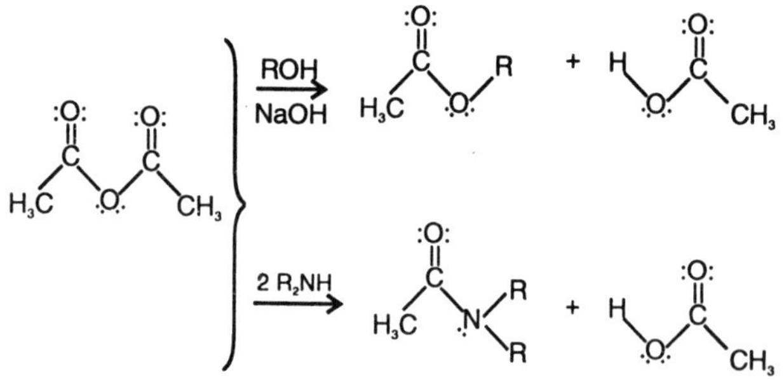

Fig. Nucleophilic substitutions of acid anhydrides.

Esters can be converted to amides but not to acid chlorides or acid anhydrides:

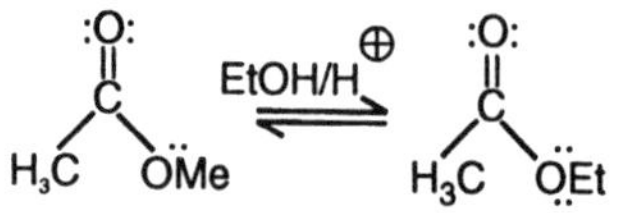

Fig. Nucleophilic substitutions of an ester.

Esters can also be converted by nucleophilic substitution from one type of ester to another and this process is called *transesterification*. For example, a methyl ester can be dissolved in ethanol in the presence of an acid catalyst and converted to an ethyl ester.

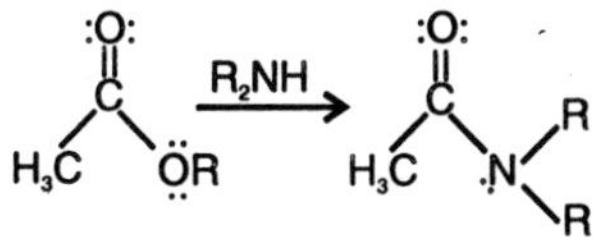

Fig. Transesterification.

The reaction is an equilibrium reaction, but if the alcohol is used as solvent, it is in large excess and the equilibrium is shifted to the desired ester. Moreover, if the alcohol to be replaced has a low boiling point, it can be distilled from the reaction as it is substituted, thus shifting the equilibrium to the desired product.

Amides are the least of the acid derivatives and cannot be converted to acid chlorides, acid anhydrides, or esters.

Hydrolysis

Reactive acid derivatives (i.e. acid chlorides and acid anhydrides) get hydrolysed by water to give the constituent carboxylic acids. The reaction is an example of nucleophilic substitution where water acts as the nucleophile. Hydrochloric acid is a by-product from the hydrolysis of an acid chloride, so pyridine is generally added to the reaction mixture to mop it up.

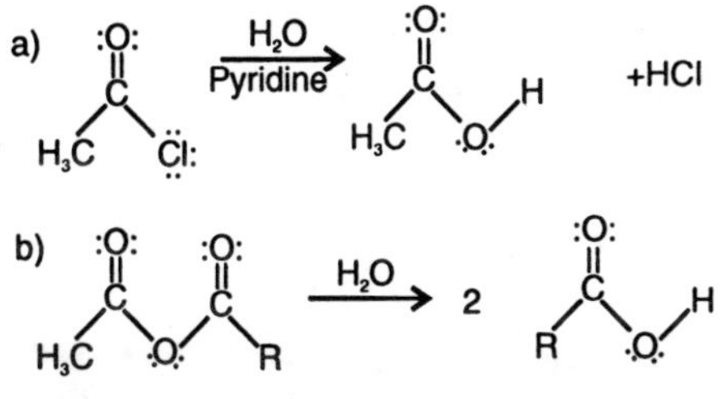

Fig. Hydrolysis of (a) an acid chloride; (b) an acid anhydride.

Esters and amides are less reactive and so the hydrolysis needs more drastic conditions using aqueous sodium hydroxide or aqueous acid with heating:

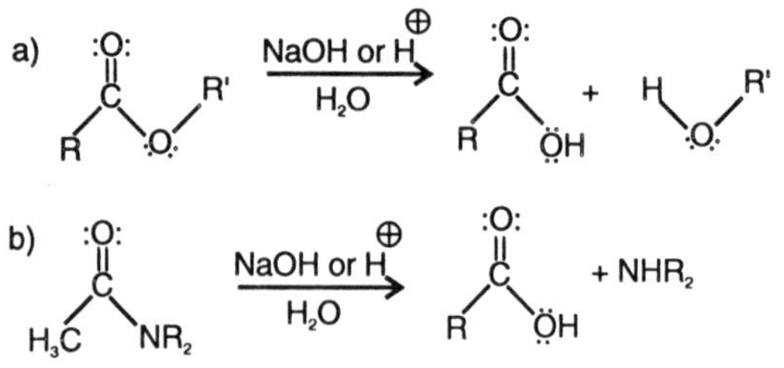

Fig. Hydrolysis of (a) esters; (b) amides.

Under basic conditions, the hydroxide ion acts as the nucleophile. For example, the mechanism of hydrolysis of ethyl acetate is shown. However, the mechanism does not stop here. The carboxylic acid which is formed reacts with sodium hydroxide to form a water soluble carboxylate ion. Moreover, the ethoxide ion that is lost from the molecule is a stronger base than water and undergoes protonation. The basic hydrolysis of an ester is also called *saponification* and produces a water soluble carboxylate ion.

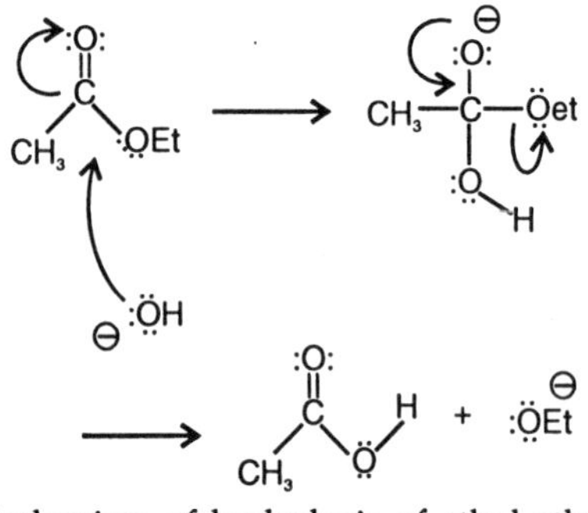

Fig. Mechanism of hydrolysis of ethyl ethanoate.

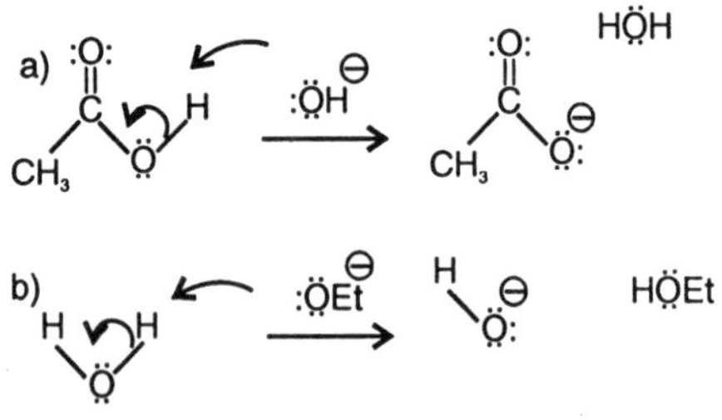

Fig. (a) Ionisation of a carboxyiic acid: (b) neutralisation of the ethoxide ion.

The same mechanism is involved in the basic hydrolysis of an amide and also results in the formation of a water soluble carboxylate ion. The leaving group from an amide is initially charged (i.e. R_2N :). However, this is a strong base and reacts with water to form a free amine and a hydroxide ion.

In the basic hydrolysis of esters and amides, the formation of carboxylate ion is irreversible and so serves to drive the reaction to completion.

To isolate carboxylic acid rather than the salt, it is essential to add acid (e.g. dilute HCl) to the aqueous solution. The acid protonates the carboxylate salt to give the carboxylic acid that (in most cases) is no longer soluble in aqueous solution and precipitates out as a solid or as an oil.

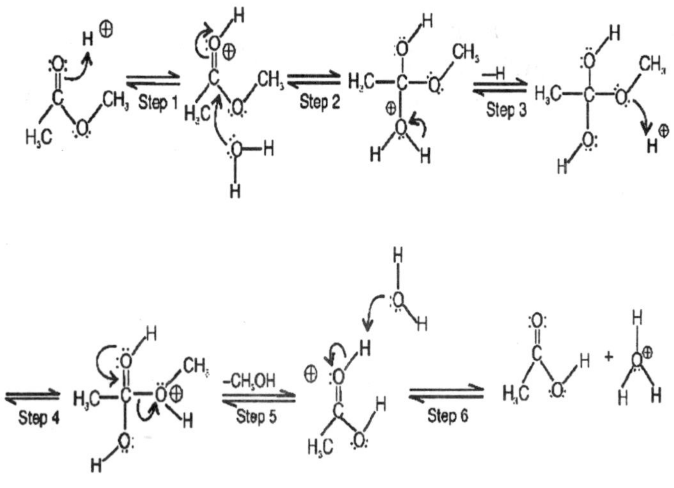

Fig. Mechanism for the acid-catalysed hydrolysis of an ester.

In the mechanism for acid-catalysed hydrolysis water acts as a nucleophile. However, water is a poor nucleophile as it gains an unfavourable positive charge when it forms a bond. Therefore, the carbonyl group has to be activated that takes place when the carbonyl oxygen is protonated by the acid catalyst (step 1). Nucleophilic attack by water in now favoured because it neutralises the unfavourable positive charge on the carbonyl oxygen (step 2).

The intermediate has a positive charge on the oxygen derived from water, but this is neutralised by losing the attached proton such that the oxygen gains the electrons in the O–H bond (step 3). Another protonation now occurs (step 4) This is necessary so as to convert a poor leaving group (the methoxide ion) into a good leaving group (methanol). The π bond can now be reformed (step 5) with loss of methanol. Finally, water can act as a base to remove the proton from the carbonyl oxygen (step 6).

The acid-catalysed hydrolysis of an ester is not as effective as basic hydrolysis because all the steps in the mechanism are reversible and there is no salt formation to pull the reaction through to products.

Therefore, it is important to use an excess of water so as to shift the equilibria to the products. In contrast to esters, the hydrolysis of an amide in acid does result in the formation of an ion. The leaving group here is an amine and since amines are basic, they will react with the acid to form a water soluble aminum ion. This is an irreversible step that pulls the equilibrium through to the products.

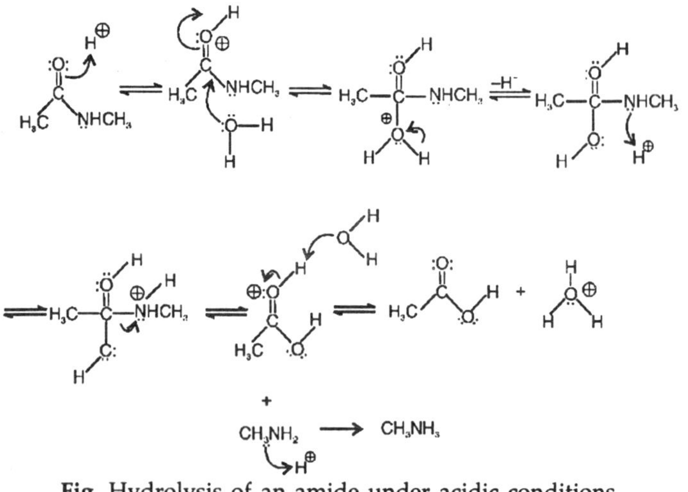

$CH_3NH_2 \longrightarrow CH_3NH_3$

Fig. Hydrolysis of an amide under acidic conditions.

In the acid-catalysed hydrolysis of an ester, only a catalytic amount of acid is needed since the protons used during the reaction mechanism are regenerated. However, with an amide, at least one equivalent of acid is required because of ionisation of the amine.

Friedel-Crafts Acylation

Acid chlorides can be treated with aromatic rings in the presence of a Lewis acid to give aromatic ketones. The reaction involves formation of an acylium ion from the acid chloride, followed by electrophilic substitution of the aromatic ring.

$$C_6H_6 + CH_3COCl \xrightarrow{AlCl_3} C_6H_5COCH_3$$

Fig. Friedel-Crafts acylation.

Grignard Reaction

Acid chlorides and esters react with two equivalents of a Grignard reagent to produce a tertiary alcohol where two extra alkyl groups are provided by the Grignard reagent:

a) $$H_3C{-}CO{-}Cl \xrightarrow[\text{b) } H_3O^{\oplus}]{\text{a) 2NMgI}} H_3C{-}C(OH)R_2$$

b) $$H_3C{-}CO{-}OR' \xrightarrow[\text{b) } H_3O^{\ominus}]{\text{a) RMgI}} H_3C{-}C(OH)R_2$$

Fig. Grignard reaction with (a) an acid chloride; and (b) an ester to produce a tertiary alcohol.

There are two reactions involved in this process. The acid chloride reacts with one equivalent of Grignard reagent in a nucleophilic substitution to produce an intermediate ketone. This ketone is also reactive to Grignard reagents and immediately reacts with a second equivalent of Grignard reagent by the nucleophilic addition mechanism.

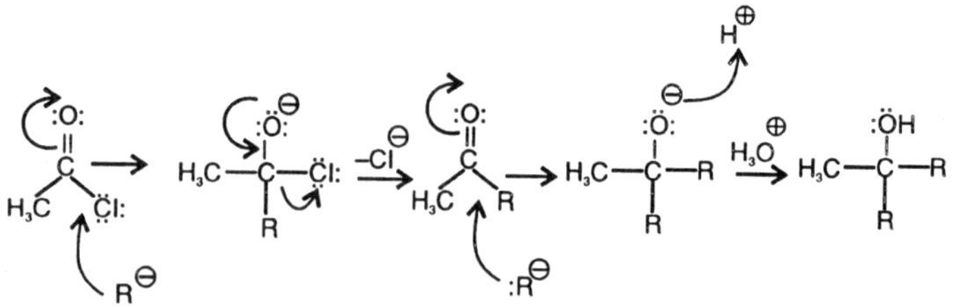

Fig. Mechanism of the Grignard reaction with an acid chloride.

Carboxylic acids reacts with Grignard reagents in an acid-base reaction forming carboxylate ion and an alkane. This has no synthetic use and it is important to protect carboxylic acids when carrying out Grignard reactions on another part of the molecule to avoid wastage of Grignard reagent.

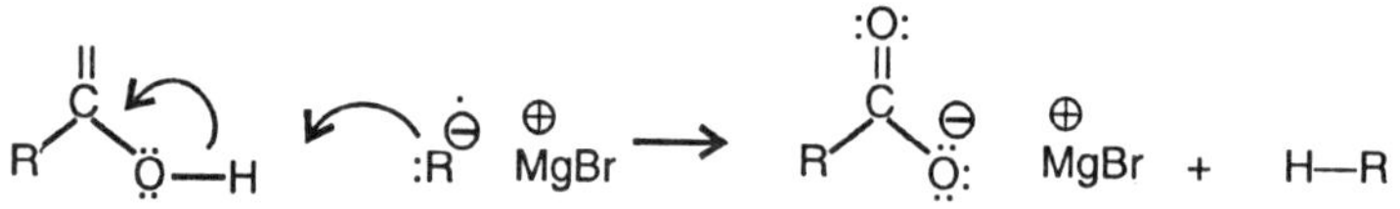

Fig. Acid-base reaction of a Grignard reagent with a carboxylic acid.

Organolithium Reactions

Esters react with two equivalents of an organolithium reagent to yield a tertiary alcohol in which two of the alkyl groups are derived from the organolithium reagent. The mechanism of the reaction is the same as that described in the Grignard reaction, i.e., nucleophilic substitution to a ketone followed by nucleophilic addition.

It is necessary to protect any carboxylic acids present when carrying out organolithium reactions since one equivalent of the organolithium reagent would be wasted in an acid-base reaction with the carboxylic acid.

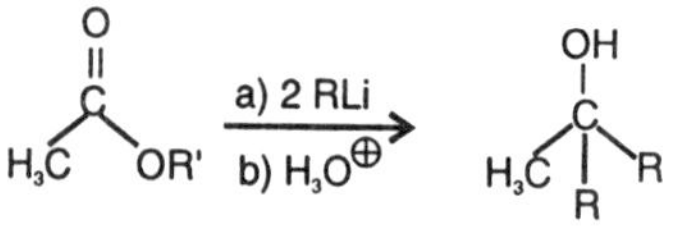

Fig. Reaction of an ester with an organolithium reagent to form a tertiary alcohol.

Organocuprate Reactions

Acid chlorides react with diorganocuprate reagents to form ketones. Like the Grignard reaction, an alkyl group displaces the chloride ion to form a ketone. However, unlike the Grignard reaction, the reaction stops at the ketone stage. The mechanism is believed to be radical based rather than nucleophilic substitution. The reaction does not occur with carboxylic acids, acid anhydrides, esters, or amides.

$$H_3C{-}C(=O){-}Cl \xrightarrow{R2Cu^{\ominus}\ Li^{\oplus}} H_3C{-}C(=O){-}R$$

Fig. Reaction of an acid chloride with a diorganocuprate reagent to produce a ketone.

Reduction

Carboxylic acids, acid chlorides, acid anhydrides and esters get reduced to primary alcohols when treated with lithium aluminium hydride (LiAlH). The reaction involves nucleophilic substitution by a hydride ion to give an intermediate aldehyde. This cannot be isolated since the aldehyde immediately undergoes a nucleophilic addition reaction with another hydride ion.

$$R{-}C(=O){-}Y \xrightarrow[\text{b) } H_3O^{\oplus}]{\text{a) } LiAlH_4} R{-}CH_2{-}OH \qquad Y = OH, Cl, OCOR', OR$$

Fig. Reduction of acid chlorides, acid anhydrides, and esters with lithium aluminium hydride.

$$H_3C{-}C(=O){-}OEt \xrightarrow[\text{Nucleophilic substitution}]{LiAlH_4} \left[H_3C{-}C(=O){-}H\right] \xrightarrow[\text{Nucleophilic addition}]{LiAlH_4} H_3C{-}CH(OH){-}H$$

1° Alcohol

Fig. Intermediate involved in the $LiAlH_4$ reduction of an ester.

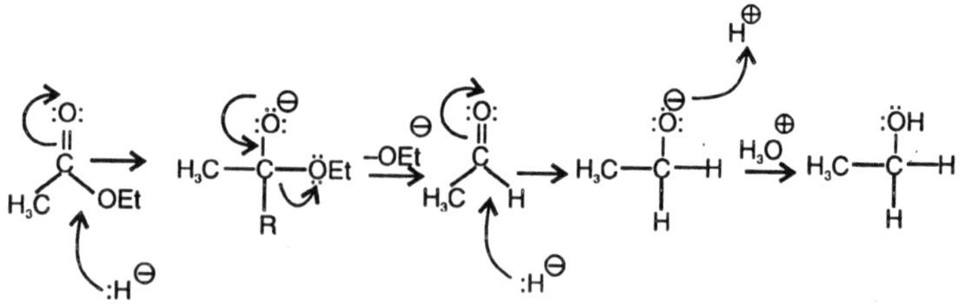

Fig. Mechanism for the $LiAlH_4$ reduction of an ester to a primary alcohol.

Amides differ from carboxylic acids and other acid derivatives in their reaction with $LiAlH_4$. Instead of forming primary alcohols, amides are reduced to amines. The mechanism involves addition of the hydride ion to form an intermediate that is converted to an organoaluminium intermediate. The difference in this mechanism is the intervention of the nitrogen's lone pair of electrons. These are fed into the electrophilic centre to eliminate the oxygen that is then followed by the second hydride addition.

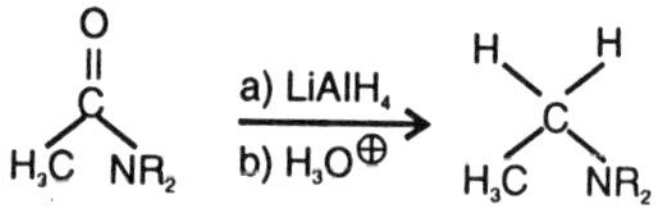

Fig. Reduction of an amide to an amine.

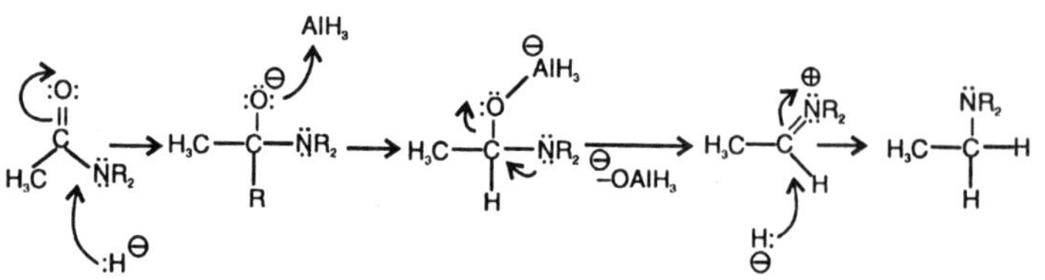

Fig. Mechanism for the $LiAlH_4$ reduction of an amide to an amine.

Although acid chlorides and acid anhydrides get converted to tertiary alcohols with $LiAlH_4$, there is little synthetic advantage in this because the same reaction can be achieved on the more readily available esters and carboxylic acids. However, since acid chlorides are more reactive than carboxylic acids, they can be treated with a milder hydride-reducing agent and this allows the synthesis of aldehydes. The hydride reagent used *(lithium of tri-tert-butoxyaluminium hydride)* contains three bulky alkoxy groups that lowers the reactivity of the remaining hydride ion. This means that the reaction stops after nucleophilic substitution with one hydride ion. Another sterically hindered hydride *reagent-diisobutylaluminium hydride (DIBAH)* can be used to reduce esters to aldehydes. Normally low temperatures are required to avoid overreduction.

$$H_3C-C(=O)-Cl \xrightarrow[\text{b) } H_3O^{\oplus}]{\text{a) } LiAlH[OC(CH_3)_3]_3} H_3C-C(=O)-H \xleftarrow[\text{b) } H_3O^{\oplus}]{\text{a) } AlH[CH_2CH(CH_3)_2]_2 \text{ (DIBAH)}} H_3C-C(=O)-OEt$$

Fig. Reduction of an acid chloride and an ester to an aldehyde.

Borane (B_2H_6) can be used as a reducing agent to convert carboxylic acids to primary alcohols. The advantage of using borane rather than $LiAlH_4$ is that the former does not reduce any nitro groups that might be present. $LiAlH_4$ reduces a nitro group (NO_2) to an amino group (NH_2).

Carboxylic acids and acid derivatives are not reduced by the milder reducing agent such as sodium borohydride (NaBH). This reagent can, therefore, be used to reduce aldehydes and ketones without affecting any carboxylic acids or acid derivatives which might be present.

Dehydration of Primary Amides

Primary amides are dehydrated to nitriles using a dehydrating agent like thionyl chloride ($SOCl_3$), phosphorus pentoxide (P_2O_5), phosphoryl trichloride ($POCl_3$), or acetic anhydride:

$$H_3C-C(=O)-NH_2 \xrightarrow[-H_2O]{SOCl_2} H_3C-C\equiv N$$

Fig. Conversion of a primary, amide to a nitrile.

The mechanism for the dehydration of an amide with thionyl chloride is shown below:

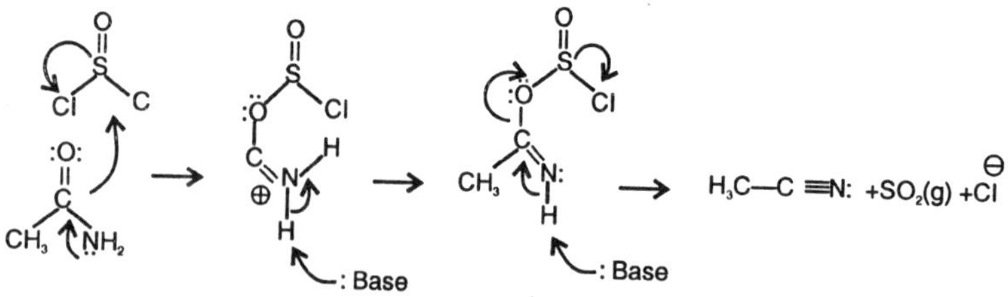

Fig. Mechanism for the dehydration of a primary amide to a nitrile.

Although the reaction is the equivalent of a dehydration, the mechanism shows that water itself is not eliminated. The reaction is driven by the loss of one sulphur dioxide as a gas.

Reactions of Enolate

Enolates

Enolate ions can be formed from aldehydes and ketones containing protons on an α-carbon. Enolate ions can also be formed from esters if they have protons on an α-carbon. Such protons are slightly acidic and can be removed on treatment with a powerful base like lithium diisopropylamide (LDA). LDA acts as a base rather than as a nucleophile since it is a bulky molecule and this prevents it attacking the carbonyl group in a nucleophilic substitution reaction.

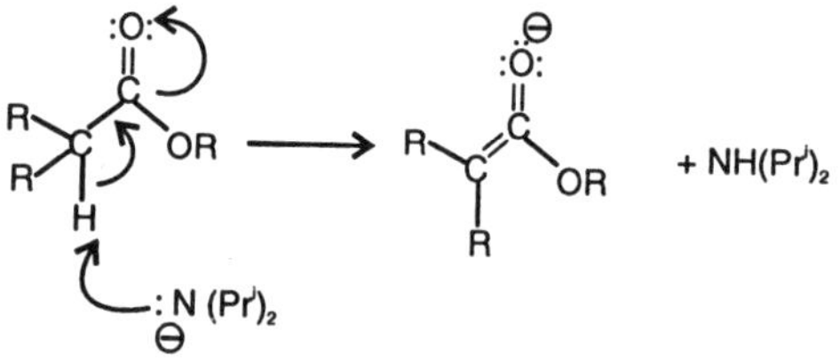

Fig. Enolate ion formation.

Fig. Formation of an enolate ion from diethyl malonate.

Formation of enolate is easier if there are two esters flanking the α-carbon since the α-proton will be more acidic. The acidic proton in diethyl malonate can be removed with a weaker base than LDA (e.g. sodium ethoxide; following fig.). The enolate ion is more stable since the charge can be delocalised over both carbonyl groups:

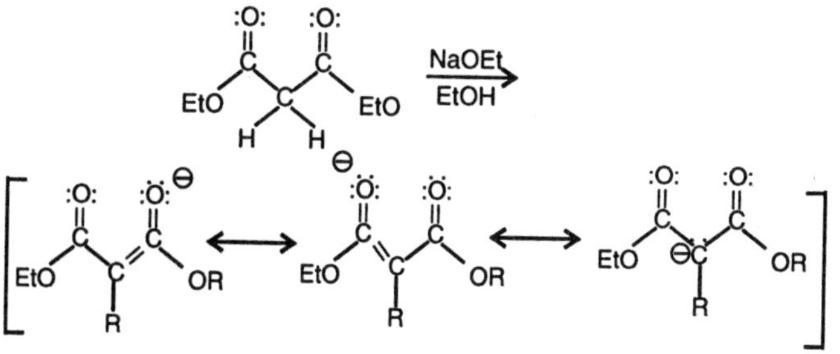

Alkylations

Enolate ions can be alkylated with alkyl halides through the S 2 nucleophilic substitution of an alkyl halide:

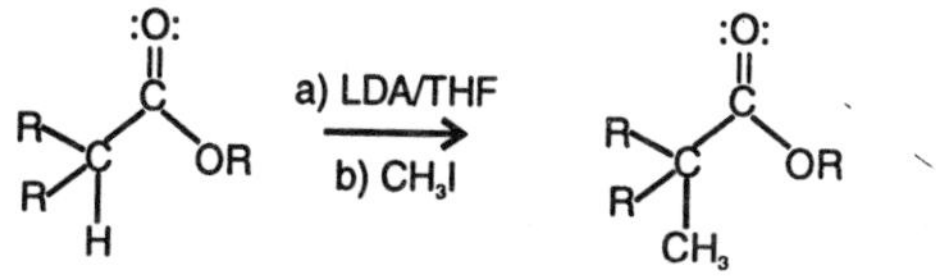

Fig. α-alkylation of an ester.

Although simple ester can be converted to their enolate ions and alkylated, the use of a molecule like diethyl malonate is far more effective. This is because the α-protons of diethyl malonate (pK_9 10-12) are more acidic than the α-protons of a simple ester like ethyl acetate (pK_a 25) and can be removed by a milder base. It is possible to predict the base needed to carry out the deprotonation reaction by considering the pK_a value of the conjugate acid for that base. If this pK_a is higher than the pK_a value of the ester, then the deprotonation reaction is possible. For example, the conjugate acid of the ethoxide ion is ethanol (pK_a 16) and so any ester having a p*Ka* less than 16 will be deprotonated by the ethoxide ion.

Therefore, diethyl malonate is deprotonated but not ethyl acetate. Moreover, the ethoxide ion is strong enough to deprotonate the diethyl malonate quantitatively such that all the diethyl malonate is converted to the enolate ion. This avoids the possibility of any competing Claisen reaction since that reaction needs the presence of unaltered ester. Diethyl malonate can be converted quantitatively to its enolate with ethoxide ion, alkylated with an alkyl halide, treated with another equivalent of base, then alkylated with a second different alkyl halide. Subsequent hydrolysis and decarboxylation of the diethyl ester yields the carboxylic acid. The decarboxylation mechanism is dependent on the presence of the other carbonyl group at the β-position.

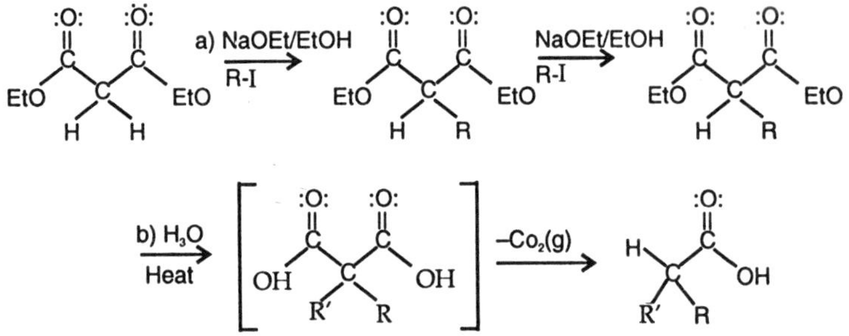

Fig. Alkylations of diethyl malonate.

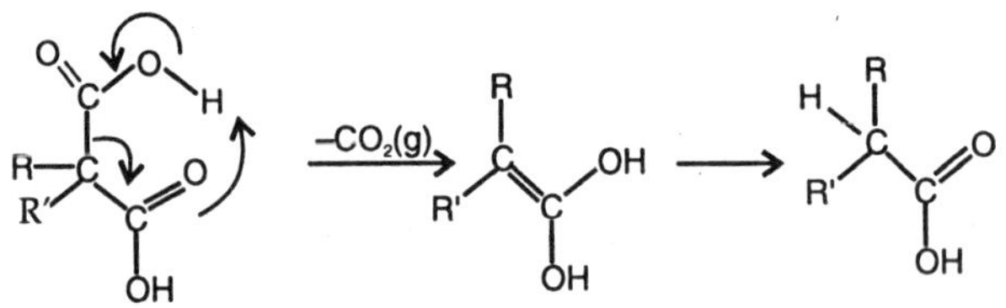

Fig. Decarboxylation mechanism.

The final product can be considered as a di-substituted ethanoic acid. Theoretically, this product could also be synthesised from ethyl ethanoate. However, the use of diethyl malonate is better because the presence of two carbonyl groups permits easier formation of the intermediate enolate ions.

Claisen Condensation

The Claisen reaction involves the condensation or linking of two ester molecules to form a β-ketoester. This reaction can be considered as the ester equivalent of the Aldol reaction The reaction involves the formation of an enolate ion from one ester molecule which then undergoes nucleophilic substitution with a second ester molecule (step 1).

The ethaoxide ion that is formed in step 2 removes an α-proton from the β-ketoester in step 3 to form a stable enolate ion and this drives the reaction to completion. The final product is isolated by protonating the enolate ion with acid.

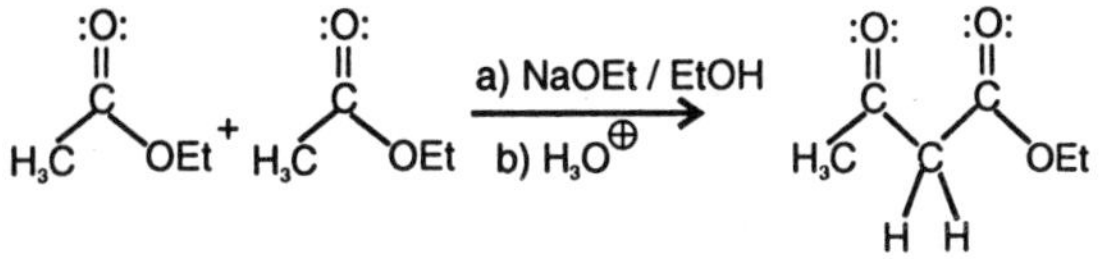

Fig. Claisen condensation.

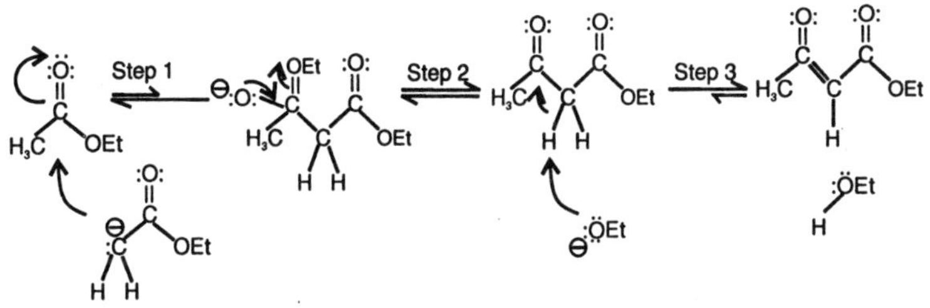

Fig. Mechanism of the Claisen condensation.

Two different esters can be used in the Claisen condensation as long as one of the esters has no α-protons and cannot form an enolate ion. β-Diketones can be synthesised from the mixed Claisen condensation of a ketone with an ester. It is better to use any ester that cannot form an enolate ion to avoid competing Claisen condensations.

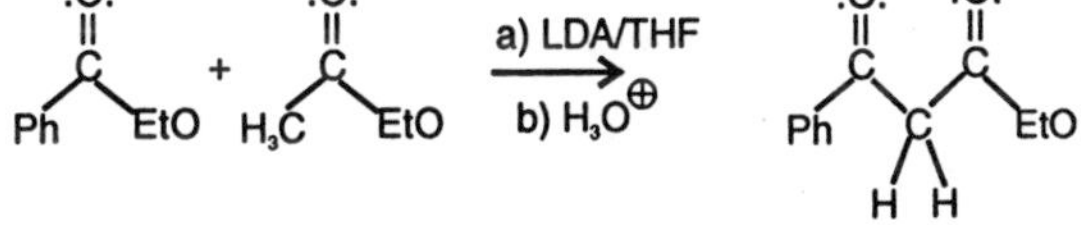

Fig. Claisen condensation of two different esters.

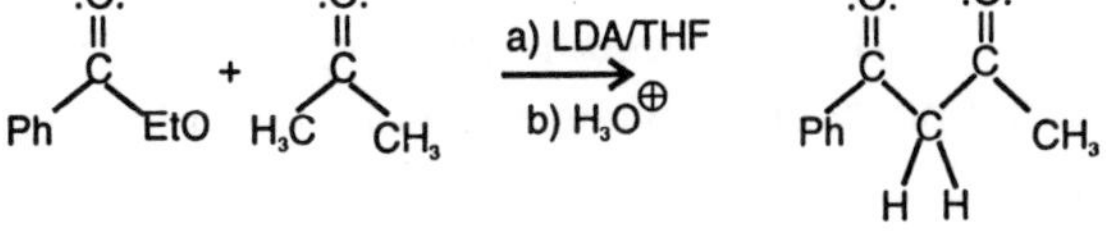

Fig. Claisen condensation of a ketone with an ester.

In both these last two examples, a very strong base is used in the form of LDA such that the enolate ion is formed quantitatively (from ethyl acetate and acetone, respectively). This avoids the possibility of self-Claisen condensation and limits the reaction to the crossed Claisen condensation.

Process of Alkyl Halides

Preparation and Physical Properties

Preparation: Alkenes when treated with hydrogen halides (HCl, HBr, and HI) or halogens (Cl_2 and Br_2) yield alkyl halides and dihaloalkanes, respectively. Another useful method of preparing alkyl halides is by treating an alcohol with a hydrogen halide (HX = HCl, HBr, or HI). The reaction works best for tertiary alcohols. Primary and secondary alcohols can be converted to alkyl halides by treating them with thionyl chloride ($SOCl_2$) or phosphorus tribromide (PBr_3).

Structure

Alkyl halides are made up of an alkyl group linked to a halogen atom (F, Cl, Br, or I) by a single (σ) bond. The carbon atom linked to the halogen atom is sp^2 hybridised and it has a tetrahedral geometry with bond angles of approximately 109°. The carbon-hydrogen bond length increases with the size of the halogen atom and this is accompanied with a decrease in bond strength. For example, C–F bonds are shorter and stronger than C–Cl bonds.

Bonding

The carbon-halogen bond is a σ bond. The bond is polar because the halogen atom is more electronegative than carbon. Due to this, halogen is slightly negative and the carbon is slightly positive. Intermolecular hydrogen bonding or ionic bonding is not possible between alkyl halide molecules and the major intermolecular bonding force consists of weak van der Waals interactions.

Properties

The polar C–X bond present in alkyl halides has a substantial dipole moment. Alkyl halides are poorly soluble in water, but are soluble in organic solvents. They have boiling points that are similar to alkanes of comparable molecular weight. Due to polarity, the carbon is an electrophilic centre and the halogen is a nucleophilic centre. Halogens are extremely weak nucleophilic centres and therefore, alkyl halides are more likely to react as electrophiles at the carbon centre.

Reactions

The important reactions of alkyl halides are, (a) nucleophilic substitution in which an attacking nucleophile replaces the halogen [fig. (a), and (b) elimination in which the alkyl halides loses HX and gets converted to an alkene fig. (b)].

a) $R_3C-\ddot{X}: \xrightarrow{:Nu^-} R_3C-Nu + :\ddot{X}:^{\ominus}$ Nucleophilic substitution

b) $H-CR_2-CR_2-\ddot{X}: \xrightarrow{:Base^{\ominus}} R_2C{=}CR_2 + H-\ddot{X}:$ Elimination

Fig. Reactions of alkyl halides.

Substitution of Nucleophilic

Definition

Due to the presence of a strongly electrophilic carbon centre alkyl halides are susceptible to nucleophilic attack, a nucleophile displaces the halogen as a nucleophilic halide ion. The reaction is called *nucleophilic substitution* and there are two types of mechanism, i.e. the S_N1 and S_N2 mechanisms. Carboxylic acids and carboxylic acid derivatives also undergo nucleophilic substitutions, but the mechanisms are totally different.

$$R-X \xrightarrow{Nu^{\ominus}} R-Nu + X^{\ominus}$$ Halide

Fig. Nucleophilic substitution.

S_N2 *Mechanism*

The reaction between methyl iodide and a hydroxide ion is an example of the S_N2 mechanism. The hydroxide ion is a nucleophile and uses one of its lone pair of electrons to form a new bond to the electrophilic carbon of the alkyl halide. Simultaneously, the C–I bond breaks. Both electrons in that bond move onto the iodine to give it a fourth lone pair of electrons and a negative charge. Since iodine is electronegative, it can stabilise this charge, so the overall process is favoured.

In the transition state for this process, the new bond from the incoming nucleophile is partially formed and the C–X bond is partially broken. The reaction centre itself (CH_3) is planar. This transition state helps to explain various other features of the S_N2 mechanism.

First, both the alkyl halide and the nucleophile are needed to form the transition state that means that the reaction rate is dependent on both components. Secondly, the hydroxide ion approaches iodomethane from one side while the iodide leaves from the opposite side.

The hydroxide and the iodide ions are negatively charged and will repel each other, so they are as far apart as possible in the transition state. Moreover, the hydroxide ion has to gain access to the reaction centre, i.e. the electrophilic carbon. There is more room to attack from the 'rear' since the large iodine atom blocks approach from the other side. Lastly from an orbital point of view, it is proposed that the orbital from the incoming nucleophile starts to overlap with the empty anti-bonding orbital of the C–X bond.

As this interaction increases, the bonding interaction between the carbon and the halogen decreases until a transition state is reached where the incoming and outgoing nucleophiles are both partially bonded. The orbital geometry requires the nucleophiles to be on opposite sides of the molecule.

Fig. S_N2 Mechanism for nucleophilic substitution.

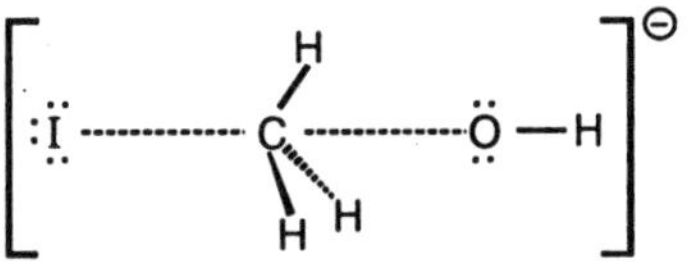

Fig. Transition state for S_N2 nucleophilic substitution.

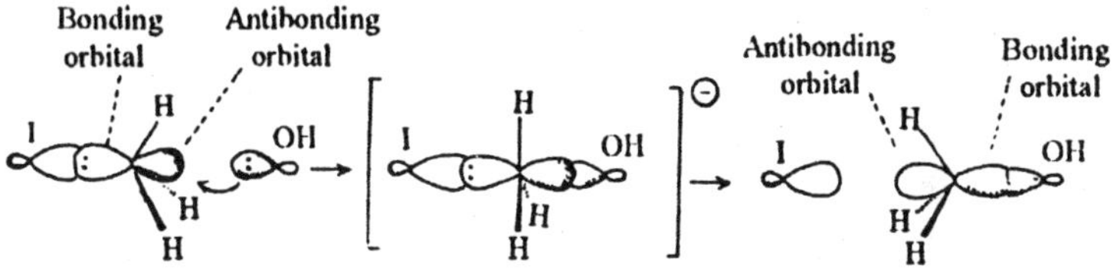

Fig. Orbital interactions in the S_N2 mechanism.

A third interesting feature about this mechanism is about the three substituents on the carbon. Both the iodide and the alcohol product are tetrahedral compounds with the three hydrogens forming an 'umbrella' shape with the carbon. However, the 'umbrella' is pointing in a different direction in the alcohol product compared to the alkyl halide. This means that the 'umbrella' has been turned inside out during the mechanism. Hence, the carbon centre has been 'inverted'. The transition state is the halfway house in this inversion.

There is no way of telling whether inversion has taken place in a molecule such as iodomethane, but proof of this inversion can be obtained by looking at the nucleophilic substitution of asymmetric alkyl halides with the hydroxide ion. Measuring the optical activity of both alkyl halide and the alcohol permits the configuration of each enantiomer to be identified. This demonstrates that inversion of the asymmetric centre occurs.

This inversion is called the *Walden Inversion* and the mechanism called *S_N2 mechanism*. The S_N stands for 'substitution nucleophilic'. The 2 signifies that the rate of reaction is second order or bimolecular and depends on both the concentration of the nucleophile and the concentration of the alkyl halide. The S_N2 mechanism is possible for the nucleophilic substitutions of primary and secondary alkyl halides, but is difficult for tertiary alkyl halides. We can draw a general mechanism to account for a range of alkyl halides and charged nucleophiles.

Umbrella Shape (open end to right)

Umbrella Shape (open end to left)

Fig. Walden inversion

The mechanism is almost the same with nucleophiles like ammonia or amines with the only difference that a salt is formed and an extra step is needed to gain the free amine. For example, consider the reaction between ammonia and 1-iodopropane. Ammonia's nitrogen atom is the nucleophilic centre for this reaction and uses its lone pair of electrons to form a bond to the alkyl halide. Due to this, the nitrogen will effectively lose an electron and will gain a positive charge.

The C–I bond is broken and an iodide ion is formed as a leaving group, which then acts as a counterion to the alkylammonium salt.

The free amine can be obtained by reaction with sodium hydroxide. This neutralise the amine to the free base that becomes insoluble in water and precipitates as a solid or as an oil.

Asymmetric centre

Asymmetric centre

Fig. Welden inversion of an asymmetric centre.

Fig. General mechanism for the S_N2 nucleophilic substitution of alkyl halides.

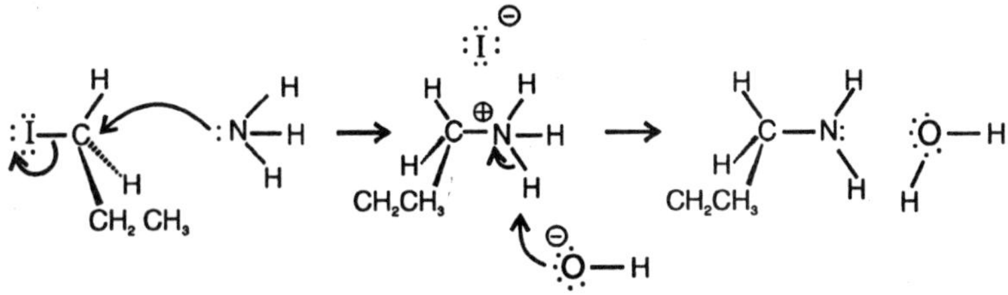

Fig. S_N2 mechanism for the reaction of 1-iodopropane with ammonia.

The reaction of ammonia with an alkyl halide is a nucleophilic substitution as far as the alkyl halide is concerned. However, the same reaction can be considered as an alkylation from the ammonia's point of view. This is because the ammonia has gained an alkyl group from the reaction.

Primary alkyl halides undergo the S_N2 reaction faster than secondary alkyl halides. Tertiary alkyl halides react extremely slowly if at all.

S_N1 Mechanism

When an alkyl is dissolved in a protic solvent like ethanol or water it gets exposed to a non-basic nucleophile (i.e. the solvent molecule).

Under these conditions, the order of reactivity to nucleophilic substitution changes dramatically from that observed in the S_N2 reaction, such that tertiary alkyl halides are more reactive then secondary alkyl halides, with primary alkyl halides not reacting at all. Thus, a different mechanism must be involved. For example, consider the reaction of 2-iodo-2-methylpropane with water. In it, the rate of reaction depends on the concentration of the alkyl halide alone and the concentration of the attacking nucleophile has no effect. Thus, the nucleophile must present if the reaction is to occur, but it does not matter whether there is one equivalent of the nucleophile or an excess. Since the reaction rate depends only on the alkyl halide, the mechanism is called the *S_N1 reaction,* where S_N stands for substitution nucleophilic and the 1 shows that the reaction is first order or unimolecular, i.e. only one of the reactants affects the reaction rate.

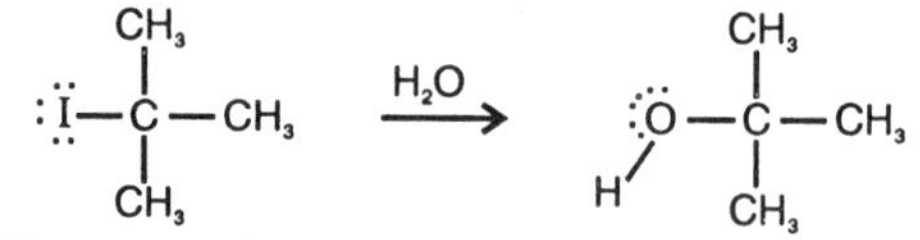

Fig. Reaction of 2-iodo-2-methylpropane with water.

There are two steps in the S_N1 mechanism. The first step is the rate determining step and it involves loss of the halide ion. The C–I bond breaks with both electrons on the bond moving onto the iodine atom to give it a fourth lone pair of electrons and a negative charge.

The alkyl portion becomes a planar carbocation in which all three alkyl groups are as far apart from each other as possible. The central carbon atom is now *sp^2* hybridised with an empty *$2p_y$* orbital. In the second step, water acts as a nucleophile and reacts with the carbocation to form an alcohol.

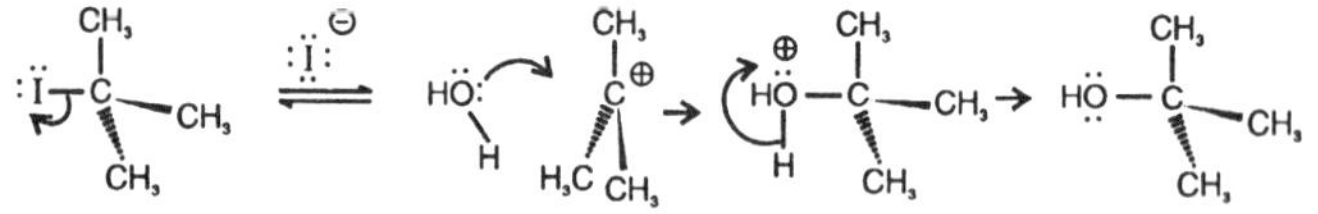

Fig. S_N1Mechanism.

In this mechanism the water molecule is coming in from the left hand side, but as the carbocation is planar, the water can attack equally well from the right hand side. Because the incoming nucleophile can attack from either side of the carbocation, so is no overall inversion of the carbon centre. This is significant when the reaction is carried out on chiral molecules. For example, if a chiral alkyl halide reacts with water by the S_N1 mechanism, both enantiomeric alcohols would be formed resulting in a racemate. However, total racemisation does not occur in S_N1 reactions. This is because the halide ion (departing from one side of the molecule) is still in the vicinity when the attacking nucleophile makes its approach. Due to this, the departing halide ion can hinder the approach of the attacking nucleophile from that particular side.

The term stereospecific indicates that the mechanism results in one specific stereochemical outcome (e.g. the S_N2 mechanism). This is distinct from a reaction which is stereoselective where the mechanism can lead to more than one stereochemical outcome, but where there is a preference for one outcome over another. Many S_N1 reactions will show a slight stereoselectivity.

CH_3 CH_3 H_3C
:I—C* $CH(CH_3)_2$ ⟶ HO—C* $CH(CH_3)_2$ + *C—OH
CH_2CH_3 CH_2CH_3 $CH(CH_3)_2$ CH_2CH_3

Fig. Racemisation of an asymmetric centre during S_N1 nucleophilic substitution.

Factors Affecting S_N2 versus S_N1 Reactions

S_N1 versus S_N2: There are two different mechanisms involved in the nucleophilic substitution of alkyl halides. When polar aprotic solvents are used, the S_N2 mechanism is preferred. Primary alkyl halides react more quickly than secondary alkyl halides, with tertiary alkyl halides hardly reacting at all. Under protic solvent conditions with non-basic nucleophiles (e.g. dissolving the alkyl halide in water or alcohol), the *S_N1* mechanism is preferred and the order of reactivity is reversed. Tertiary alkyl halides are more reactive than secondary alkyl halides and primary alkyl halides do not react at all.

There are various factors that determine if substitution will be S_N1 or S_N2 and they also control the rate at which these reactions occur. These include the nature of the nucleophile and the type of solvent is used. The reactivity of primary, secondary, and tertiary alkyl halides is controlled by electronic and steric factors.

Solvent

The S_N2 reaction is suitable in polar aprotic solvents (i.e. solvents with a high dipole moment, but with no O–H or N–H groups). These include solvents like acetonitrile (CH_3CN) or dimethylformamide (DMF). These solvents are polar enough to dissolve the ionic reagents needed for nucleophilic substitution, but they do so by solvating the metal cation rather than the anion. Anions are solvated by hydrogen bonding and because the solvent is incapable of hydrogen bonding, the anions remain unsolvated. Such 'naked' anions retain their nucleophilicity and react more strongly with electrophiles.

Polar protic solvents like water or alcohols can also dissolve ionic reagents but they solvate both the metal cation and the anion. Thus, the anion is 'caged' in by solvent molecules. Thus, stabilises the anion, makes it less nucleophilic and makes it less likely to react by the S_N2 mechanism. Due to this, the S_N1 mechanism becomes more important.

The S_N1 mechanism is specially favoured when the polar protic solvent is also a non-basic nucleophile. Therefore, it is most likely to take place when an alkyl halide is dissolved in water or alcohol. Protic solvents are bad for the S_N2 mechanism because they solvate the nucleophile, but they are good for the S_N1 mechanism. This is because polar protic solvents can solvate and stabilise the carbocation intermediate. If the carbocation is stabilised, the transition state leading to it will also be stabilised and this determines whether the S_N1 reaction is favoured or not. Protic solvents will also solvate the nucleophile by hydrogen bonding, but unlike the S_N2 reaction, this does not affect the reaction rate since the rate of reaction is independent of the nucleophile.

Non polar solvents are of no use in either the S_N1 or the S_N2 reaction as they cannot dissolve the ionic reagents needed for nucleophilic substitution.

Nucleophilicity

The relative nucleophilic strengths of incoming nucleophiles will affect the rate of the S_N2 reaction with stronger nucleophiles reacting faster. A charged nucleophile is stronger than the corresponding uncharged nucleophile (e.g. alkoxide ions are stronger nucleophiles than alcohols). Nucleophilicity is also related to base strength when the nucleophilic atom is the same (e.g. $RO^- > HO^- > RCO_2^- > ROH > H_2O$). In polar aprotic solvents, the order of nucleophilic strength for the halides is $F^- > Cl^- > Br^- > I^-$.

Because the rate of the S_N1 reaction is independent of the incoming nucleophile, the nucleophilicity of the incoming nucleophile is not so important.

Leaving Group

The nature of the leaving group is important to both the S_N1 and S_N2 reactions the better the leaving group, the faster the reaction. In the transition states of both reactions, the leaving group has gained a partial negative charge and the better that can be stabilised, the more stable the transition state and the faster the reaction. Therefore, the best leaving groups are those that form the most stable anions. This is also related to basicity in that the more stable the anion, the weaker the base. Iodide and bromide ions are stable ions and weak bases, and prove to be good leaving groups. The chloride ion is less stable, more basic and a poorer leaving group. The fluoride ion is a very poor leaving group and thus alkyl fluorides do not undergo nucleophilic substitution. The need for a stable leaving group explains why alcohols, ethers, and amines do not undergo nucleophilic substitutions since they would involve the loss of a strong base (e.g. RO^- or R_2N^-).

Alkyl Halides — S_N2

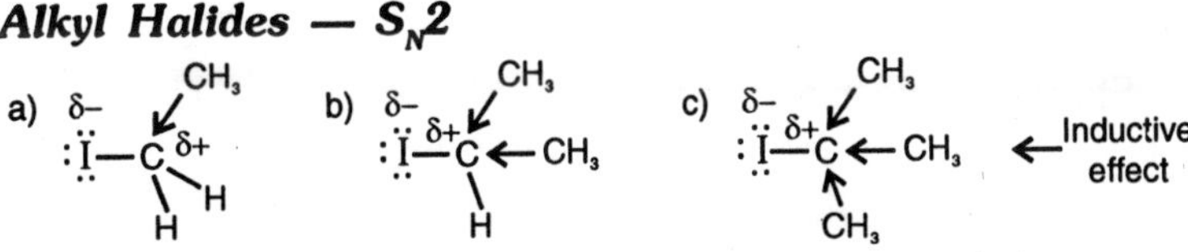

Fig. (a) Iodoethane; (b) 2-iodopropane; (c) 2-iodo-2-methylpropane.

There are two factors that affect the rate at which alkyl halides undergo the S_N2 reaction. These are electronic and steric. To illustrate why different alkyl halides react at different rates in the S_N2 reaction let us compare a primary, secondary, and tertiary alkyl halide.

Alkyl groups have an inductive, electron-donating effect that tends to lower the electrophilicity of the neighbouring carbon centre. Lowering the electrophilic strength means that the reaction centre will be less reactive to nucleophiles. Therefore, tertiary alkyl halides will be less likely to react with nucleophiles than primary alkyl halides, since the inductive effect of three alkyl groups is greater than one alkyl group.

Steric factors also play a role in making the S_N2 mechanism difficult for tertiary halides. An alkyl group is a bulky group compared to a hydrogen atom, and can therefore act like a shield against any incoming nucleophile. A tertiary alkyl halide has three alkyl shields compared to the one alkyl shield of a primary alkyl halide. Therefore, a nucleophile is more likely to be deflected when it approaches a tertiary alkyl halide and fails to reach the electrophilic centre.

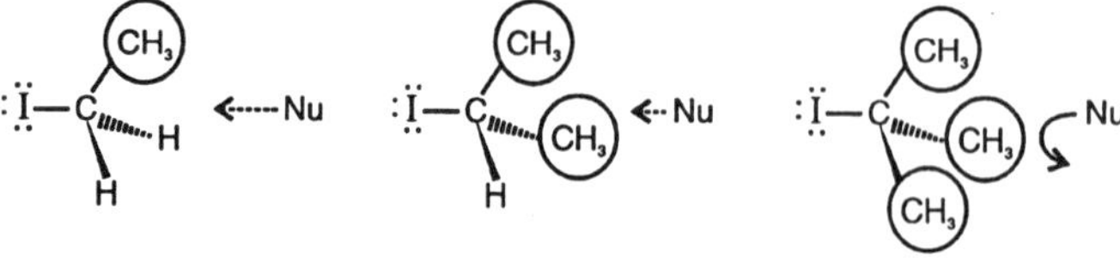

Fig. Steric factors affecting nucleophilic substitution.

Alkyl Halider — S_N1

Steric and electronic factors also play role in the rate of the S_N1 reaction because the steric bulk of three alkyl substituents makes it very difficult for a nucleophile to reach the electrophilic carbon centre of tertiary alkyl halides, these structures undergo nucleophilic substitution by the S_N1 mechanism. In this mechanism, the steric problem is relieved because loss of the halide ion creates a planar carbocation where the alkyl groups are much further apart and where the carbon centre is more accessible. Formation of the carbocation also relieves steric strain between the substituents.

Electronic factors also help in the formation of the carbocation because the positive charge can be stabilised by the inductive and hyperconjugative effects of the three alkyl groups:

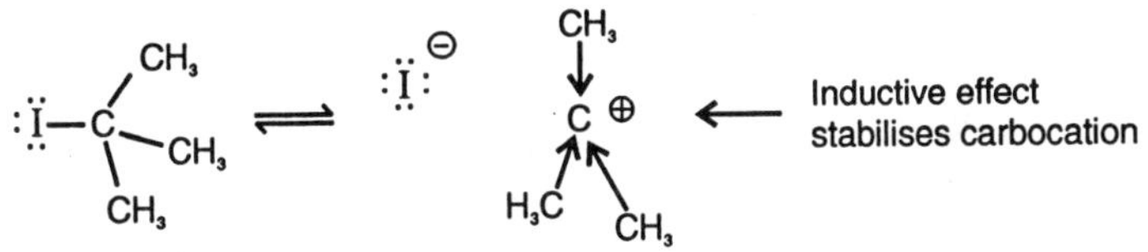

Inductive effect stabilises carbocation

Fig. Inductive effects stabilising a carbocation.

Both the inductive and hyperconjugation effects are greater when there are three alkyl groups connected to the carbocation centre than when there are only one or two. Therefore, tertiary alkyl halides are more likely to produce a stable coarbocation intermediate than primary or secondary alkyl halides.

Since the reaction rate is determined by how well the transition state of the rate determining step stabilised. In a situation in which a high energy intermediate is formed (i.e. the carbocation), the transition state leading to it will be closer in character to the intermediate than the starting material. Therefore, any factor that stabilises the intermediate carbocation also stabilises the transition state and consequently increases the reaction rate.

Determining the Mechanism

Generally the nucleophilic substitution of primary alkyl halides will occur via the S_N2 mechanism, whereas nucleophilic substitution of tertiary alkyl halides will occur by the S_N1 mechanism. Generally secondary alkyl halides are more likely to react by the S_N2 mechanism, but it is not possible to predict this with certainty.

The only way to find out for certain is to try out the reaction and see whether the reaction rate depends on the concentration of both reactants (S_N2) or whether it depends on the concentration of the alkyl halide alone (S_N1).

If the alkyl halide a chiral, the optical rotation of the product could be measured to see whether it is a pure enantiomer or not. If it is, the mechanism is S_N2. If not, it is S_N1.

Process of Elimination

Definition

Alkyl halides having a proton attached to a neighbouring β-carbon atom can undergo an elimination reaction to produce an alkene and a hydrogen halide. This reaction is the reverse of the electrophilic addition of a hydrogen halide to an alkene. There are two mechanisms by which this elimination can occur. These are E2 mechanism and the E1 mechanism.

The E2 reaction is the most effective for the synthesis of alkenes from alkyl halides and can be used on primary, secondary, and tertiary alkyl halides. The E1 reaction is not so useful from a synthetic point of view and occurs in competition with the S_N1 reaction of tertiary alkyl halides. Primary and secondary alkyl halides do not generally react by this mechanism.

Fig. Elimination of an alkyl halide.

Susceptible B-protons

An alkyl halide can undergo an elimination reaction if it has a susceptible proton situated on a β-carbon, i.e. the carbon next to the C–X group. This proton is lost during the elimination reaction along with the halide ion. In some respects, there is similarity here between alkyl halides and carbonyl compounds.

Alkyl halides can have susceptible protons at the β-position whilst carbonyl compounds can have acidic protons at their α-position. By comparing both structures, it can be seen that the acidic/ susceptible proton is attached to a carbon neighbouring an electrophilic carbon.

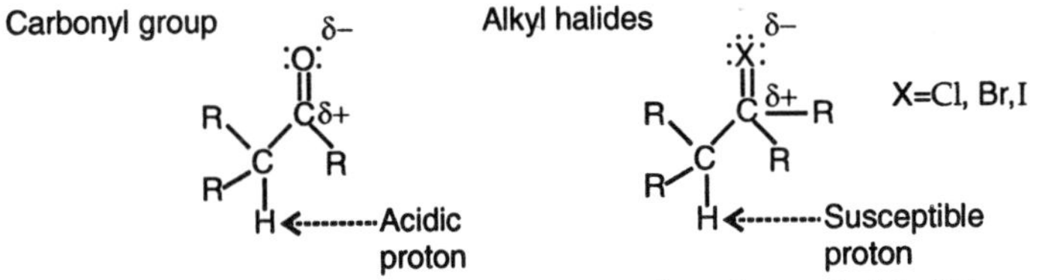

Carbonyl group Alkyl halides X=Cl, Br,I

Acidic proton Susceptible proton

Fig. Comparison of carbonyl compound and an alkyl halide.

E2 Mechanism

The E2 mechanism is a concerted mechanism and involves both the alkyl halide and the nucleophile. Due to this, the reaction rate depends on the concentration of both reagents and is called *second order* (E2 = Elimination second order). To illustrate the mechanism, we shall look at the reaction of 2-bromopropane with a hydroxide ion given below:

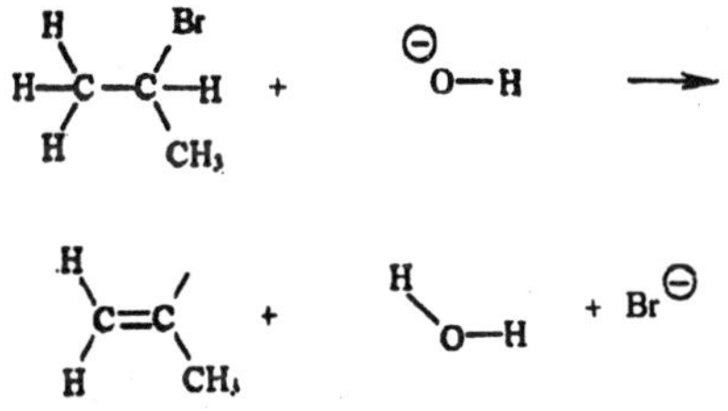

Fig. Reaction of 2-bromopropane with the hydroxide ion.

The mechanism involves the hydroxide ion forming a bond to the susceptible proton. As the hydroxide ion forms its bond, the C–H bond breaks.

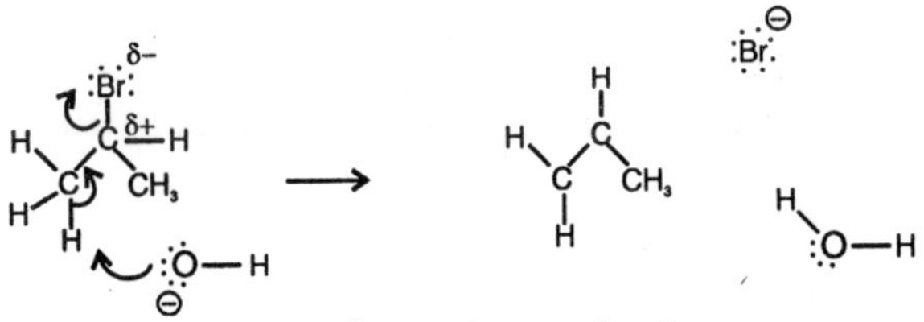

Fig. E2 Elimination mechanism.

Both electrons in that bond could move onto the carbon, but there is a neighbouring electrophilic carbon that attracts the electrons and so the electrons move in to form a π bond between the two carbons. Simultaneously as this π bond is formed, the C–Br bond breaks and both electrons end up on the bromine atom that is lost as a bromide ion.

Circled atoms are in one plane   Antiperiplanar arrangement

Fig. Relative geometry of the atoms involved in the E2 elimination mechanism.

The E2 elimination is stereospecific, with elimination taking place in an anti-periplanar geometry. The diagrams given below show that the four atoms involved in the reaction are in plane with the H and Br on opposite sides of the molecule.

The reason for this stereospecificity can be explained using orbital diagrams. In the transition state of this reaction, the C–H and C–Br σ bonds are in the process of breaking. As they do so, the sp^3 hybridised orbitals which were used for these σ bonds are changing into *p* orbitals that begins to interact with each other to form the eventual π bond. For all this to happen in the one transition state, an anti-periplanar arrangement is essential.

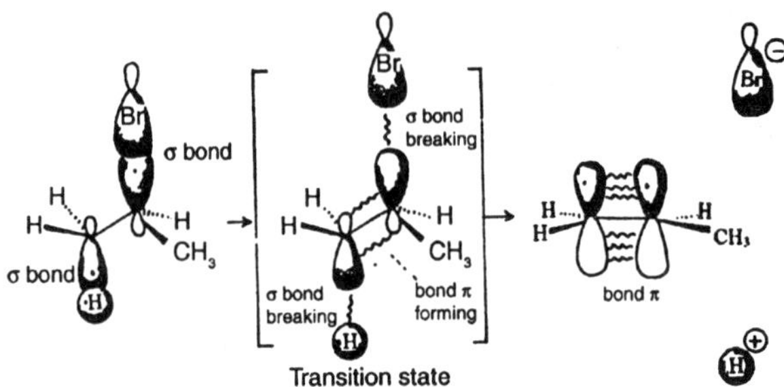

Fig. Orbital diagram of the E2 elimination process.

E1 Mechanism

The E1 mechanism generally occurs when an alkyl halide is dissolved in protic solvent where the solvent can act as a non-basic nucleophile. These are the same conditions for the S_N1 reaction and so both these reactions generally take place simultaneously forming a mixture of products. For example, the E1 mechanism is the reaction of 2-iodo-2-methyl-butane with methanol:

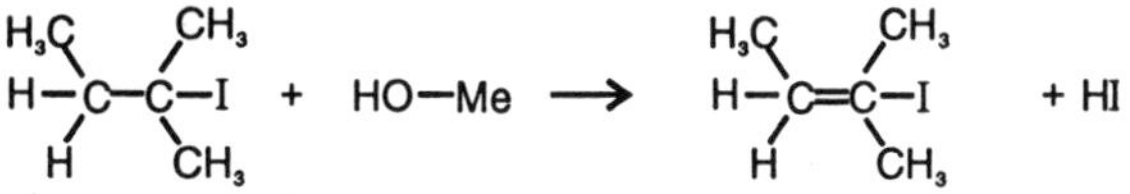

Fig. Elimination reaction of 2-iodo-2-methyl-butane.

There are two stages to the mechanism. The first stage is exactly the same as described for the S_N1 mechanism and that is cleavage of the C–X bond to form a planar carbocation intermediate in which the positive charge is stabilised by the three alkyl groups surrounding it. In the second stage, the methanol forms a bond to the susceptible proton on the β-carbon. The C–H bond breaks and both electrons are used to form a π bond to neighbouring carbocation. The first step of the reaction mechanism is the rate-determining step and as this is dependent only on the concentration of the alkyl halides, the reaction is first order (E1 = elimination first order). There is no stereospecificity involved in this reaction and a mixture of isomers can be obtained with the more stable (more substituted) alkene being favoured.

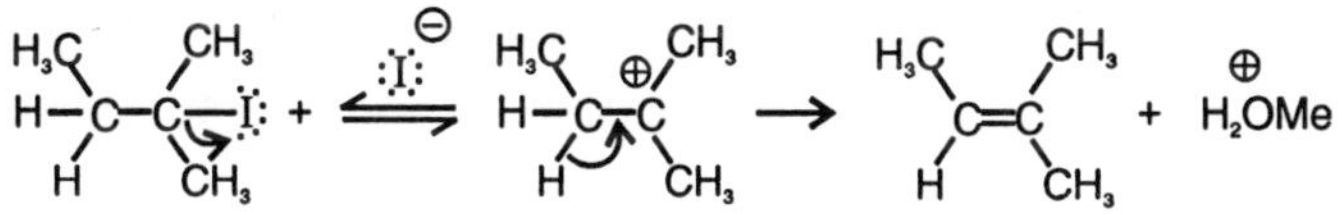

Fig. The El mechanism.

E2 versus E1

The E2 elimination occurs with a strong base (like a hydroxide or ethoxide ion) in a protic solvent (like ethanol or water). The E2 reaction is more common than the E1 elimination and more useful. All types of alkyl halide can undergo the E2 elimination and the method is useful for preparing alkenes.

The conditions that favour E1 are the same which that favour the S_N1 reaction (i.e. a protic solvent and a non-basic nucleophile). Therefore, the E1 reaction normally only takes place with tertiary alkyl halides and will be in competition with the S_N1 reaction.

Derivatives and Meaning of Carboxylic Acid

Structure and Properties

Structure: The structures of derivatives of carboxylic acids are derived from the parent carboxylic acid structure, The four common types of acid derivative are acid chlorides, acid anhydrides, esters, and amides. These functional groups contain a carbonyl group (C=O) in which both atoms are sp^2 hybridised. The carbonyl group along with the two neighbouring atoms is planar with bond angles of 120°.

The carbonyl group along with the attached carbon chain is known as *carboxylic acids* and carboxylic acid derivatives differ in atoms/groups attached to the acyl group (i.e. Y=Cl, OCOR, OR, NR_2, or OH). In all, these can founds the atom in Y that is directly attached to the carbonyl group is a hetero atom (Cl, O, or N). This distinguishes carboxylic acids and their derivatives from aldehydes and ketones in which the corresponding atom is hydrogen or carbon. This is important with respect to the type of reactions that carboxylic acids and their derivatives undergo. The carboxylic acid group (COOH) is generally called *carboxyl group*.

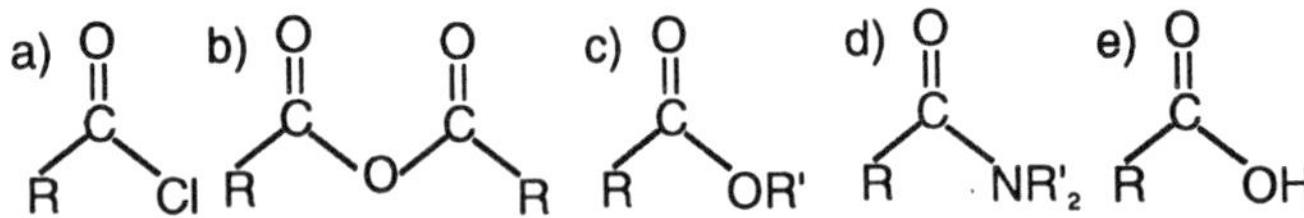

Fig. (a) Acid chloride; (b) acid anhydride; (c) ester; (d) amide; (e) carboxylic acid.

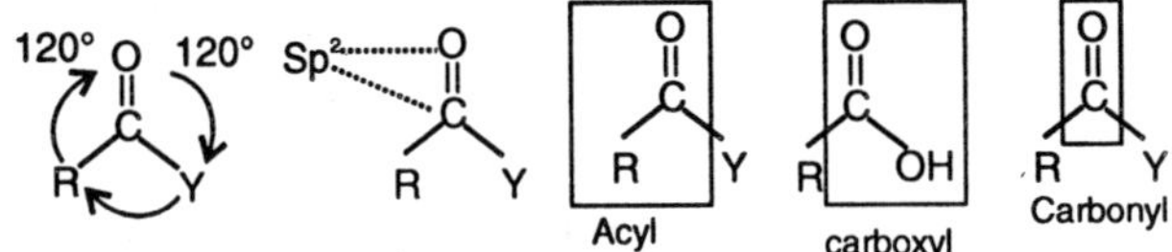

Fig. Structure of the functional group.

Bonding

The bonds in the carbonyl C=O group consist of a strong σ bond and a weaker πbond. As oxygen is more electronegative than carbon, the carbonyl group is polarised in such a way that the oxygen is slightly negative and the carbon is slightly positive, so oxygen can act as a nucleophilic centre and carbon can act as an electrophilic centre.

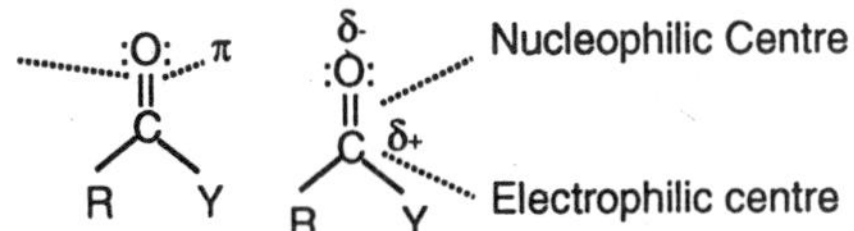

Fig. Bonding and properties.

Properties

Carboxylic acids and their derivatives are polar molecules because of the polar nature of carbonyl group and the presence of a heteroatom in the group Y. Carboxylic acids can associate with each other as dimers through the formation of two intermolecular hydrogen bonds, due to this, the carboxylic acids have higher boiling points than alcohols of comparable molecular weight. The low molecular weight carboxylic acids are soluble in water because of this hydrogen bonding.

H —bond

H —bond

Fig. Intermolecular H-bonding.

However, as the molecular weight of the carboxylic acid increases, the hydrophobic character of the alkyl portion outweighs the polar character of the functional group and thus at higher molecular weight carboxylic acids are insoluble in water.

Primary amides and secondary amides also have a hydrogen that is capable of hydrogen bonding (i.e. RCONHR', $RCONH_2$), due to this, these compounds also have higher boiling points for as compared to aldehydes and ketones of similar molecular weight. Acid chlorides, acid anhydrides, esters, and tertiary amides are polar, but due to lack a hydrogen atom that is capable of participating in hydrogen bonding, they have lower boiling points than carboxylic acids or alcohols of similar molecular weight, and similar boiling points to comparable aldehydes and ketones.

Carboxylic acids are weak acids in aqueous solution, forming an equilibrium between the free acids and the carboxylic ion. In the presence of a base like sodium hydroxide or sodium hydrogen carbonate, they ionise to form water-soluble salts and this provides a method of separating carboxylic acids from other organic compounds.

Reactions

Carboxylic acids and carboxylic acid derivatives commonly react with nucleophiles in a reaction called *nucleophilic substitution*. These reaction involves replacement of one nucleophile with another. Nucleophilic substitution is possible because the displaced nucleophile contains an electronegative heteroatom (Cl, O, or N) that is capable of stabilising a negative charge.

$$RC(=O)Y \xrightarrow{:Nu^{\ominus}} RC(=O)Nu + :Y^{\ominus}$$

$Y = Cl, OCOR, OR, NR_2$

Carboxylic acid derivatives

Fig. Nucleophilic substitution.

Substitution of Nucleophilic

Definition

Nucleophilic substitutions reactions are those reactions in which the substitution of one nucleophile for another is involved. Alkyl halides, carboxylic acids, and carboxylic acid derivatives undergo nucleophilic substitution. However, the mechanisms involved for alkyl halides are quite different from those involved for carboxylic acids and their derivatives. The reaction of a methoxide ion with ethanoyl chloride is a nucleophilic substitution reaction. In it one nucleophile (the methoxide ion) substitutes another nucleophile Cl^-.

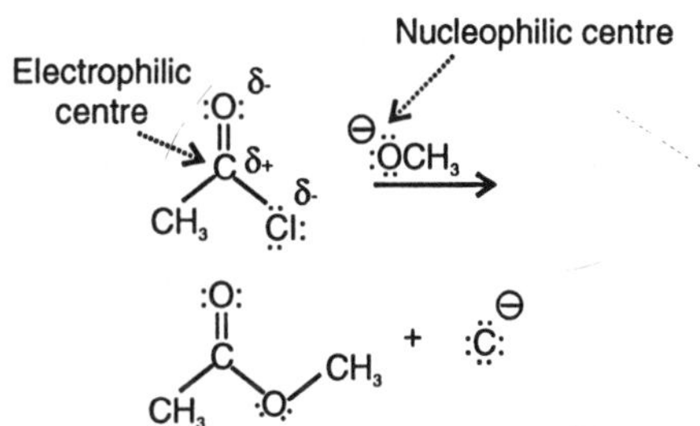

Fig. Nucleophilic substitution

Mechanism: Charged Nucleophiles

The methoxide ion uses of its lone pairs of electrons to form a bond to the electrophilic carbonyl carbon of the acid chloride. Simultaneously, the relatively weak π bond of the carbonyl group breaks and both of the π electrons move onto the carbonyl oxygen to give it a third lone pair of electrons and a negative charge. This is exactly the same first step involved in nucleophilic addition to aldehydes and ketones. However, with an aldehyde or a ketone, the tetrahedral structure is the final product. With carboxylic acid derivatives, the lone pair of electrons on oxygen return to reform the carbonyl π bond (step 2). As this happens, the C–Cl σ bond breaks with both electrons moving onto the chlorine to form a chloride ion that departs the molecule.

This explains how the products are formed, but why should the C–Cl σ bond break in preference to the C–OMe σ bond or the C–CH_3 σ bond can be explained by looking at the leaving groups which would be formed from these processes. The leaving groups would be a chloride ion, a methoxide ion and a carbanion, respectively.

The chloride ion is the best leaving group as it is the most stable. This is because chlorine is more electronegative than oxygen or carbon and can stabilise the negative charge. This same mechanism is involved in the nucleophilic substitutions of all other carboxylic acid derivatives and a general mechanism can be drawn as follow.

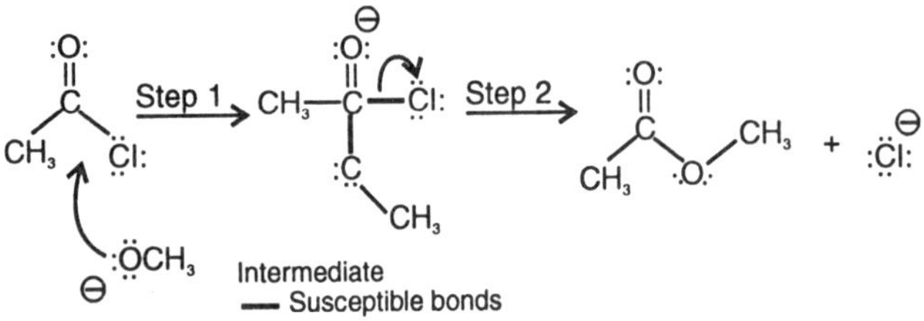

Fig. Mechanism of the nucleophilic substitution.

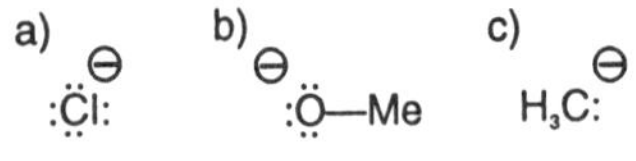

Fig. Leaving groups; (a) chloride; (b) methoxide; (c) carbanion.

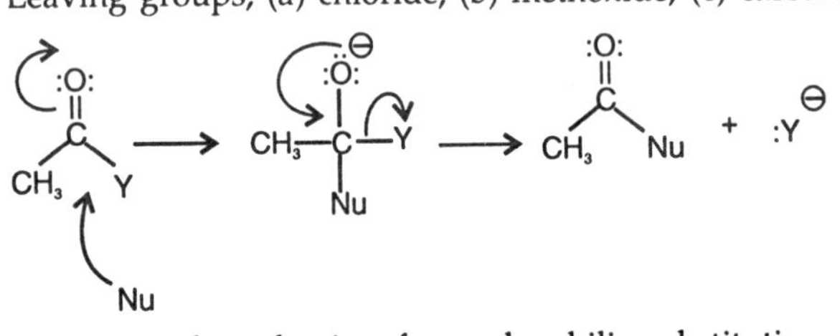

Fig. General mechanism for nucleophilic substitution.

Mechanism: Neutral Nucleophiles

Acid chlorides are quite reactive an *a* react with uncharged nucleophiles. For example, ethanoyl chloride react with methanol to form an ester. Oxygen is the nucleophilic centre in methanol and uses one of its lone pairs of electrons to form a new bond to the electrophilic carbon of the acid chloride. As this new bond forms, the carbonyl π bond breaks and both electrons move onto the carbonyl oxygen to give it a third pair of electrons and a negative charge (step 1).

The methanol oxygen gains a positive charge as it has effectively lost an electron by sharing its lone pair with carbon in the new bond. A positive charge on oxygen is not very stable and so the second stage in the mechanism is the loss of a proton. Both electrons in the O–H bond move onto the oxygen to restore a second lone pair of electrons and thus neutralise the charge. Methanol can help the process by acting as a base. The final stage in the mechanism is the same as before. The carbonyl π bond is reformed and as this happens, the C–Cl σ bond breaks with both electrons ending up in the departing chloride ion as a fourth lone pair of electrons. Finally, the chloride anion can remove a proton from $CH_3OR,^+$ to form HCl and methanol.

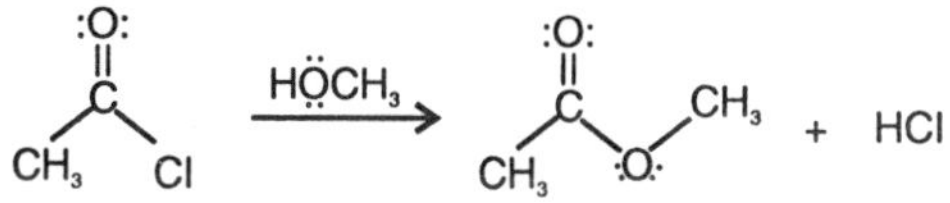

Fig. Ethanoyl chloride reacting with methanol to form methyl ethanoate.

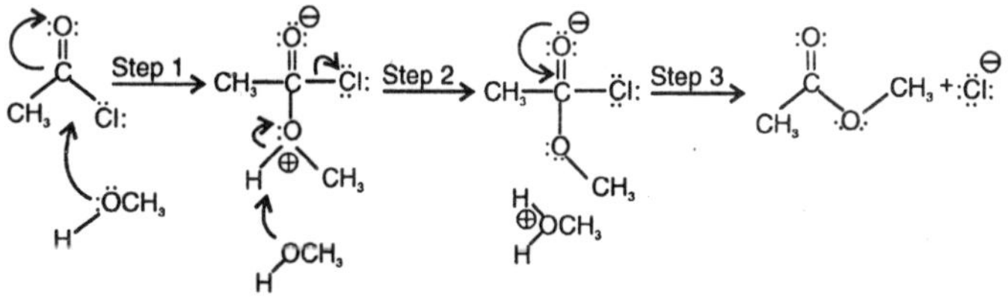

Fig. Mechanism for the reaction of an alcohol with an acid chloride.

The above mechanism is essentially the same mechanism involved in the reaction of ethanoyl chloride with sodium methoxide, the only difference being that we have to remove a proton during the reaction mechanism.

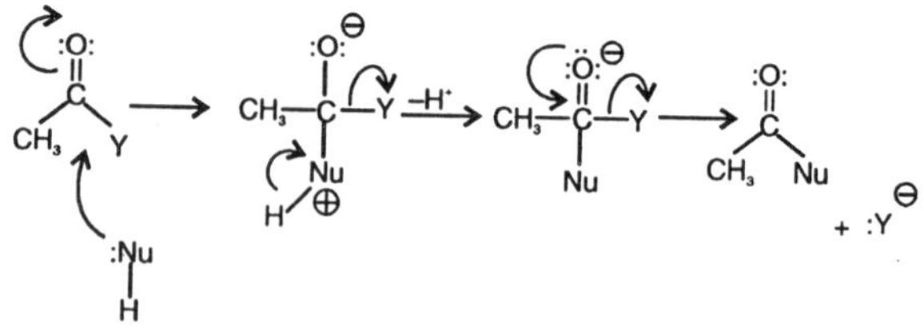

Y=Cl, OCOR, OR, NR_2

Fig. General mechanism for the nucleophilic substitution of a neutral nucleophile with a carboxylic add derivatives.

The same mechanism is true for nucleophilic substitutions of other carboxylic acid derivatives with neutral nucleophiles. In practice, acids or bases are generally added to improve yields.

Addition vs Substitution

Carboxylic acid derivatives undergo nucleophilic substitution whereas aldehydes and ketones undergo nucleophilic addition. This is because nucleophilic substitution of a ketone or an aldehyde would generate a carbanion or a hydride ion, respectively. These ions are unstable and highly reactive, so they are only formed with difficulty. Furthermore, C–C and C–H σ bonds are easily broken. Therefore, nucleophilic substitutions of aldehydes or ketones are not feasible.

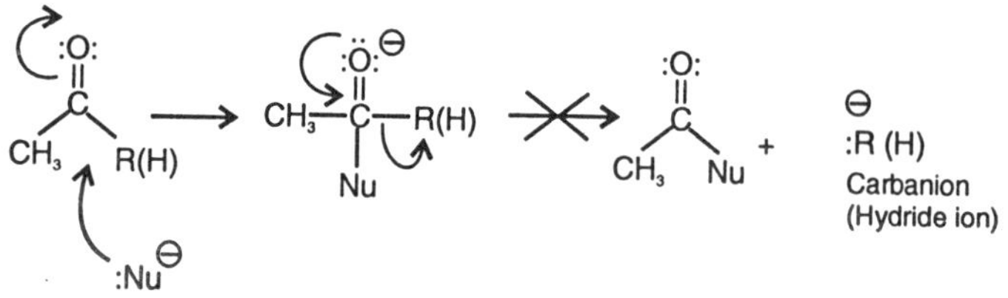

Fig. Unfavourable formation of an unstable carbanion hydride ion.

Process of Reactivity

Reactivity Order

Acid chlorides can be converted to acid anhydrides, esters, or amides. These reactions are possible because acid chlorides are the most reactive of the four carboxylic acid derivatives. Nucleophilic substitutions of the other acid derivatives are more limited because they are less reactive. For example, acid anhydrides can be used to synthesise esters and amides, but cannot be used to synthesise acid chlorides.

The possible nucleophilic reactions for each carboxylic acid derivative depends on its reactivity with respect to the other acid derivatives. Reactive acid derivatives can be converted to less reactive (more stable) acid derivatives, but not the other way round. For example, an ester can be converted to an amide, but not to an acid anhydride.

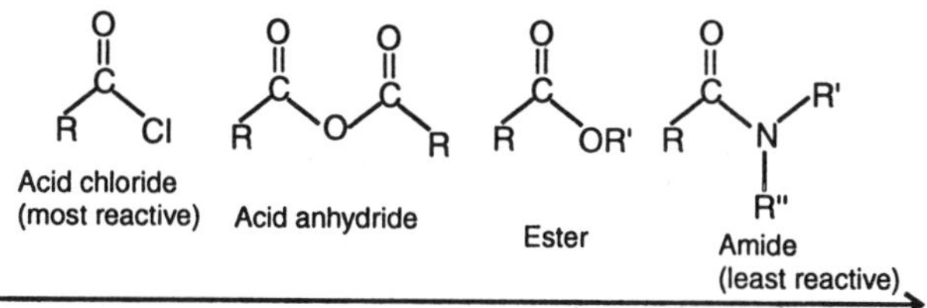

Fig. Relative reactivity of carboxylic acid derivatives.

Electronic Factors

To understand this difference in reactivity of various acid derivatives look at the first step in the nucleophilic substitution mechanism (involving the addition of a nucleophile to the electrophilic carbonyl carbon) which is the rate-determining step. Therefore, the more electrophilic this carbon is, the more reactive it will be. The nature of Y has a significant effect in this respect:

electrophilic centre

y=Cl, OCOR, OR, NR_2

Fig. The electrophilic centre of a carboxylic acid derivative.

Y is linked to the acyl group by an electronegative heteroatom (Cl, O, or N) that makes the carbonyl carbon more electrophilic. The extent to which this happens depends on the electronegativity of Y. If Y is strongly electronegative (e.g.

chlorine), it has a strong electron-withdrawing effect on the carbonyl carbon making it more electrophilic and more reactive to nucleophiles. Because chlorine is more electronegative than oxygen, and oxygen is more electronegative than nitrogen, acid chlorides are more reactive than acid anhydrides and esters, while acid anhydrides and esters are more reactive than amides.

The electron-withdrawing effect of Y on the carbonyl carbon is an inductive effect. With amides, there is an important resonance contribution that decreases the electrophilicity of the carbonyl carbon. The nitrogen has a lone pair of electrons that can form a bond to the neighbouring carbonyl carbon. As this new bond is formed, the weak π bond breaks and both electrons move onto oxygen to give it a third lone pair of electrons and a negative charge. Because the nitrogen's lone pair of electrons is being fed into the carbonyl group, the carbonyl carbon becomes less electrophilic and is less prone to attack by an upcoming nucleophile.

Theoretically, this resonance could also take place in acid chlorides, acid anhydrides, and esters to give resonance structures. However, the process is much less important because oxygen and chlorine are less nucleophilic than nitrogen. In these structures, the positive charge ends up on an oxygen or a chlorine atom.

These atoms are more electronegative than nitrogen and less able to stabilise a positive charge. These resonance structures might occur to a small extent with esters and acid anhydrides, but are far less likely in acid chlorides. This tend also matches the trend in reactivity.

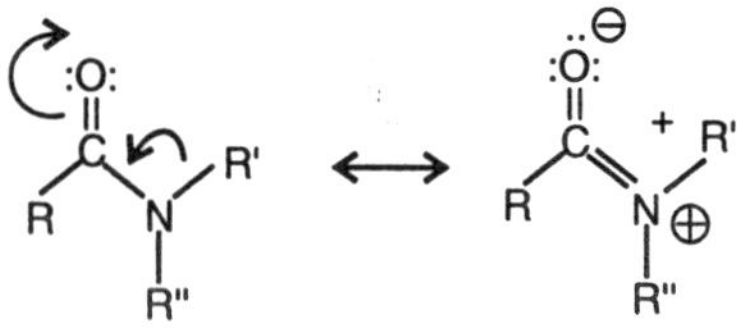

Fig. Resonance contribution in an amide.

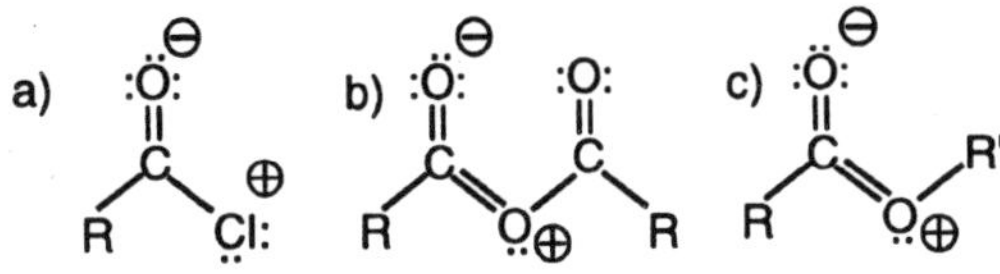

Fig. Resonance structures for (a) an acid chloride; (b) an acid anhydride (c) an ester

Although the resonance effect is weak in esters and acid anhydrides, it explain why acid anhydrides are more reactive than esters. Acid anhydrides have two carbonyl groups and so resonance can occur with either carbonyl group. Due to this, the lone pair of the central oxygen is 'split' between both groups that means that the resonance effect is split between both carbonyl groups.

This means that the effect of resonance at any one carbonyl group is diminished and it will remain strongly electrophilic. With an ester, there is only one carbonyl group and so it experiences the full impact of the resonance effect. Therefore, its electrophilic strength will be diminished relative to an acid anhydride.

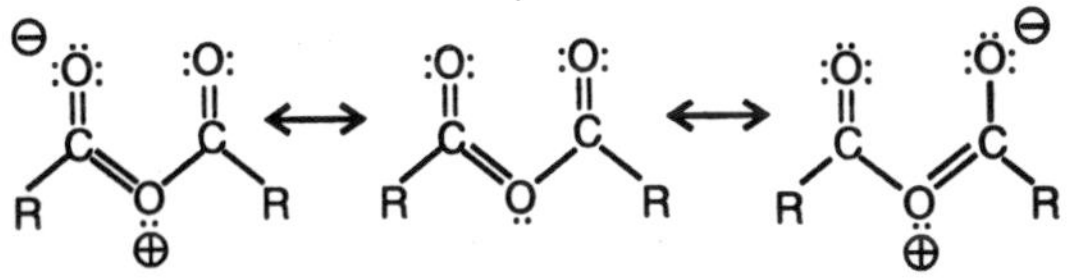

Fig. Resonance structures for an acid anhydride.

Steric Factors

Steric factors can also play a part in the reactivity of acid derivatives. For example, a bulky group attached to the carbonyl group can hinder the approach nucleophiles and hence lower reactivity. The steric bulk of the nucleophile can also have an influence in slowing down the reaction.

For example, acid chloride react faster with primary alcohols than they do with secondary or tertiary alcohols. This allows selective esterification if a molecule has more than one alcohol group present:

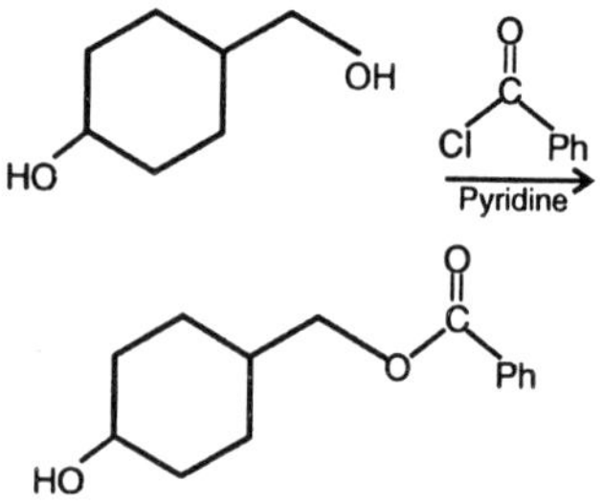

Fig. Selective esterification of a primary alcohol.

Carboxylic Acids

The nucleophilic substitution of carboxylic acids is complicated due to the fact that an acidic proton is present. Since most nucleophiles can act as bases, the reaction of a carboxylic acid with a nucleophile results in an acid-base reaction rather than nucleophilic substitution.

However, carboxylic acids can undergo nucleophilic substitution if they are activated before the reaction.

Process of Oxidation and Reduction

Alkenes: Oxidation and Reduction

Alkenes to Alkanes

Alkenes can be converted to alkanes by their reaction with hydrogen over a finely divided metal catalyst such as palladium, nickel, or platinum. This is an addition reaction, as it involves the addition of hydrogen atoms to each end of the double bond. It is also called a *catalytic hydrogenation* or a *reduction reaction.*

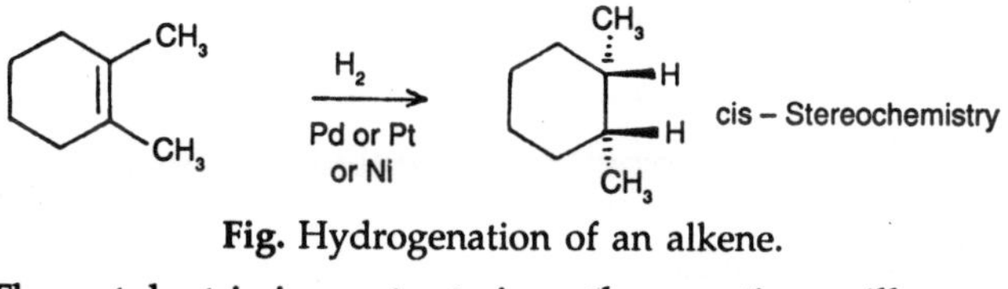

CH_3, CH_3 $\xrightarrow[\text{Pd or Pt or Ni}]{H_2}$ CH_3, H, H, CH_3 cis – Stereochemistry

Fig. Hydrogenation of an alkene.

The catalyst is important since the reaction will not occur at room temperature in its absence. This is because hydrogenation has a high free energy of activation (ΔG_1). The role of the catalyst is to bind the alkene and the hydrogen to a common surface such that they can react more easily. This results in a much lower energy of activation (ΔG_2) allowing the reaction to proceed in the much milder conditions.

The catalyst itself is unchanged after the reaction and can be used in small quantity.

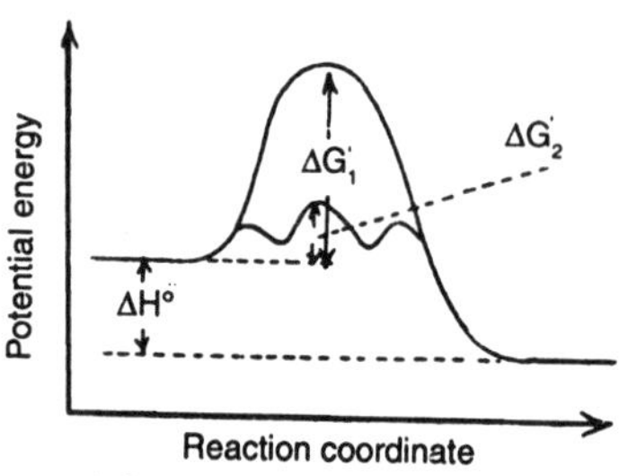

Fig. Graph of potential energy versus reaction coordinate for an uncatalysed and a catalysed hydrogenation reaction of an alkene.

Both the hydrogen and the alkene are bound to the catalyst surface before the hydrogen atoms are transferred, which means that both hydrogens are added to the same side of the double bond–*syn-addition. Note that the hydrogen molecule is split once it has been added to the catalyst.*

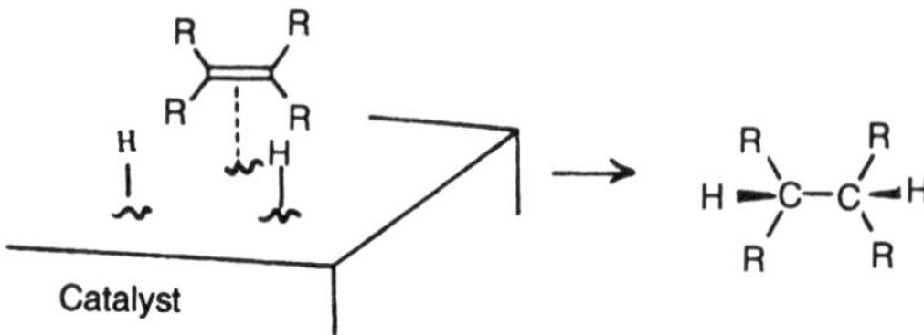

Fig. Binding of alkene and hydrogen to catalytic surface.

Alkenes to Aldehydes and Ketones

Considering that an alkene on oxidation with ozone that proceeds with the formation of an initial ozonide which then rearranges to an isomeric ozonide. This second ozonide is unstable and potentially explosive and so it is not generally isolated. Instead, it is reduced with zinc and water leading to the formation of two separate molecules:

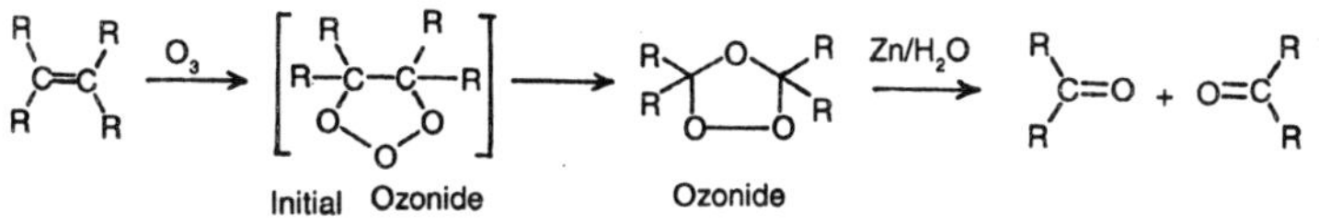

Initial Ozonide   Ozonide

Fig. Ozonolysis of an alkene.

The alkene is split across the double bond to form two carbonyl compounds. These will be ketones or aldehydes depending on the substituents present. For example, 3-methyl-2-pentene gives an aldehyde and a ketone:

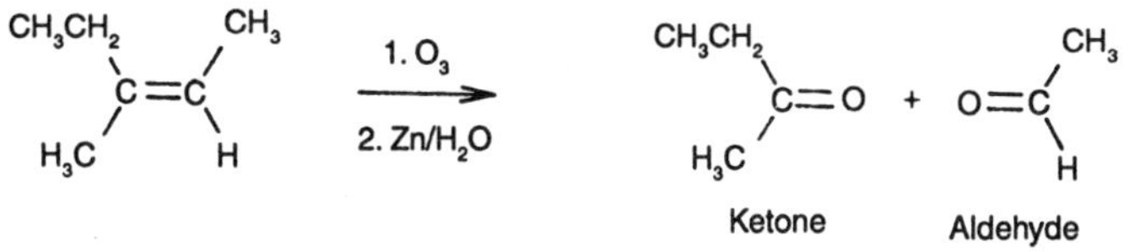

Ketone   Aldehyde

Fig. Ozonolysis of 3-methyl-2-pentene.

Alkenes to Carboxylic Acids and Ketones

Alkenes can be oxidatively cleaved with hot permanganate solution to give carboxylic acids and/or ketones. The products obtained depend on the substituents present on the alkene.

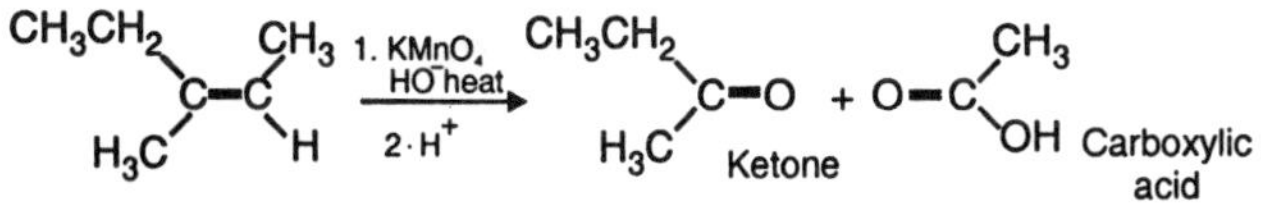

Ketone   Carboxylic acid

Fig. Oxidative cleavage of 3-methyl-2-pentene.

Alkenes to 1, 2-diols

The reaction of alkenes with osmium tetroxide (OsO_4) is an example of an oxidation reaction. In this case the alkene is not split, but, a 1,2-diol is obtained which is also called a *glycol*. The reaction involves the formation of a cyclic intermediate where the osmium reagent is attached to one face of the alkene.

On treatment with sodium bisulphite, the intermediate is cleaved such that the two oxygen atoms linking the osmium remain attached. Due to this both the alcohols being added to the same side of the double bond, i.e. *syn*-hydroxylation.

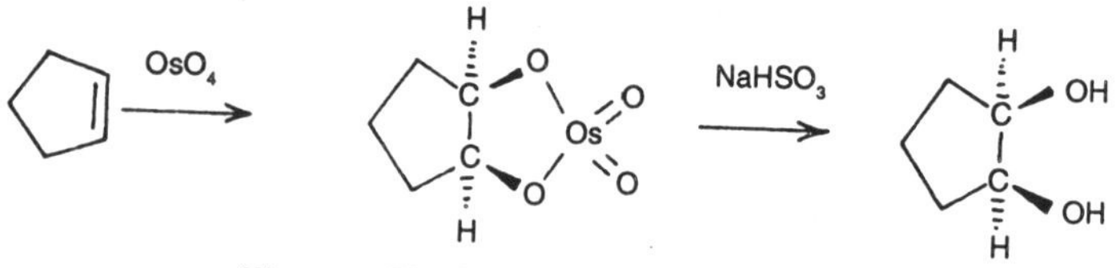

Fig. syn-Hydroxylation of an alkene.

The same reaction can also be carried out using cold alkaline potassium permanganate ($KMnO_4$) followed by treatment with aqueous base. It is important to keep the reaction cold since potassium permanganate can cleave the diol by further oxidation.

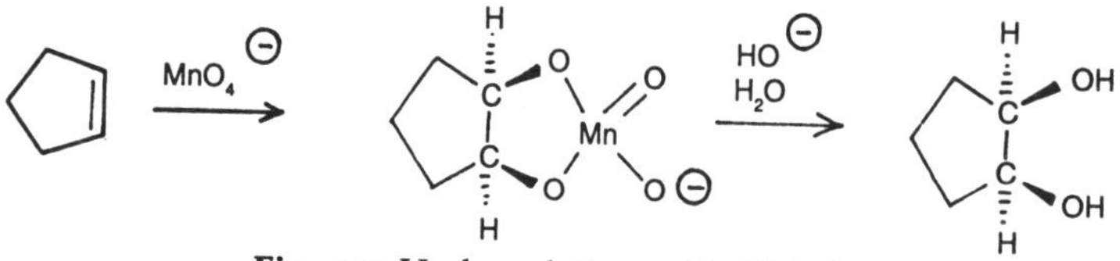

Fig. syn-Hydroxylation with $KMnO_4$

The reaction works better with osmium tetroxide. However, this is a highly toxic and expensive reagent and has to be handled with care.

Anti-hydroxylation of the double bond can also be achieved by forming an epoxide, then carrying out an acid-catalysed hydrolysis.

Alkenes to Expoxides

When an alkene is treated with a peroxyacid (RCO_3H) it forms a epoxide. m-Chloroperoxybenzoic acid is one of the most commonly used peroxyacids for this reaction. The reaction is unusual because there is no carbocation intermediate, and it involves a one-step process without intermediates.

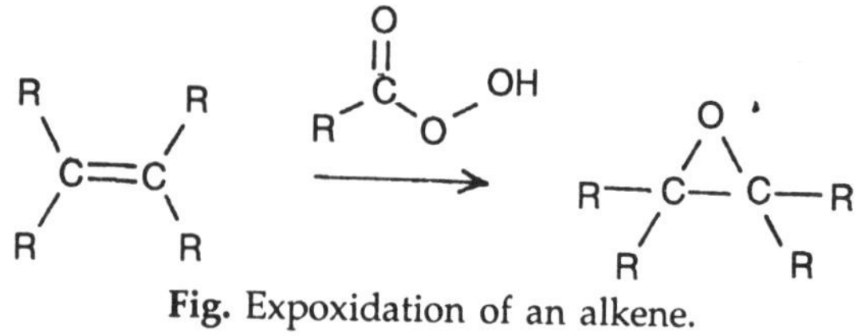

Fig. Expoxidation of an alkene.

Alkenes Hydroboration

Reaction

Alcohols can be generated from alkenes by reaction with diborane (B_2H_6 or BH_3), followed by treatment with hydrogen peroxide. The first part of the reaction involves the splitting of a B–H bond in BH_3 with the hydrogen joining one end of the alkene and the boron joining the other. Each of the B–H bonds is split in this way such that each BH_3 molecule reacts with three alkenes to give an organoborante intermediate where boron is liked to three alkyl groups. This can then be oxidised with alkaline hydrogen peroxide to produce the alcohol.

$$3\ R_2C{=}CR_2 \xrightarrow{BH_3} (H{-}CR_2{-}CR_2)_3B \xrightarrow{H_2O_2/NaOH} 3H{-}CR_2{-}CR_2{-}OH$$

Fig. Hydroboration of an alkene.

With unsymmetrical alkenes, the least substituted alcohol is obtained and so the organoborane reaction is complementary to the electrophilic addition reaction with aqueous acid. Steric factors appear to play a role in controlling this preference with the boron atom preferring to approach the least sterically hindered site. Electronic factors also play a role as described in the mechanism below:

$$(H_3C)_2C{=}CH_2 \xrightarrow{1)\ BH_3\quad 2)\ H_2O_2/H_2O} \underset{\text{I}}{(H_3C)_2CH{-}CH_2{-}OH} \quad \left(\underset{\text{II}}{HO{-}C(CH_3)_2{-}CH_3}\right)$$

Fig. Hydroboration of 2-methylpropane to give a primary alcohol (I) The tertiary alcohol (II) is not obtained.

Mechanism

The mechanism involves the alkene π bond interacting with the empty *p* orbital of boron to form a π complex. One of BH_3's hydrogen atom is then transferred to one end of the alkene as boron itself forms a o bond to the other end. This takes place through a four-centred transition state where the alkene's π bond and the B–H bond are partially broken, and the eventual C–H and C–B bonds are partially formed. There is an imbalance of electrons in the transition state which results in the boron being slightly negative and on of the alkene carbons being slightly positive.

The carbon best able to handle this will be the most substituted carbon and so the boron will end up on the least substituted carbon. (Note that boron has six valence electrons and is electrophilic. Therefore, the addition of boron to the least substituted position actually follows Markovnikov's rule.)

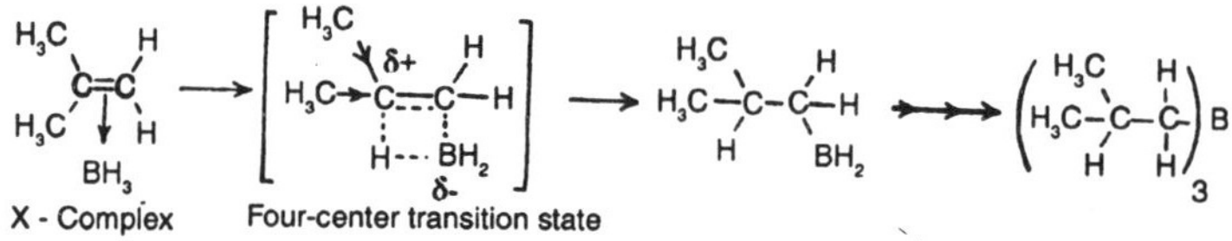

X - Complex   Four-center transition state

Fig. Mechanism of hydroboration.

Since subsequent oxidation with hydrogen peroxide replaces the boron with a hydroxyl group, the eventual alcohol will be on the least substituted carbon. Furthermore, the addition of the boron and hydrogen atoms takes place such that they are on the same side of the alkenes. This is called *syn-addition.*

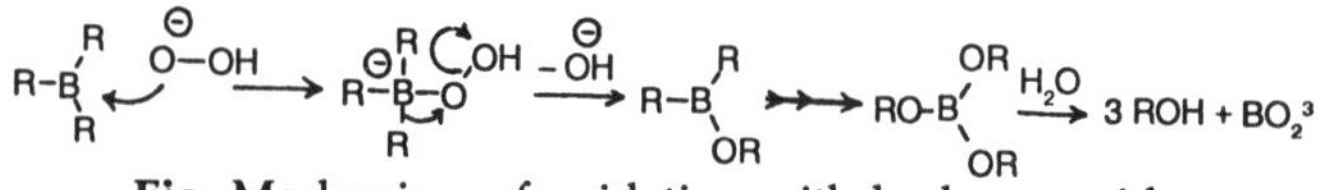

$R_2B{-}R + {}^{\ominus}O{-}OH \longrightarrow \ldots \xrightarrow{-\,{}^{\ominus}OH} \ldots \longrightarrow RO{-}B(OR)_2 \xrightarrow{H_2O} 3\,ROH + BO_3^{3-}$

Fig. Mechanism of oxidation with hydroperoxide.

The mechanism of oxidation involves addition of the hydroperoxide to the electron deficient boron to form an unstable intermediate which then rearranges such that an alkyl group migrates from the boron atom to the neighbouring oxygen and expels a hydroxide ion. This process is then repeated for the remaining two hydrogens on boron and the final trialkyl borate $B(OR)_3$ can then be hydrolysed with water to give three molecules of alcohol plus a borate ion.

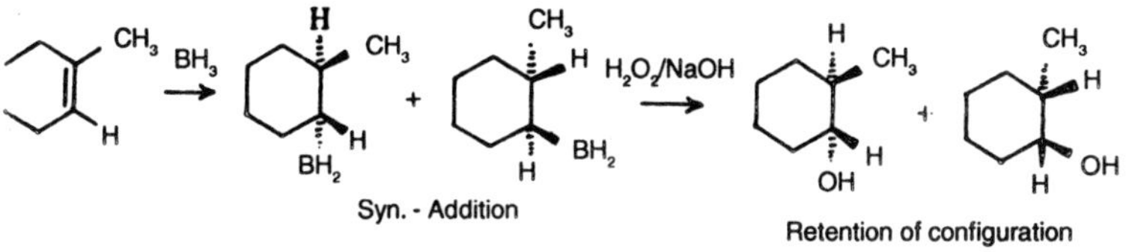

BH_3 $H_2O_2/NaOH$

Syn. - Addition

Retention of configuration

Fig. Stereochemical aspects of hydroboration.

The mechanism of oxidation takes place with retention of stereochemistry at the alcohol's carbon atom and so the overall reaction is stereospecific. Note that the reaction is stereospecific such that the alcohol group is *trans* to the methyl group in the product. However, it is not enantiospecific and both enantiomers are obtained in equal amounts (a racemate).

Alkynes: Electrophilic Addition

Additions to Symmetrical Alkynes

Alkynes give electrophilic addition reactions with the same reagents as in case of alkenes (e.g. halogens and hydrogen halides). Since there are two *n* bonds in alkynes, it is possible for the reaction to go once or twice depending on the amount of reagent added. For example, reaction of 2-butyne with one equivalent of bromine gives an (E)-dibromoalkene (fig. a). With two equivalents of bromine, the initially formed (E)-dibromoalkene reacts further to give a tetrabromoalkane (fig. b).

a) 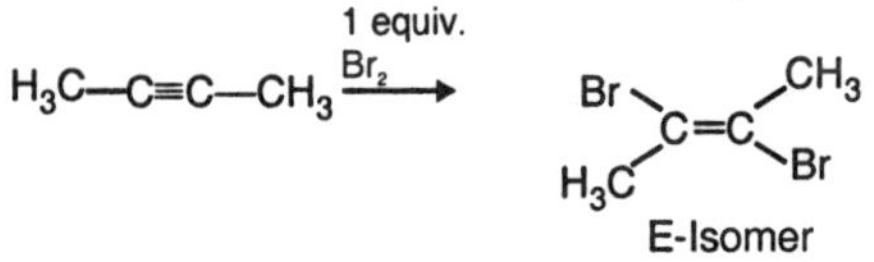 $H_3C{-}C{\equiv}C{-}CH_3 \xrightarrow[1\ equiv.]{Br_2}$ E-Isomer

b) 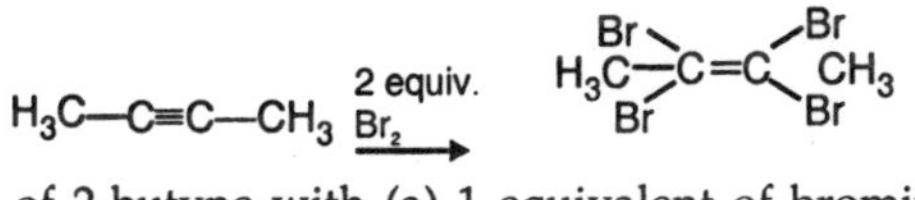 $H_3C{-}C{\equiv}C{-}CH_3 \xrightarrow[2\ equiv.]{Br_2}$

Fig. Reaction of 2-butyne with (a) 1 equivalent of bromine; (b) 2 equivalents of bromine.

When an alkyne is treated with one equivalent of HBr gives a bromoalkene (fig. a). If two equivalents of hydrogen bromide are present the reaction can go twice to give a geminal dibromoalkane where both bromine atoms are added to the same carbon (fig. b).

a) $H_3C—C{\equiv}C—CH_3 \xrightarrow[\text{HBr}]{\text{1 equiv.}}$ $(H_3C)(H)C{=}C(Br)(CH_3)$

b) $H_3C—C{\equiv}C—CH_3 \xrightarrow[\text{HBr}]{\text{2 equiv.}}$ $H_3C—CH_2—CBr_2—CH_3$

Fig. Reaction of 2-butyne with (a) 1 equivalent of HBr; (b) 2 equivalents of HBr.

These addition reactions are similar to the addition reactions of alkenes. However, the reaction is much slower for an alkyne, because alkynes are less reactive. Alkynes can be expected to be more nucleophilic, as they are more electron rich in the vicinity of the multiple bond, that is, six electrons in a triple bond as compared to four in a double bond. However, electrophilic addition to an alkyne involves the formation of a vinylic carbocation. This carbocation is much less stable than the carbocation intermediate formed during electrophilic addition to an alkene.

$$H_3C — C \equiv C — CH_3 \xrightarrow{\text{HBr}}$$

$$H_3C — \overset{\oplus}{C}{=}C(CH_3)(H) \xrightarrow{\text{HBr}} (H_3C)(Br)C{=}C(CH_3)(H)$$

Fig. Electrophilic addition to an alkyne via a vinylic carbocation (I).

Because this low reactivity, alkynes react slowly with aqueous acid and mercuric sulphate has to be added as a catalyst. The product that might be expected from this reaction is a diol:

$$H_3C—C{\equiv}C—CH_3 \xrightarrow[\text{HgSO}_4]{\text{H}^+/\text{H}_2\text{O}} \underset{\text{Intermediate}}{(H_3C)(HO)C{=}C(H)(CH_3)} \xrightarrow[\text{HgSO}_4]{\text{H}^+/\text{H}_2\text{O}} \underset{\text{Diol}}{H_3C—C(OH)_2—CH(H)(CH_3)}$$

Fig. Reaction of 2-butyne with aqueous acid and mercuric sulphate.

Actually, a diol is not formed. The intermediate (an enol) undergoes acid-catalysed rearrangement to give a ketone. This process is called a *keto-enol tautomerism.*

$$H_3C(HO)C{=}CH(CH_3) \xrightarrow{H^{\oplus}} H_3C{-}C({=}O){-}CH_2{-}CH_3$$

Enol Ketone

Fig. Keto-enol tautomerism.

Tautomerism is used to describe the rapid interconversion of two different isomeric forms (tautomers). In this case the keto and enol tautomers of a ketone. The keto tautomer is by far the dominat species for a ketone and the enol tautomer is generally present in only very small amounts (typically 0.0001 per cent). Therefore, as soon as the enol is formed in the above reaction, it rapidly tautomerised to the keto form and further electrophilic addition does not occur.

Additions to Terminal Alkynes

When a terminal alkyne is treated with an excess of hydrogen halide the halogens both end up on the more substituted carbon. This is in accordance with the Markovnikov's rule which states that the additional hydrogens end up on the carbon which already has the most hydrogens. The same rule applies for the reaction with acid and mercuric sulphate which means that a ketone is formed after keto-enol tautomerism instead of an aldehyde.

$$H_3C{-}C{\equiv}C{-}H \xrightarrow{HBr} H_3C(Br)C{=}CH_2 \xrightarrow{HBr} H_3C{-}CBr_2{-}CH_3$$

Intermediate

Fig. Reaction of propyne with HBr.

$$H_3C{-}C{\equiv}C{-}H \xrightarrow[HgSO_4]{H^+/H_2O} H_3C(HO)C{=}CH_2 \xrightarrow[HgSO_4]{H^+/H_2O} H_3C{-}C({=}O){-}CH_3$$

Intermediate

Fig. Reaction of propyne with aqueous acid and mercuric sulphate.

Alkynes Reduction

Hydrogenation

Alkynes react with hydrogen gas in the presence of a metal catalyst and the process called *hydrogenation.* It is an example of a reduction reaction. With a fully active catalyst like platinum metal, two molecules of hydrogen are added to produce an alkane.

$$R-C\equiv C-R \xrightarrow[Pt]{2\,H_2} H-\underset{H}{\overset{R}{C}}-\underset{H}{\overset{R}{C}}-H$$

Fig. Reduction of an alkyne to an alkene.

The reaction involves the addition of one molecule of hydrogen to form na alkene intermediate which then reacts with a second molecule of hydrogen to form the alkane. With less active catalysts, it is possible to stop the reaction at the alkene stage. *In particular, (Z)-alkenes can be synthesised from alkynes by reaction with hydrogen gas and Lindlar's catalyst.* This catalyst consists of metallic palladium deposited on calcium carbonate which is then treated with lead acetate and quinoline. The later treatment 'poisons' the catalyst in such a way that the alkyne reacts with hydrogen to give an alkene, but doe not react further. Since the starting materials are absorbed onto the catalyst surface, both hydrogens are added to the same side of the molecule to produce the (Z) isomer.

The same result can be achieved with nickel boride (Ni_2B) by using the P-2 catalyst.

$$R-C\equiv C-R \xrightarrow[H_2]{\text{Lindlar's Catalyst or P-2 Catalyst}} (R)(H)C=C(R)(H)$$

Fig. Reduction of alkyne to a (Z)-alkyne.

Dissolving Metal Reduction

$$R-C\equiv C-R \xrightarrow[\text{2) } NH_4Cl]{\text{1) Na or Li, } NH_3} (R)(H)C=C(H)(R)$$

Fig. Reduction of an alkyne to a *E-alkene.*

Reduction of an alkyne to an (E)-alkene can be achieved by treating the alkyne with lithium or sodium metal in ammonia at low temperatures. This is called *dissolving metal reduction.*

$$R{-}C{\equiv}C{-}R \xrightarrow{+e^{\ominus}} \text{Radical anion} \xrightarrow{H{-}NH_2} \text{Vinylic radical} \longrightarrow \text{Vinylic anion} \xrightarrow{H{-}NH_2} \text{(E)-RHC=CHR}$$

Fig. Mechanism for the dissolving metal reduction of an alkyne.

In this reaction, the alkali metal donates its valence electron to the alkyne to produce a radical anion. This removes a proton from ammonia to produce a *vinylic radical* that receives an electron from a second alkali metal to produce a *trans-vinylic anion.* This anion then removes a proton from a second molecule of ammonia and forms the *trans* or (E)-alkene. Only half curly arrows are used in the mechanism because this is a radical reaction *involving the movement of single electrons.*

Terminal Alkynes: Alkylation

Terminal Alkynes

A terminal alkyne is an alkyne that has a hydrogen substituents. This hydrogen substituent is acidic and can be removed with strong base (e.g. sodium amide $NaNH_2$) to produce an alkynide. This is an acid-base reaction.

$$R{-}C \equiv C{-}H$$

Fig. Terminal alkyne.

$$\underset{\text{Acid}}{R{-}C \equiv C{-}H} + \underset{\text{Base}}{:\overset{\ominus}{N}H_2} \longrightarrow \underset{\text{Alkynide}}{R{-}C \equiv \overset{\ominus}{C}:} + :NH_3$$

Fig. Reaction of a terminal alkyne with a strong base.

Alkylation

Once the alkynide is formed, it can be treated with an alkyl halide to form more complex alkynes. This reaction is called an *alkylation* and is an example of *nucleophilic substitution.*

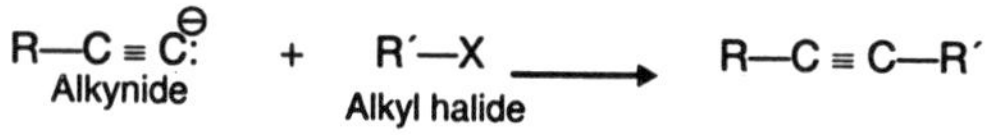

R—C ≡ C:$^{\ominus}$ (Alkynide) + R′—X (Alkyl halide) ⟶ R—C ≡ C—R′

Fig. Reaction of an alkynide ion with an alkyl halide.

This reaction works best with primary alkyl halides. When secondary or tertiary alkyl halides are used, the alkynide reacts like a base and this results in elimination of hydrogen halide from the alkyl halide to produce an alkene:

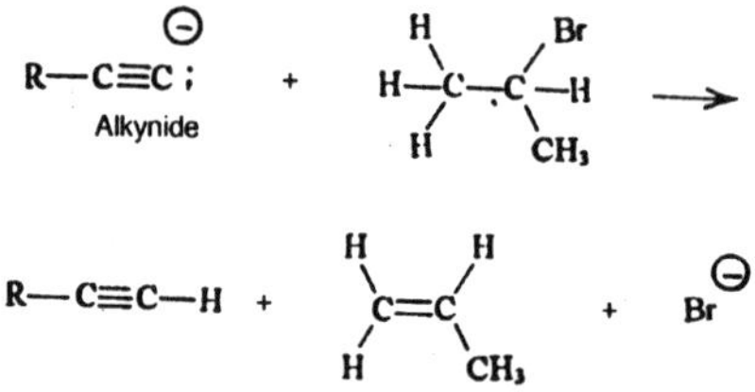

R—C≡C:$^{\ominus}$ (Alkynide) + CH_2H—CH(Br)—CH_3 ⟶

R—C≡C—H + H_2C=CH—CH_3 + $Br^{\ominus}$

Fig. Elimination of HBr.

Dienes, which are Conjugated

Structure

A conjugate diene is made up of two alkene units separated by a single bond (fig. a). Dienes are separated by more than one single bond known as *non-conjugated dienes* (fig. b).

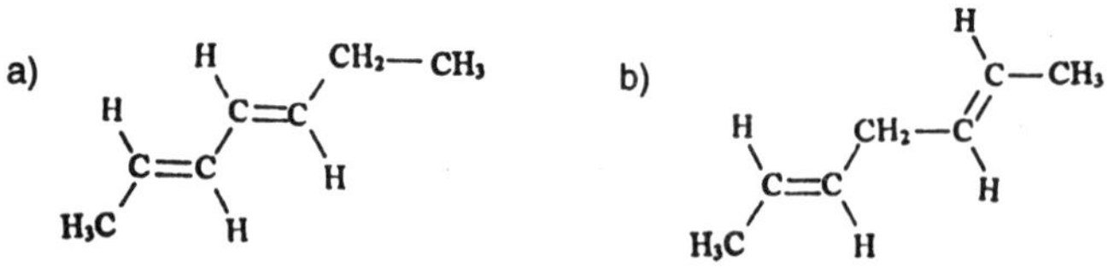

a) H_3C—CH=CH—CH=CH—CH_2—CH_3

b) H_3C—CH=CH—CH_2—CH=CH—CH_3

Fig. (a) Conjugated diene; (b) non-conjugated diene.

Bonding

A conjugated diene does not behave like two isolated alkenes. For example, the length of the 'single' bond connecting the two alkene units is slightly shorter than expected for a typical single bond (1.48Å). This shows that there is a certain amount of double-bond character present in this bond. Two sp^2 hybridised carbons rather than two sp^3 hybridised carbons. Therefore, an sp^2 hybridised orbital from each carbon is used for the single bond. Since this hybridised orbital ha more s-character than an sp^3 hybridised orbital, the bond is expected to be shorter. An alternative explanation is that the π orbitals of the two alkene systems can overlap to produce the partial double-bond character:

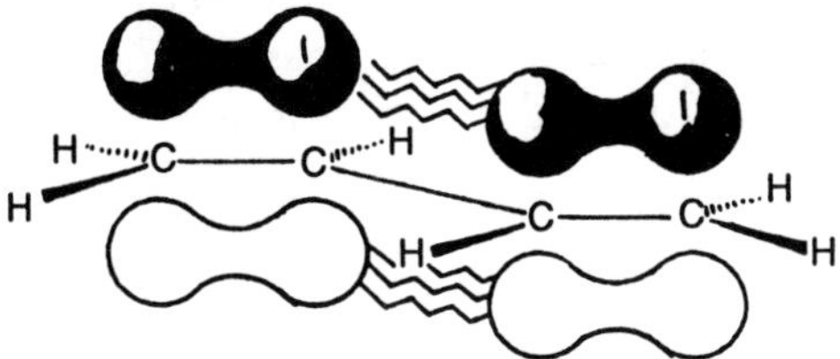

Fig. π-Orbital overlap.

Electrophilic Addition

The reactions of a conjugate diene reflect the fact that a conjugated diene should be viewed as a functional group in its own right, rather than as two separate alkenes. Electrophilic addition to a conjugated diene results in a mixture of two possible products arising from 1,2-addition and 1,4 addition:

a) Br_2 → 1,2-Addition + 1,4-Addition

b) HBr → 1,2-Addition + 1,4-Addition

Fig. Electrophilic addition to a conjugated diene of (a) bromine and (b) HBr.

In 1,2-addition, new atoms have been added to each end of one of the alkene units. This is the normal electrophilic addition of an alkene with which we are familiar. In 1,4-addition, new atoms have been added to each end of the entire diene system. Furthermore, the double bond remaining has shifted position (isomerised) to the 2, 3-position.

The mechanism of 1, 4-addition starts off in the same way as a normal electrophilic addition. We shall consider the reaction of a conjugated diene with hydrogen bromide as an example. One of the alkene units of the diene uses its *n* electrons to form a bond to the electrophilic hydrogen of HBr. The H–Br bond breaks at the same time to produce a bromide ion. The intermediate carbocation produced has a double bond next to the carbocation centre and is called an *allylic carbocation.*

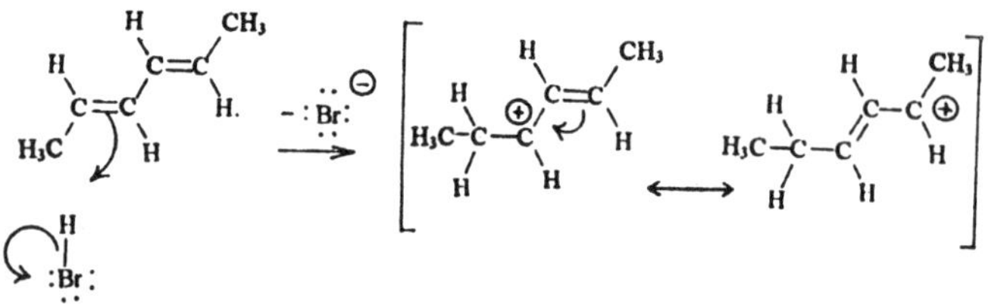

Fig. Mechanism of 1,4-addition-first step.

This system is now set up for resonance involving the remaining alkene and the carbocation centre, resulting in delocalisation of the positive charge between positions 2 and 4. Due to this delocalisation, the carbocation is stabilised and this in turn explains two features of this reaction. First of all, the formation of two different products is now possible since the second stage of the mechanism involve the bromide anion attacking either at position 2 or at position 4.

Secondly, it explains why the alternative 1, 2-addition product is not formed. The intermediate carbocation required for this 1, 2-addition cannot be stabilised by resonance. Therefore, the reaction proceeds through the allylic carbocation instead.

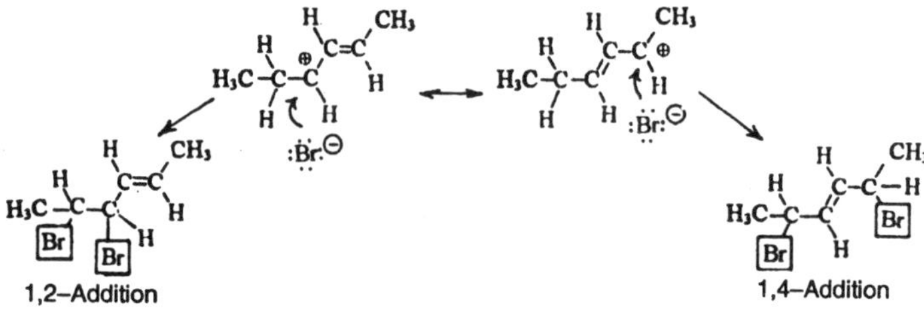

1,2–Addition

1,4–Addition

Fig. Mechanism of 1, 2- and 1, 4-addition - second step.

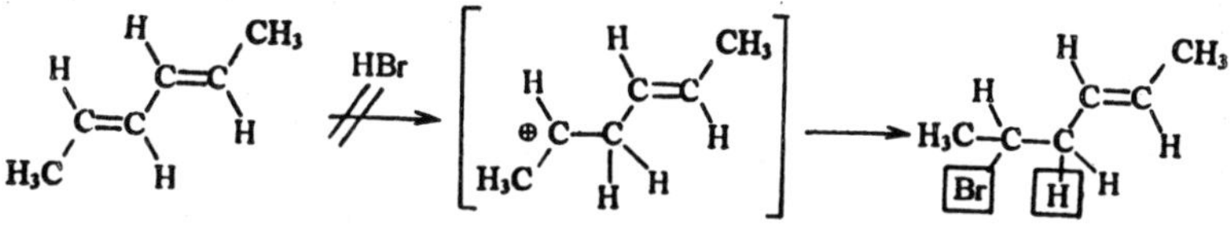

Fig. Unfavoured reaction mechanism.

Diets-Alder Cycloaddition

The diels-Alder cycloaddition reaction is an important reaction by which six-numbered rings can be synthesised. It involves a *conjugated diene* and an *alkene*. The alkene is called a *dienophile* (diene-lover) and generally has an electron-withdrawing group linked to it (e.g. a carbonyl group or a nitrile). The *mechanism is concerted with* new bonds being formed simultaneously as old bonds are being broken. No intermediates are involved.

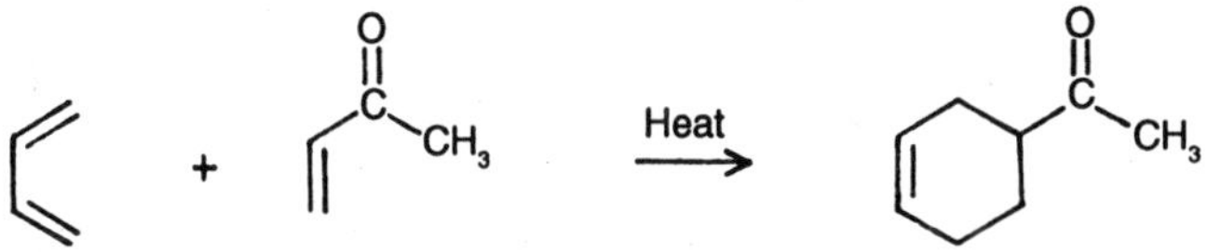

Fig. Diels-Alder cycloaddition.

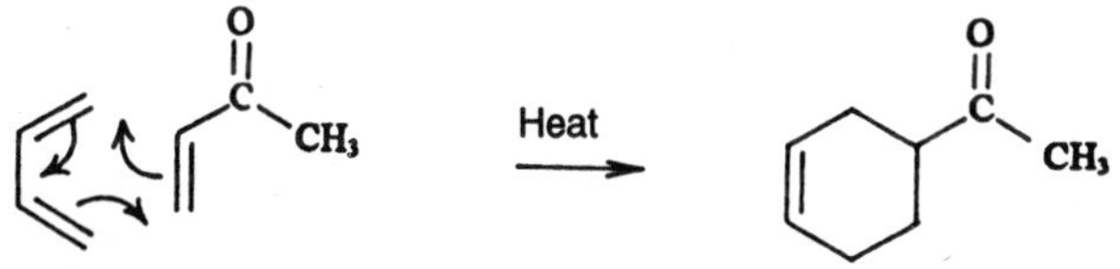

Fig. Mechanism of Diels-Alder cycloaddition.

Process of Aromaticity

Definition

Originally term aromatic was used for benzene-like structures because of the distinctive aroma of such compounds. Aromatic compounds undergo certain distinctive reactions that set them apart from other functional groups. They are *highly unsaturated compounds,* but unlike alkenes and alkynes, they are *relatively unreactive* and tend to undergo reactions that involve a retention of their unsaturation. Benzene is a six-numbered ring structure having three formal double bonds (fig. a). However, the six π electrons involved are not localised between any two carbon atoms. Instead, they are delocalised around the ring that results in an increased stability. Because of this, benzene is often written with a circle in the centre of the ring to signify the delocalisation of the six π electrons (fig. b). Reactions that disrupt this delocalisation are not favoured as it means a loss of stability, so benzene undergoes reactions in which the aromatic ring system is retained. All six carbon atoms in benzene are sp^2 hybridised, and the molecule itself is cyclic and planar.

The planarity is essential if the $2p$ atomic orbitals on each carbon atom are to overlap and result in delocalisation.

Fig. Representations of benzene.

Huckel Rule

An aromatic molecule must be *cyclic* and *planar with sp^2 hybridised atoms* (i.e. conjugated), but it must also obey the Huckel rule. It states, that *the ring system must have $4n + 2$ π electrons* where n = 1, 2, 3, *etc*. Therefore, ring systems which have 6, 10, 14, ... π electrons are aromatic. Benzene fits the Huckel rule as it has six π electrons. Cyclooctatetraene has eight π electrons and *does not obey the Huckel rule.* Although all the carbon atoms in the ring are sp^2 hybridised, cyclooctateraene reacts like a conjugated alkene.

It is not planar, the π electrons are not delocalised and the molecule is made up of alternating single and double bonds (fig. a). However, the 18-membered cyclic system (fig. b) obeys the Huckel rule (n = 4) and is a planar molecule with aromatic properties and a delocalised π system.

Fig. (a) Cyclooctatetraene; (b) 18-membered aromatic ring.

We can also possible to get aromatic ring. The *cyclopentadienyl anion* and the *cycloheptatrienyl cation* are both aromatic. Both are cyclic and planar, containing six π electrons, and all the atoms in the ring are sp^2 hybridised.

Fig. (a) Cyclopentadienyl anion; (b) cycloheptatrienyl cation.

Bicyclic and polycyclic systems can also be aromatic.

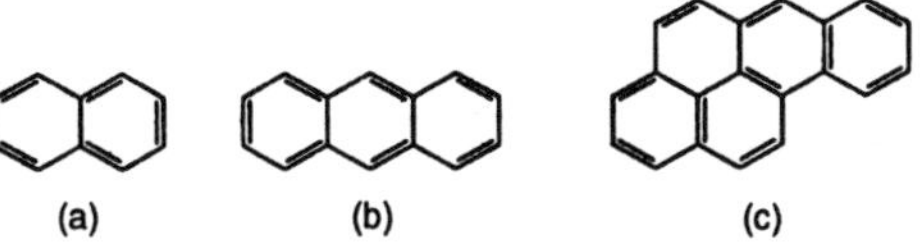

Fig. (a) Naphthalene; (b) anthracene; (c) benzo[a]pyrene.

Aromatic Compounds: Properties and Formation

Preparation

It is difficult to synthesise aromatic compounds in the laboratory from scratch and most aromatic compounds are prepared from benzene or other simple aromatic compounds (e.g. toluene and naphthalene). These in turn are isolated from natural sources like coal or petroleum.

Properties

Many aromatic compounds have a characteristic aroma and they burn with a smoky flame. That are hydrophobic, non-polar molecules and will dissolve in organic solvents. They are soluble in water. Aromatic molecules can interact with each other through *intermolecular bonding by van der Waals interactions* (fig. a). However, induced dipole interactions are also possible with alkyl ammonium ions or metal ions where the positive charge of the cation induces a dipole in the aromatic ring such that the face of the ring is slightly negative and the edges are slightly positive (fig. b). This results in the cation being sandwiched between two aromatic rings.

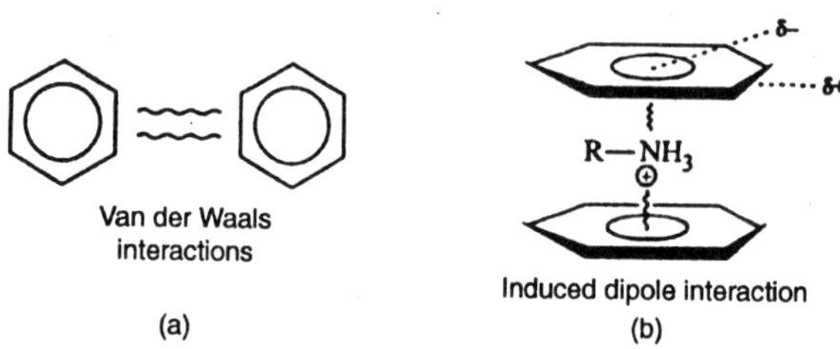

Fig. Intermolecular bonding involving aromatic rings.

Aromatic compounds are generally stable and they do not react like alkenes. They prefer to undergo reactions in which the stable aromatic ring is retained. The most common type of reaction for aromatic rings is electrophilic substitution, but reduction is also possible.

Benzene's Electrophilic Substitution

Definition

Aromatic rings undergo electrophilic substitution, for example the bromination of benzene (Following fig). The reaction involves an electrophile (Br^+) replacing another electrophile (H^+) with the aromatic ring remaining intact. Therefore, one electrophile replaces another and the reaction is called an *electrophilic substitution.*

$$C_6H_5{-}H + Br^{\oplus} \longrightarrow C_6H_5{-}Br + H^{\oplus}$$

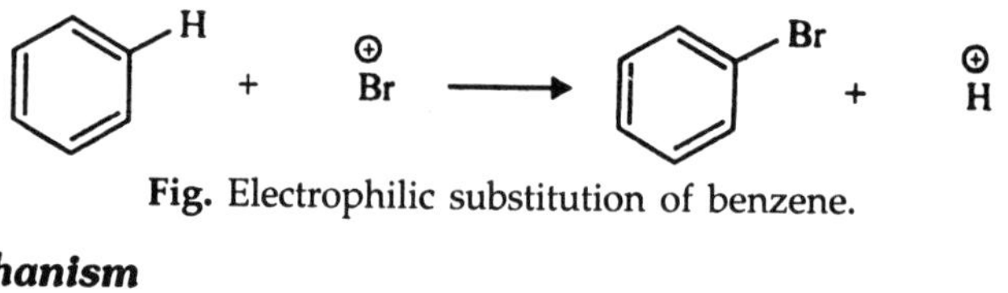

Fig. Electrophilic substitution of benzene.

Mechanism

In the mechanism of electrophilic substitution reactions the aromatic ring acts as a nucleophile and it provides two of its π electrons to form a bond to Br^+. The aromatic ring now has lost one of its formal double bonds which results in a positively charged carbon atom. This first step in the mechanism is the same as that described for the electrophilic addition to alkenes, and so the positively charged intermediate here is equivalent to the carbocation intermediate in electrophilic addition.

However, in step 2, the mechanism of electrophilic addition and electrophilic substitution differ. Whereas the carbocation intermediate from an alkene reacts with a nucleophile to form an addition product, the intermediate from the aromatic ring loses a proton. The C–H σ bond breaks and the two electrons move into the ring to reform the π bond, thus regenerating the aromatic ring and neutralising the positive charge on the carbon. This is the mechanism undergone in all electrophilic substitutions. The only difference is the nature of the electrophile.

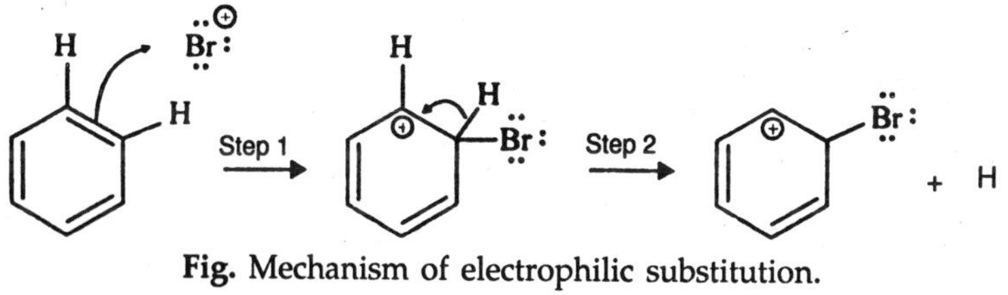

Fig. Mechanism of electrophilic substitution.

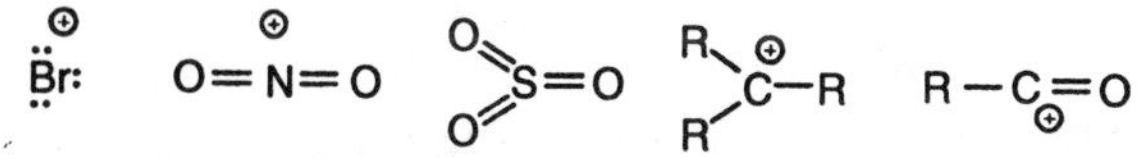

Fig. Examples of electrophiles used in electrophilic substitution.

Intermediate Stabilisation

The rate-determining step in the electrophilic substitution is *the formation of the positively charged intermediate,* and so the rate of the reaction is determined by the energy level of the transition state leading to that intermediate. The transition state resembles the intermediate in character and so any factor stabilising the intermediate also stabilises the transition state and lowers the activation energy needed for the reaction.

Therefore, electrophilic substitution is more likely to occur if the positively charged intermediate can be stabilised. Stabilisation is possible if the positive charge can be spread amongst different atoms, i.e. by delocalisation. The process by which this can occurs is called *resonance*:

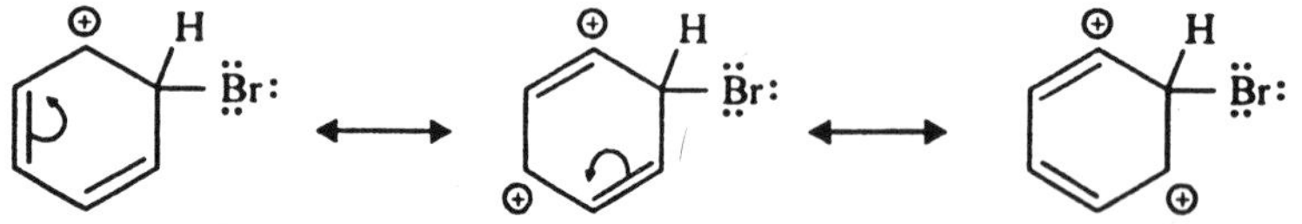

Fig. Resonance stabilisation of the charged intermediate.

The resonance process involves two π electrons shifting their position round the ring to provide the 'top' carbon with a fourth bond and thus neutralise its positive charge. In the process, another carbon in the ring is left short of bonds and gains the positive charge. This process can be repeated such that the positive charge is spread to a third carbon. The structures drawn in figure given above are called *resonance structures* (Canonical forms).

Halogenation

The stable aromatic ring means that aromatic compounds are less reactive than alkenes to electrophiles. For example, an alkene will react with Br_2, but an aromatic ring will not. Therefore, we have to activate the aromatic ring (i.e. make it a better nucleophile) or activate the Br_2 (i.e. make it a better electrophile) if we want a reaction to take place. The electron-donating substituents on an aromatic ring increase the nucleophilicity of the aromatic ring. A Br_2 molecule can be activated to make it a better electrophile by adding a Lewis acid like $FeCl_3$, $FeBr_3$, or $AlCl_3$ to the reaction medium.

These compounds contain a central atom (iron or aluminium) that is strongly electrophilic and does not have a full valence shell of electrons. Due to this, the central atom can accept a lone pair of electrons, even from a weakly nucleophilic atom like a halogen. For example bromine uses a lone pair of electrons to form a bond to the Fe atom in $FeBr_3$ and becomes positively charged. Bromine is now activated to act as an electrophile and will react more easily with a nucleophile (the aromatic ring) by the normal mechanism for electrophilic substitution.

An aromatic ring can be chlorinated in a similar manner, making use of Cl_2 in the presence of $FeCl_3$.

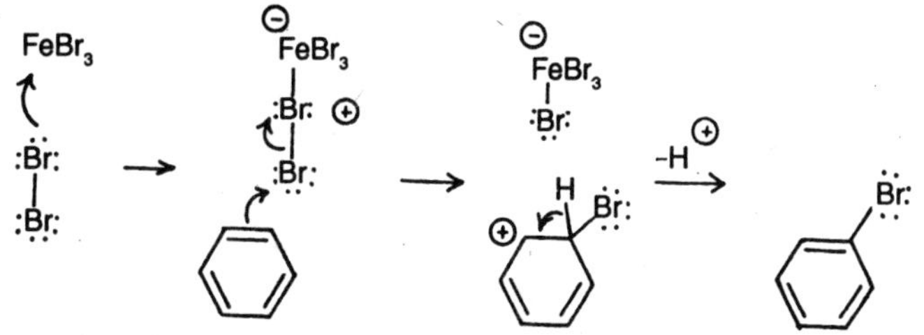

Fig. Mechanism by which a Lewis acid activates bromine towards electrophilic substitution.

Friedel-Crafts Alkylation and Acylation

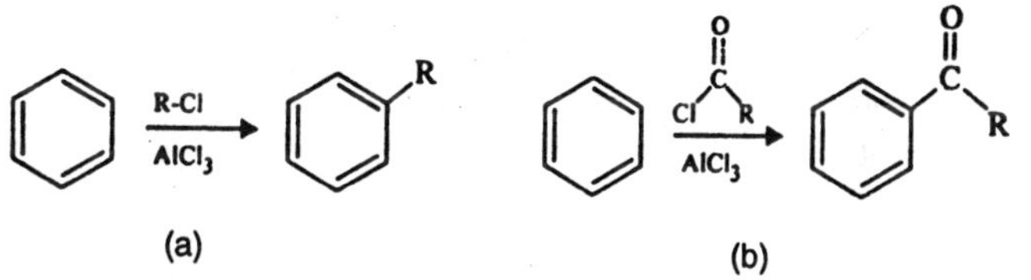

Fig. (a) Friedel-Crafts alkylation; (b) Friedel-Crafts acylation.

Friedel-Crafts alkylation and acylation are two other examples of electrophilic substitution requiring the presence of a Lewis acid.

These are particularly important as they allow the synthesis of larger organic molecules by adding alkyl (R) or acyl (RCO) side chains to an aromatic ring.

The example of Friedel-Crafts alkylation is the reaction of benzene with 2-chloropropane. The Lewis acid ($AlCl_3$) promotes the formation of the carbocation needed for the reaction and does so by accepting a lone pair of electrons from chlorine to form an unstable intermediate that fragments to give a carbocation and $AlCl_4^-$.

Once the carbocation is formed it reacts as an electrophile with the aromatic ring by the electrophilic substitution mechanism already described.

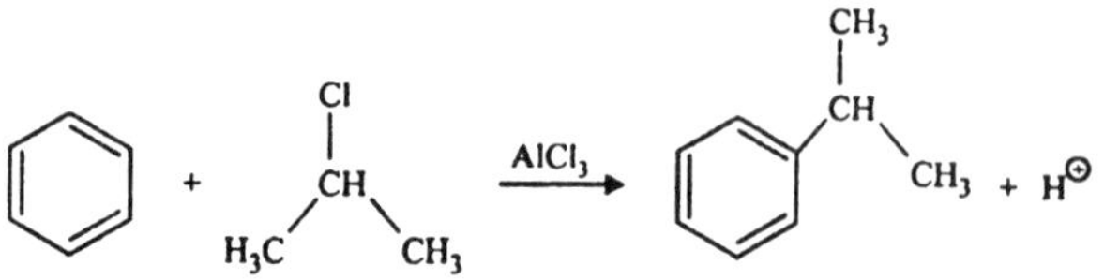

Fig. Friedel-Crafts reaction of benzene with 2-chloropropane.

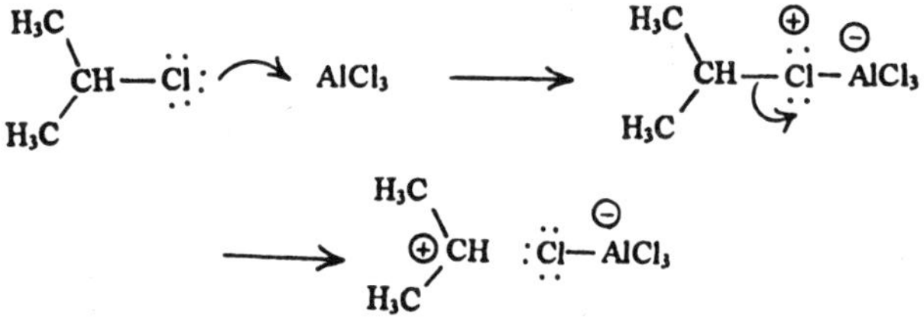

Fig. Mechanism of carbocation formation.

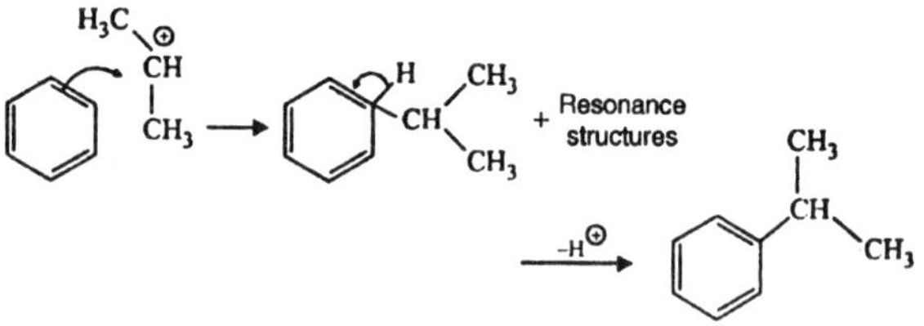

Fig. Mechanism for the Friedel-Crafts alkylation.

Following are the limitations to the Friedel-Crafts alkylation. For example, the reaction of 1-chlorobutane with benzene gives two products with only 34 per cent of the desired product. This is because of the fact that the primary carbocation that is generated can rearrange to a more stable secondary carbocation where a hydrogen (and the *two* sigma *electrons* making up the C–H bond) 'shift' across to the neighbouring carbon atom. This is called a *hydride shift* and it occurs because the secondary carbocation is more stable than the primary carbocation. Such rearrangements limit the type of alkylations that can be carried out by the Friedel-Crafts reaction.

$$C_6H_6 + Cl-CH_2CH_2CH_2CH_3 \xrightarrow{AlCl_3} C_6H_5CH_2CH_2CH_2CH_3\ (34\%) + C_6H_5CH(CH_3)CH_2CH_3\ (66\%)$$

Fig. Friedel-Crafts reaction of 1-chlorobutane with benzene.

$$\overset{\oplus}{H_2C}-CH(H)CH_2CH_3 \longrightarrow H_3C-\overset{\oplus}{C}HCH_2CH_3$$

Fig. Hydride shift.

Keeping this fact in mind, we can make structures such 1-butylbenzene in good yield. In the Friedel-Crafts acylation of benzene with butanoyl chloride instead of 1-chlorobutane, the necessary 4-C skeleton is linked to the aromatic ring and no rearrangement occurs. The carbonyl group can then be removed by reducing it with hydrogen over a palladium catalyst to give the desired product.

$$C_6H_6 \xrightarrow[AlCl_3]{CH_2CH_3CH_2COCl} C_6H_5COCH_2CH_2CH_3 \xrightarrow{H_2/Pd} C_6H_5CH_2CH_2CH_2CH_3$$

Fig. Synthesis of 1-butylbenzene by Friedel-Crafts acylation and reduction.

The mechanism of the Friedel-Crafts acylation is the same as the Friedel-Crafts alkylation. It involves an acylium ion instead of a carbocation. Like Friedel-Crafts alkylation, a Lewis acid is needed to generate the acylium ion (R–C = O)\ but unlike a carbocation the acylium ion does not rearrange since there is resonance stabilisation from the oxygen:

Friedel-Crafts alkylations can also be done using alkenes instead of alkyl halides. A Lewis acid is not needed, but a mineral acid is required. Treatment of the alkene with the acid leads to a carbocation that can then react with an aromatic ring by the electrophilic substitution mechanism already described.

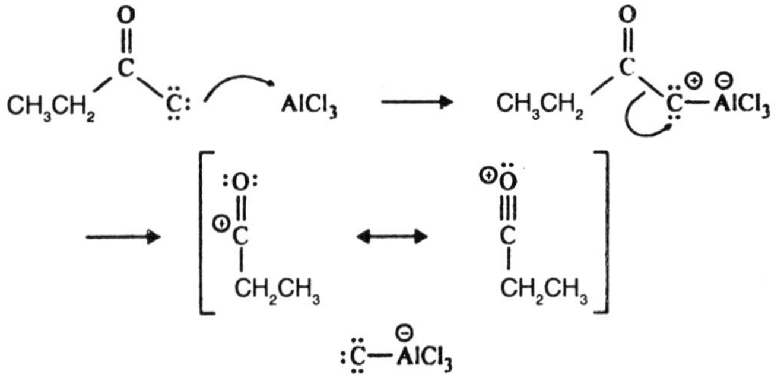

Fig. Generation of the acylium ion.

For an alkene, this is another example of electrophilic addition where a proton is attached to one end of the double bond and a phenyl group is added to the other.

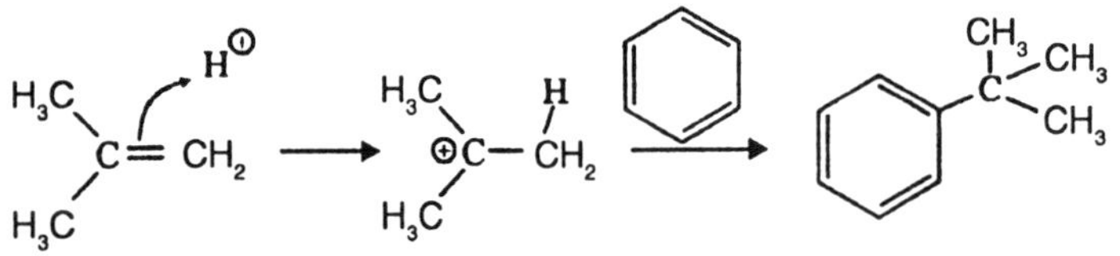

Fig. Friedel-Crafts alkylation of benzene with an alkene.

Friedel-Crafts reactions can also be done with alcohols in the presence of mineral acid.

The acid leads to the elimination of water from the alcohol resulting in the formation of an alkene that can then be converted to a carbocation:

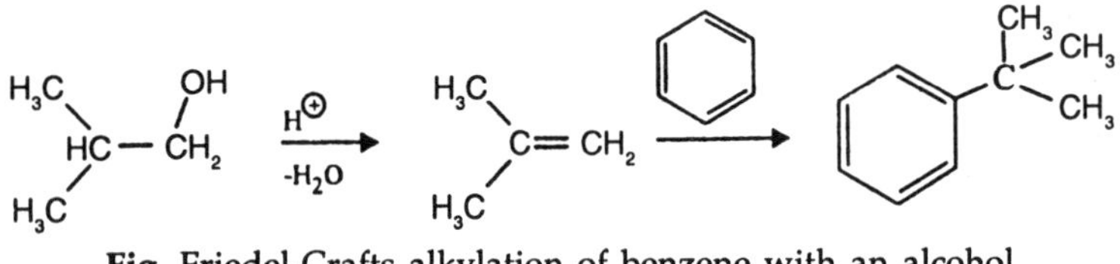

Fig. Friedel-Crafts alkylation of benzene with an alcohol.

Sulphonation and Nitration

Sulphonation and nitration are electrophilic substitutions that involve strong electrophiles and do not need the presence of a Lewis acid:

Fig. (a) Sulphonation of benzene; (b) nitration of benzene.

In Sulphonation, the electrophile is sulphur tetroxide (SO_3) that is generated under the acidic reaction conditions. Protonation of an OH group generates a protonated intermediate (I). As the oxygen gains a positive charge it *becomes a good leaving group* and water is lost from the intermediate to give sulphur trioxide. Although sulphur trioxide has no positive charge, it is a strong electrophile.

This is because the sulphur atom is bonded to three electronegative oxygen atoms that are all 'pulling' electrons from the sulphur, and making it electron deficient (i.e. electrophilic). During electrophilic substitution, the aromatic ring forms a bond to sulphur and one of the π bonds between sulphur and oxygen is broken.

Both electrons move to the more electronegative oxygen top form a third lone pair and produce a negative charge on that oxygen. This finally gets neutralised when the third lone pair of electrons is used to form a bond to a proton.

Fig. Generation of sulphur trioxide.

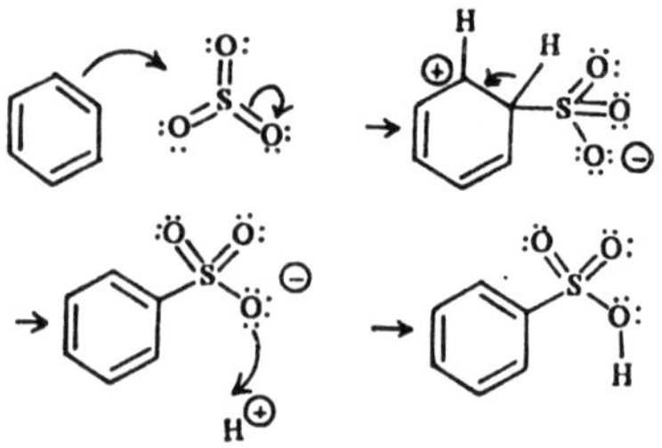

Fig. Sulphonation of benzene.

In nitration, sulphuric acid acts as an acid catalyst for the formation of a nitronium ion (NO_2^+) that is generated from nitric acid by a similar mechanism to that used in the generation of sulphur trioxide from sulphuric acid:

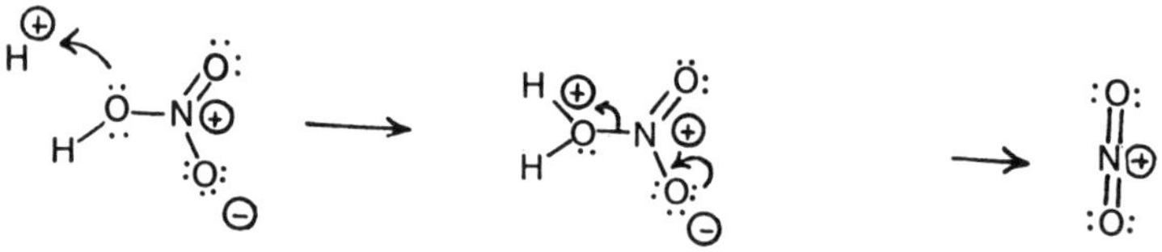

Fig. Generation of the nitronium ion.

The mechanism for the nitration of benzene is similar to sulphonation. As the aromatic ring forms a bond to the electrophilic nitrogen atom, a π bond between N and O breaks and both electrons move onto the oxygen atom. Unlike sulphonation, this oxygen keeps its negative charge and does not pick up a proton. This is because it acts as a counterion to the neighbouring positive charge on nitrogen.

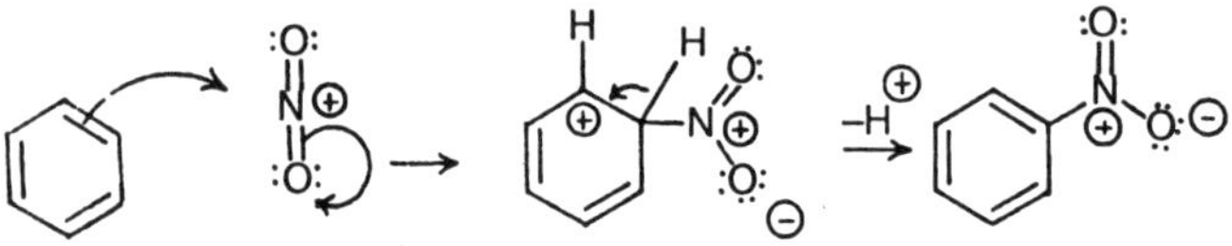

Fig. Nitration of benzene.

Formation of Mono-substituted Bonzenes

Functional Group Transformations

Some substituents cannot be introduced directly into an aromatic ring by electrophilic substitution. These include the following groups: $-NH_2$, $-NHR$, NR_2, $NHCOCH_3$, CO_2H, CN, OH. Although these groups cannot be added directly into the aromatic ring they can be obtained by transforming a functional group that can be added directly by electrophilic substitution. Some of the most important transformations are shown:

NO_2 —Sn/HCl→ NH_2

R —$KMnO_4$, R = alkyl→ CO_2H

RC=O —Pd / H_2→ CH_2R

Fig. Functional group transformations of importance in aromatic chemistry.

Nitro ($-NO_2$), *alkyl* (–*R*) *and acyl* (*RCO*–) groups can readily be added by electrophilic substitution and can then be converted to amino, carboxylic acid, and alkyl groups, respectively. Once the amino and carboxylic acid groups have been obtained, they can be further converted to a large range of other functional groups like secondary and tertiary amines, amides, diazonium salts, halides, nitriles, esters, phenols, alcohols, and ethers.

Synthetic Planning

In planning, the synthesis of an aromatic compound, it is best to work backwards from the products and to ask what it could have been synthesised from a process known as *retrosynthesis*. To illustrate this, consider the synthesis of an aromatic ester. An ester functional group cannot be attached directly by electrophilic substitution, so the synthesis must involve various steps. The ester can be prepared from an acid chloride which can be synthesised from a carboxylic acid. Alternatively, the ester can able made directly from the carboxylic acid by treating it with an alcohol and an acid catalyst. Either way, benzoic acid is required to synthesise the ester.

Carboxylic acids cannot be added directly to aromatic rings either, so we have to look for a different functional group that can be added directly, then transformed to a carboxylic acid. A carboxylic acid group can be obtained from the oxidation of a methyl group. Methyl groups can be added directly by *Friedel-Crafts alkylation.*

Therefore, a possible synthetic route would be as shown in the figure given below.

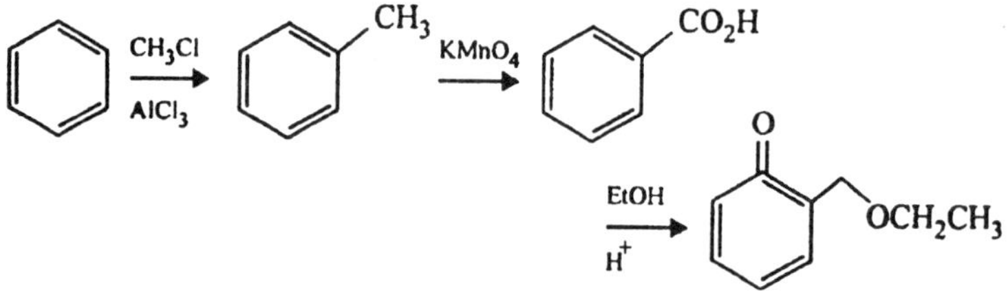

Fig. Possible synthesis of an aromatic ester.

The problem with this route is the possibility of poly-methylation in the first step. This is likely since the product (toluene) will be more reactive than the starting material. This problem can be overcome by using an excess of benzene.

Again consider the synthesis of an aromatic amine. The alkylamine group cannot be applied to an aromatic ring directly and so must be obtained by modifying another functional group. The alkyamine group could be obtained by alkylation of an amino group (NH_2). An amino group cannot be directly applied to an aromatic ring either.

However, an amino group could be obtained by reduction of a nitro group. A nitro group can be applied directly to an aromatic ring. Thus, the overall synthesis would be nitration followed by reduction, followed by alkylation.

Fig. Possible synthetic of an aromatic amine.

There are two methods of converting aniline ($PhNH_2$) to the final product. Alkylation is the direct method, but sometimes acylation followed by reduction gives better yields. This is because it is sometimes difficult to control the alkylation to only one alkyl group.

Fig. Possible synthetic route to an aromatic ether.

Lastly let us consider the synthesis of an aromatic ether. Here an ethoxy group is attached to the aromatic ring. The ethoxy group cannot be added directly to an aromatic ring, so we have to find a way of obtaining it from another functional group. Alkylation of a phenol group would give the desired ether, but a phenol group cannot be added directly to the ring ether. However, we can obtain the phenol from an amino group, that in turn can be obtained from a nitro group. The nitro group can be added directly to the ring and so the synthesis involves a nitration, conversion of the amino group to a diazonium salt, hydrolysis, and finally an alkylation.

Mono-substituted Aromatic Rings: Electrophilic Addition

Ortho, Meta and Para Substitution

An aromatic compounds containing a substituent can undergo electrophilic substitution at three different positions relative to the substituent. For example in the bromination of toluene, three different products are possible depending on where the bromine enters the ring. These products have the same molecular formula and are therefore constitutional isomers. The aromatic ring is said to be di-substituted and the three possible isomers are referred to as being *ortho, meta, and para.*

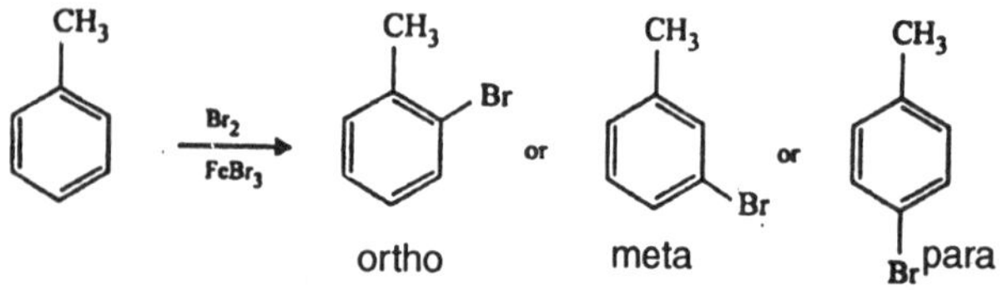

CH_3 $\xrightarrow[FeBr_3]{Br_2}$ ortho or meta or para

Fig. *Ortho, meta* and *para* isomers of bromotoluene.

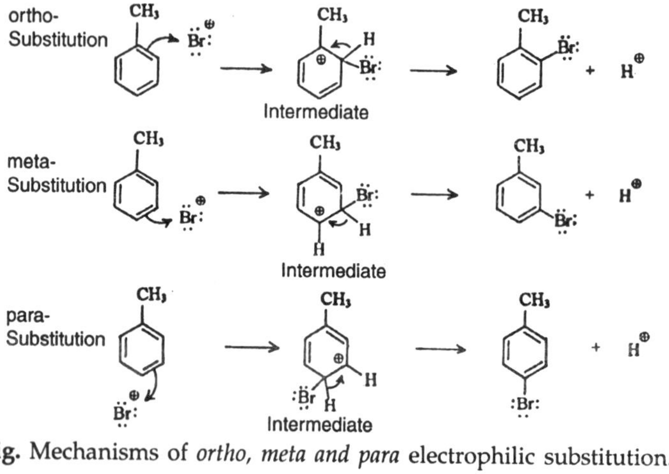

ortho-Substitution → Intermediate → + $H^{\oplus}$

meta-Substitution → Intermediate → + $H^{\oplus}$

para-Substitution → Intermediate → + $H^{\oplus}$

Fig. Mechanisms of *ortho, meta and para* electrophilic substitution.

Substituent Effect

Out of the three possible isomers arising from the bromination of toluene; only two (the *ortho* and *para)* are formed in significant quantity. Moreover, the bromination of toluene goes at a faster rate than the bromination of benzene.

This is so because the methyl substituent can affect the rate and the position of further substitution. A substituent can either activate or deactivate the aromatic ring towards electrophilic substitution and does so through inductive or resonance effects. A substituent can also direct the next substitution so that it goes mainly *ortho/para* or mainly *meta.*

The substituents can be classified into four groups depending on the effect they have on the rate and the position of substitution:

(i) Activating groups which direct *ortho/para* by inductive effects.

(ii) Deactivating groups which direct *meta* by inductive effects.

(iii) Activating groups which direct *ortho/para* by resonance effects.

(iv) Deactivating groups which direct *meta* by resonance effects.

There are no substituents which activate the ring and are *meta* directly.

Reaction Profile

For explaining the reasons behind the substituent effect, we shall have to consider the reaction profile of electrophilic substitution with respect to the relative energies of starting material, intermediate, and product. The energy diagram illustrates the reaction pathway for the bromination of benzene. The first stage in this mechanism which is also the rate-determining step is the formation of the carbocation.

This is endothermic and proceeds through a transition state which needs an activation energy ($\Delta G^{\#}$). The magnitude of $\Delta G^{\#}$ determines the rate at which the reaction will occur and this in turn is determined by the stability of the transition state.

The transition state resembles the carbocation intermediate and so any factor that stabilises the intermediate also stabilises the transition state and favours the reaction. Thus, we can consider the stability of relative carbocations to determine which reaction is more favourable.

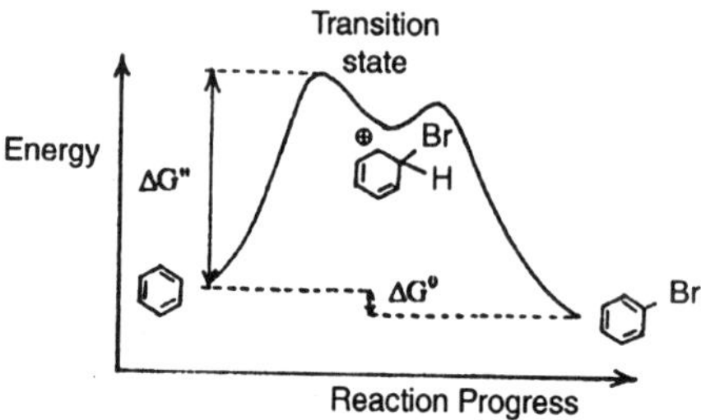

Fig. Energy diagram for electrophilic substitution.

Activating Groups—Inductive ortho/para Directing

A methyl substituent is an inductive activating group and let us consider again the bromination of toluene. To explain the directing properties of the methyl group, we look more closely at the mechanisms involved in generating the *ortho, meta,* and *para* isomers. The preferred reaction pathway will be the one that goes through the most stable intermediate. Since a methyl group directs *ortho* and *para,* the intermediate involved in these reaction pathways are more stable than the intermediates involved in *meta* substitution. The relevant intermediates and their resonance structures are shown below:

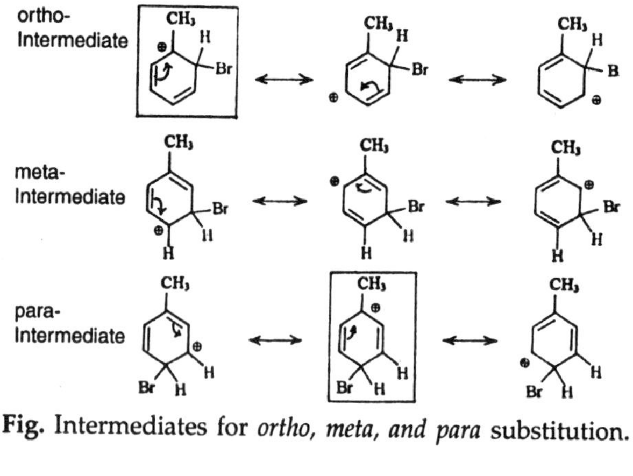

Fig. Intermediates for *ortho, meta, and para* substitution.

Now if we compare all the resonance structures, we can spot one *ortho* and one *para* resonance structure (boxed) where the positive charge is positioned immediately next to the methyl substituent. An alkyl group can stabilise a neighbouring positive charge by an inductive, electron-donating effect that results in some of the positive charge being spread over the alkyl group.

This is an additional stabilising effect that is only possible for the intermediates arising from *ortho* and *para* substitution. There is no such equivalent resonance structure for the *meta* intermediate and so that means that the *ortho* and *para* intermediates experience an increased stability over the *meta*, that results in a preference for these two substitution pathways.

Similarly, toluene will be more reactive than benzene. The electron-donating effect of the methyl group into the aromatic ring makes the ring inherently more nucleophilic and more reactive to electrophiles, as well as providing extra stabilisation of the reaction intermediate. Thus, alkyl groups are activating groups and are *ortho, para directing*.

CH_3 c. H_2SO_4 / HNO_3 → CH_3, NO_2 + CH_3, NO_2 + CH_3, NO_2

ortho(58%) meta(4%) para(38%)

Fig. Nitration of toluene.

Consider the nitration of toluene. The amount of *meta* substitution is very small as expected and there is a preference for the *ortho* and *para* products. The formation of more *ortho* substitution compared to *para* substitution is due to the fact that there are two *ortho* sites on the molecule to one *para* site and so there is double the chance of *ortho* attack to *para* attack. Based on pure statistics it would be expected that the ratio of *ortho* to *para* attack to be 2:1. In fact, the ratio is closer to 1.5:1. In other words, there is less *ortho* substitution than expected.

This is because the *ortho* sites are immediately 'next door' to the methyl substituents and the size of the substituent tends to inference with *ortho* attack—a steric effect. The significance of the steric effect will vary according to the size of the alkyl substituent. The larger the substituent, the more *ortho* attack will be hindered.

Deactivating Groups—Inductive in Directing

Alkyl groups are activating groups and direct substitution to the *ortho, para* positions. Electron withdrawing substitutions have the opposite effect. They deactivate the ring, make the ring less nucleophilic and less likely to react with an electrophile.

The electron-withdrawing effects also destabilises the reaction intermediate and makes the reaction more difficult. This destabilisation is more pronounced in the intermediates arising from *ortho/para* attack and so *meta* attack is favoured.

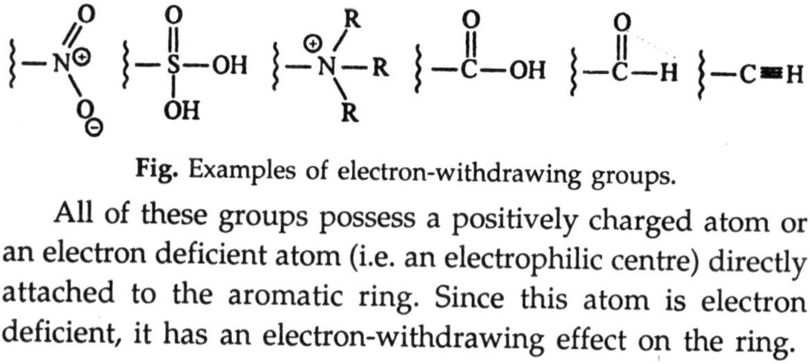

$$-NO_2 \quad -SO_2OH \quad -\overset{\oplus}{N}R_3 \quad -COOH \quad -CHO \quad -C\equiv H$$

Fig. Examples of electron-withdrawing groups.

All of these groups possess a positively charged atom or an electron deficient atom (i.e. an electrophilic centre) directly attached to the aromatic ring. Since this atom is electron deficient, it has an electron-withdrawing effect on the ring.

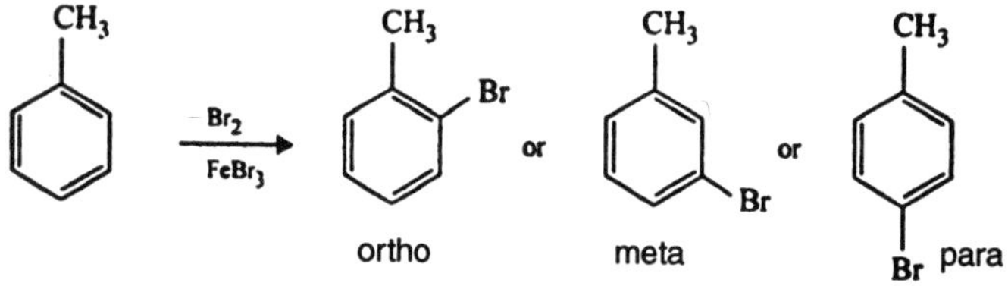

$$C_6H_5CH_3 \xrightarrow[FeBr_3]{Br_2} \text{ortho-}BrC_6H_4CH_3 \text{ or meta-}BrC_6H_4CH_3 \text{ or para-}BrC_6H_4CH_3$$

ortho meta para

Fig. Bromination of nitrobenzene.

Deactivating groups make electrophilic substitution more difficult but the reaction will proceed under more forcing reaction conditions.

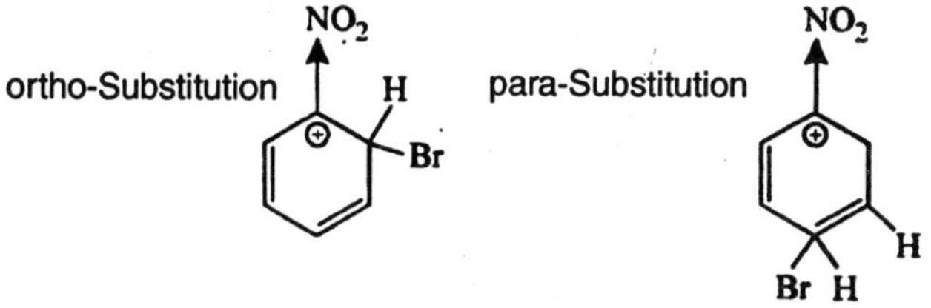

ortho-Substitution NO_2 H Br para-Substitution NO_2 H Br H

Fig. Destabilising resonance structures for the intermediate arising from *ortho* and *para* substitution.

However, substitution is now directed to the *meta* position. This can be explained by comparing all the possible resonance structures arising form *ortho, meta* and *para* attack. For example, consider the bromination of nitrotoluene. Of all the possible resonance structures arising from *ortho, meta,* and *para* attack, there are two specific resonance structures (arising from *ortho* and *para* attack) in which the positive charge is placed directly next to the electron-withdrawing nitro group. Due to this, these resonance structures are generally destabilised. This does not take place with any of the resonance structures arising from *meta* attack and so *meta* attack is favoured.

Activating Groups—Resonance Ortho/para Directing

Phenol group activates the aromatic ring by resonance effects and it directs substitution to the *ortho* and *para* positions. In phenol, an electronegative oxygen atom is next to the aromatic ring. As oxygen is electronegative so it should have an electron-withdrawing inductive effect and it might be expected to deactivate the ring. The fact that the phenolic group is a powerful activating group is because of the fact that oxygen is electron rich and can also act as a nucleophile, feeding electrons into the ring through a resonance process. For example, consider the nitration of phenol:

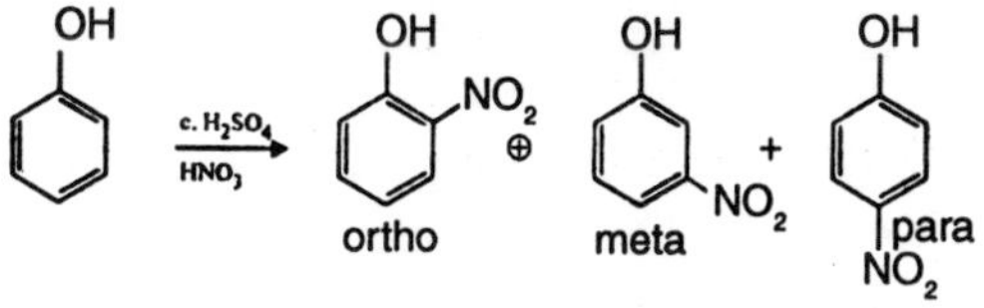

OH $\xrightarrow[HNO_3]{c.\ H_2SO_4}$ OH NO_2 ortho ⊕ OH meta NO_2 + OH para NO_2

Fig. Nitration of phenol.

There are three resonance structures for the intermediate formed in each form of electrophilic substitution, but there are two crucial ones to consider, arising from *ortho* and *para* substitution. These resonance structures have the positive charge next to the OH substituent. If oxygen only had an inductive effect, these resonance structures would be highly unstable. However, oxygen can act as a nucleophile and so can use one of its lone pairs of electrons to form a new π bond to the neighbouring electrophilic centre. This results in a fourth resonance structure where the positive charge is moved out of the ring and onto the oxygen atom. Delocalising the charge like this further stabilises it and makes the reaction proceed more easily.

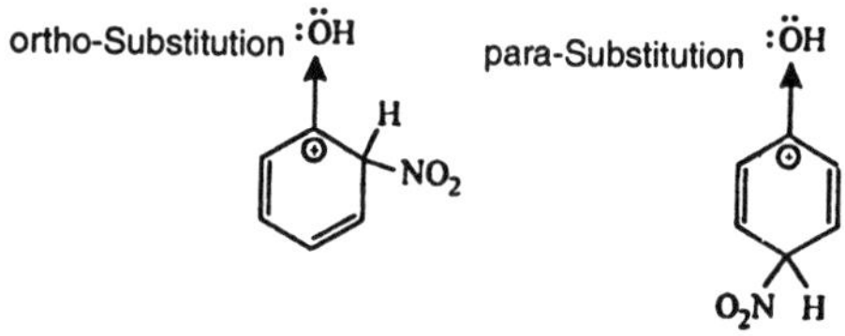

ortho-Substitution :ÖH H NO_2 para-Substitution :ÖH O_2N H

Fig. Resonance structures for the intermediates arising from *ortho* and *para* substitution.

:ÖH H NO_2 ⊕ ÖH H NO_2

:ÖH O_2N H ⊕ ÖH O_2N H

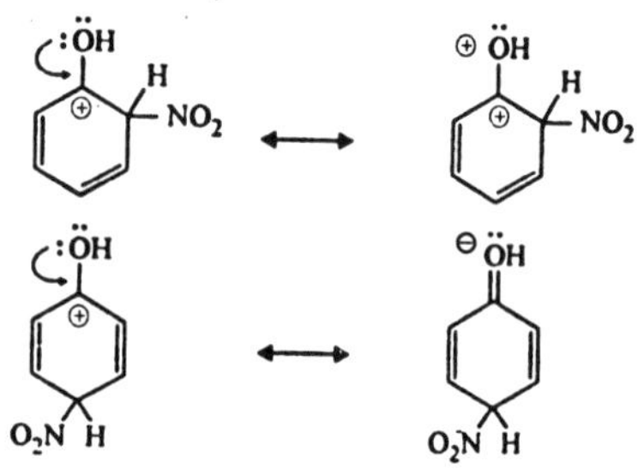

Fig. Resonance interactions between the aromatic ring and oxygen.

As none of the resonance structures arising from *meta* attack places the positive charge next to the phenol group, this fourth resonance structure is not available to the *meta* intermediate and so *meta* attack is not favoured. Thus, the phenol group is an activating group that is *ortho, para* directing due to resonance effects. This resonance effect is more important than any inductive effect that the oxygen might have.

The same is true for the following substituents: alkoxy (–OR), esters (–OCOR), amines (–NH_2, –NHR, –NR_2) and amides (–NHCOR). In all these cases, there is either a nitrogen or an oxygen next to the ring. Both these atoms are nucleophilic and have lone pairs of electrons which can be used to form an extra bond to the ring. The ease with which the group can do this depends on the nucleophilicity of the attached atom and how well it can cope with a positive charge.

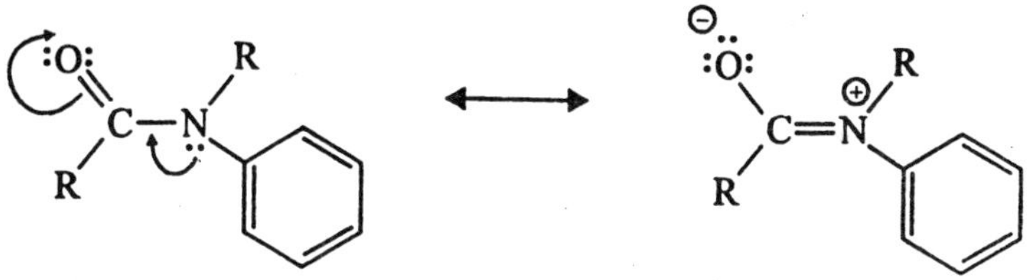

Fig. Amide resonance.

Nitrogen is more nucleophilic than oxygen since it is better able to cope with the resulting positive charge. Therefore, amine substituents are stronger activating groups than ethers. On the other hand, an amide group is a weaker activating group since the nitrogen atom is less nucleophilic. This is because the nitrogen's lone pair of electrons is pulled towards the carbonyl group and is less likely to form a bond to the ring. This property of amides can be quite useful.

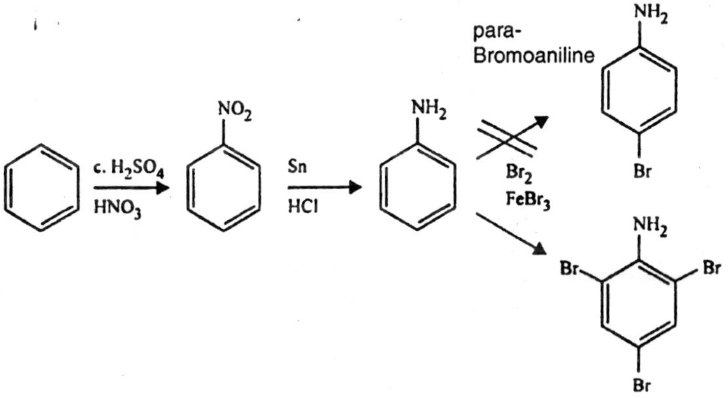

Fig. Bromination of aniline.

For example, if we want to make *para*-bromoaniline by brominating aniline. Theoretically, this reaction scheme should give the desired product. In practice, the NH_2 group is such a strong activating group that the final bromination goes three times to give the tri-bromianted product rather than the mono-brominated product.

To lower the activation of the amino group, we can convert it to the less activating amide group. The bromination then only goes once. We also find that the bromination reaction is more selective for the *para* position than for the *ortho* position.

This is because the amide group is bulkier than the NH_2 group and tends to shield the *ortho* positions from attack. Once the bromination has been completed the amide can be converted back to the amino group by hydrolysis.

NO_2 NH_2

c.H_2SO_4 / HNO_3 → ; Sn / HCl →

$NHCOCH_3$ $NHCOCH_3$ NH_2

CH_3COCl → ; Br_2 / $FeBr_3$ → ; $H^{\oplus}$ / H_2O →

Br Br

Fig. Synthesis of *para*-bromoaniline.

Deactivating Groups—Resonance Ortho/Para Directing

The last group of aromatic substituents are the halogen substituents that deactivate the aromatic ring and that direct substitution to the *ortho* and *para* positions. These are perhaps the most difficult to understand as they deactivate the ring by one effect, but direct substitution by a different effect. The halogen atom is strongly electronegative and so we would expect it to have a strong electron-withdrawing inductive effect on the aromatic ring. This would make the aromatic ring less nucleophilic and less reactive to electrophiles. It would also destabilise the required intermediate for electrophilic substitution. Halogens are also poorer nucleophiles and so any resonance effects they might have are less important than their conductive effects.

However, if halogen atoms are deactivating the ring due to inductive effects, they should not direct substitution to the *meta* position like other electron-withdrawing groups. Consider the nitration of bromobenzene. There are three resonance structures for each of the three intermediates leading to these products, but the crucial ones to consider are those which position a positive charge next to the substituent.

These occur with *ortho* and *para* substitution, but not *meta* substitution. These are the crucial resonance structures as far as the directing properties of the substituents is concerned. If bromine acts inductively, it will destabilise these intermediates and direct substitution to the *meta* position. However, since bromine directs *ortho/para* and so it must be stabilising the *ortho/para* intermediates rather than destabilising them. The bromine can stabilise the neighbouring positive charge only by resonance in the same way as a nitrogen or oxygen atom.

Fig. Nitration of bromobenzene.

Thus, the bromine acts as a nucleophile and donates one of its lone pairs to form a new bond to the electrophilic centre beside it. A new π bond is formed and the positive charge is moved onto the bromine atom. This resonance effect is weak since the halogen atom is a much weaker than oxygen or nitrogen and is less capable of stabilising a positive charge.

However, it is significant enough to direct substitution to the *ortho* and *para* positions.

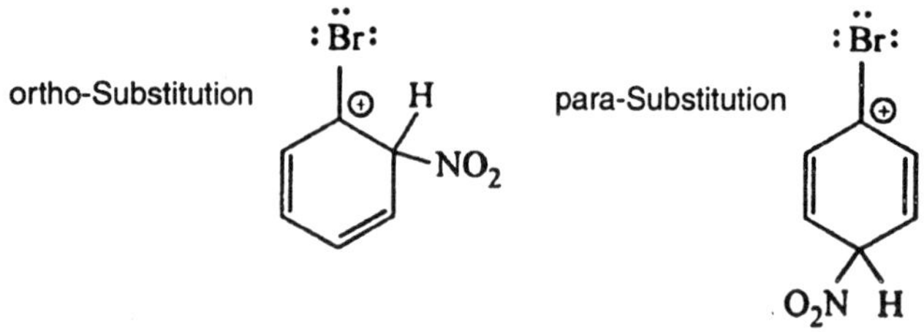

ortho-Substitution :Br: H NO_2 para-Substitution :Br: O_2N H

Fig. Crucial resonance structures for *ortho* and *para* substitution.

In case of halogen substituents, the inductive effect is more important than the resonance effect in deactivating the ring. However, once electrophilic substitution does occur, resonance effects are more important than inductive effects in directing substitution.

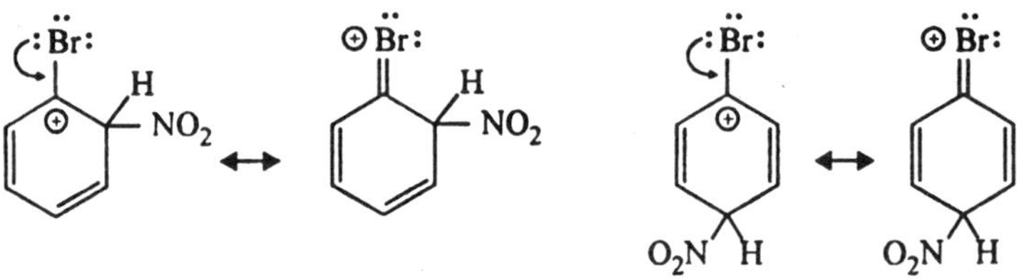

:Br: H NO_2 ↔ ⊕Br: H NO_2 :Br: O_2N H ↔ ⊕Br: O_2N H

Fig. Resonance interactions involving bromine.

Alkynes and Alkenes

Alkenes

Preparation of Alkenes: In the laboratory alkenes can be obtained by

(i) *passing alcohol vapours overheated alumina* (Al_2O_3) at about 700K.

$$\underset{\text{Ethyl alcohol}}{CH_3CH_2OH} \xrightarrow{Al_2O_3\ 650K} CH_2{=}CH_2 + H_2O$$

Dehydration may also be affected by heating in the presence of strong acids like H_2SO_4, H_3PO_4 or p-toluene sulphuric acid. The process is called *"acid-catalysed dehydration"*.

(ii) *Dehydrohalogenation of alkyl halides*

$$C_2H_5 + \text{Alc. KOH} \longrightarrow \underset{CH_2}{\overset{CH_2}{\|}} + KCl + H_2O$$

(iii) *Dehalogenation of vicinal dihalides*

$$\underset{\times}{R-\underset{|}{CH}}-\underset{\times}{\underset{|}{CH}}-R + \underset{\text{(zinc dust)}}{Zn} \xrightarrow{\text{Alcohol}} \underset{\text{alkene}}{RCH=CHR} + Zn\ X_2$$

Vincinal dihalide

$$CH_2-CH_2 + Zn \xrightarrow{\text{Alcohol}} CH_2=CH_2 + ZnBr_2$$

The dibromide itself is usually prepared from the same alkene and so the reaction is not particularly useful for the synthesis of alkenes. It is useful, however, in protection strategy. During a lengthy synthesis, it may be necessary to protect a double bond so that it does not undergo any undesired reactions. Bromine can be added to form the dibromide and removed later by denomination in order to restore the functional group.

$$R_2C(Br)-C(Br)R_2 \xrightarrow[\text{NaI, acetone}]{\text{Zn/HO Ac or}} R_2C=CR_2$$

Fig. Synthesis of an alkene from a vicinal dibromide.

(iv) *Partial hydrogenation of alkynes*

Partial reduction of alkynes with sodium in liquid ammonia or bycatalytic hydrogenation.

$$\underset{\text{Ethyne}}{CH\equiv CH} + H_2 \xrightarrow{\text{Pd, 575 K}} \underset{\text{Ethene}}{CH_2\equiv CH_2}$$

Preparation of Alkynes

Ethyne, $CH\equiv CH$ commonly known as *acetylenes* in the first number of this series and is prepared in the laboratory by the action of water on calcium carbide (CaC_2).

$$CaC_2 + 2H_2O \rightarrow \underset{\text{Ethyne}}{CH\equiv CH} + Ca(OH)_2$$

Higher alkynes can be synthesised from alkenes through a two-step process which involves the electrophilic addition of bromine to form a vicinal dibromide then dehydrohalogenation with strong base. The second stage involves the loss of two molecules of hydrogen bromide and so two equivalents of base are required.

$$\text{RHC=CHR} \xrightarrow[\text{CCl}_4]{\text{Br}_2} \text{R–CHBr–CHBr–R} \xrightarrow[\text{heat}]{\text{NaNH}_2} \text{R–C}\equiv\text{C–R} + 2\text{HBr}$$

Fig. Synthesis of an alkyne from an alkene.

Higher homologues of acetylene can also be obtained from acetylene itself. Acetylene is converted into acetylide and the acetylide is reacted with alkyl halide to get a higher homologue.

$$\underset{\text{Ethyne}}{CH \equiv CH} + \underset{\text{Sodamide}}{NaH_2} \xrightarrow{\text{Liquid Ammonia}} \underset{\text{Sod. acetylide}}{HC \equiv \bar{C}\overset{+}{Na}}$$

Characteristics of Alkynes and Alkenes

Properties of Alkenes

The functional group in alkenes is (> C = C<). The alkene functional group ($R_2C = CR_2$) is planar in shape with bond angles of 120°. The two carbon atoms involved in the double bond are both sp^2 hybridised. Each carbon has three sp^2 hybridised orbitals which are used for *a* bonds while the *p* orbital is used for a π bond. Thus, the double bond is made up of one a bond and one π bond:

Fig. Structure of an alkene functional group.

Strength of C = C Bond

The C = C bond is stronger (152 kcal mol^{-1}) and shorter (1.33Å) than a C–C single bond (88 kcal mol^{-1} and 1.54. A, respectively). A C = C bond contains one σ bond and one π bond, with the π bond being weaker than the σ bond. This is important with respect to the reactivity of alkenes.

Bond rotation is not possible for a C = C double bond since this would require the σ bond to be broken. Therefore, isomers of alkenes are possible depending on the relative position of the substituents. These can be defined as the *cis* or *trans,* but are more properly defined as (Z) or (E).

Alkenes are defined as mono, di, tri, or tetrasubstituted depending on the number of substituents which are present. The more substituents which are present, the more stable the alkene.

Physical Properties

The physical properties of alkenes are similar to those of alkanes. They are relatively *non-polar,* dissolve in non-polar solvents and are not soluble in water.

Since only weak van der Waals interactions are possible between unsaturated molecules such as alkenes, so they have comparatively *low boiling points.*

Properties of Alkynes

In them the functional group is (–C≡C–). The alkyne functional group consists of a carbon-carbon triple bond and is linear in shape with bond angles of 180°.

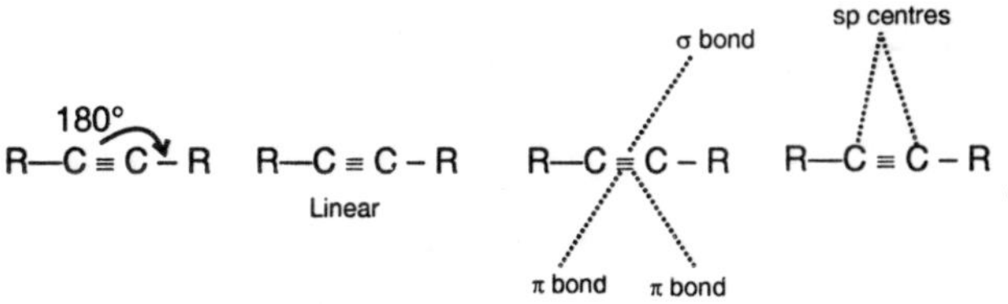

Fig. Structure of an alkyne functional group.

The two carbon atoms involved in the triple bond are *sp* hybridised, such that each carbon atom has two *sp* hybridised orbitals and two *p* orbitals. The *sp* hybridised orbitals are used for two π bonds while the *p* orbitals are used for two π bonds.

Thus, the triple bond is made up of one σ bond and two π bonds.

The bond length of a carbon triple bond is 1.20 Å and the bond strength is 200 kcal mol^{-1}. The π bonds are weaker than the *a* bond. The presence of the π bonds explains why alkynes are more reactive than alkanes.

Physical Properties of Alkynes

Alkynes have physical properties similar to alkanes. They are relatively non-polar, dissolve in non-polar solvents and are not very soluble in water. Only weak van der Waals interactions are possible between unsaturated molecules such as alkynes, and so these structures have low boiling points compared to other functional groups.

Nucleophilicity

Alkenes and alkynes are *nucleophilic* and they react with electrophiles in a reaction called *electrophilic addition*. The nucleophilic centre of the alkene or alkyne is the double bond or triple bond. These are areas of high electron density due to the bonding electrons. The specific electrons which are used to form bonds to attacking electrophiles are those involving in *n* bonding.

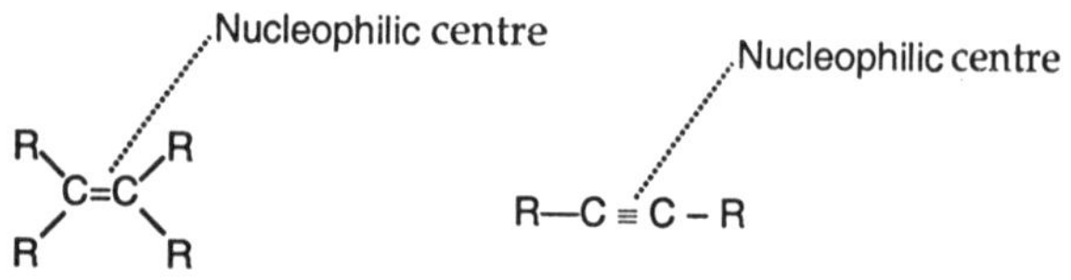

Fig. Nucleophilic centres of an alkene and an alkyne.

Symmetrical Alkenes: Electrophilic Addition

Reactions

Many of the reactions which alkenes undergo take place by a mechanism known as *electrophilic addition*. In these reactions, the π bond of the double bond has been used to form a bond to and incoming electrophile and is no longer present in the product. Furthermore, a new substituents has been added to each of the carbon atoms.

$$R_2C{=}CR_2 \xrightarrow{X_2} X{-}CR_2{-}CR_2{-}X \qquad R_2C{=}CR_2 \xrightarrow{HX} H{-}CR_2{-}CR_2{-}X$$

Fig. Electrophilic additions.

Symmetrical and Unsymmetrical Alkenes

Now we shall look at the electrophilic addition of symmetrical alkenes. A symmetrical alkene is an alkene that has the same substituents at each end of the double bond (fig. a). Unsymmetrical alkenes have different substituents at each end of the double bond (fig. b).

a) $(H_3C)_2C{=}C(CH_3)_2 \qquad H(H_3C)C{=}C(CH_3)H$

b) $(H_3C)_2C{=}CH_2 \qquad (H_3C)_2C{=}C(H)CH_3$

Fig. (a) Symmetrical alkenes; (b) unsymmetrical alkenes.

Hydrogen Halide

Alkenes react with hydrogen halides (HCl, HBr, and HI) to produce an alkyl halide. The hydrogen halide molecule gets split and the hydrogen atom adds to one end of the double bond while the halogen atom adds to the other, e.g. the addition of HBr with 2,3-dimethyl-2-butene. In this reaction, the alkene acts as a nucleophile.

It has an electron-rich double bond containing four electrons, two of which make up a strong a bond and two of which make up a weaker π bond. The *double bond* can be considered as a *nucleophilic centre.* Hydrogen bromide has a polar H-Br bond and so the *hydrogen is an electrophilic centre* and the *bromine is a nucleophilic centre.* Since, halogen atoms are extremely weak nucleophilic centre. So this molecule is a more likely to react as an electrophile through its electrophilic hydrogen.

$(H_3C)_2C{=}C(CH_3)_2 + \overset{\delta+}{H}{-}\overset{\delta-}{Br} \longrightarrow (H_3C)_2CH{-}CBr(CH_3)_2$

Nucleophilic centre

Electrophilic centre

Fig. Reaction of HBr with 2,3-dimethyl-2-butene.

Various Steps Involved in Electrophilic Addition Reactions

The first step of electrophilic addition is the one in which alkene acts as a nucleophile and uses its two 71 electrons to form a new bond to the hydrogen of HBr. As a new bond is formed, the H–Br bond breaks since hydrogen can form one bond. Both electrons in that bond end up on the bromine atom to produce a bromide ion. Since the electrons form the π bond have been used for the formation of a new σ bond, the π bond is no longer present. Because of this, the 'left hand' carbon is left with only three bonds and becomes positively charged. This is called a *carbocation* since the positive charge is on a carbon atom.

This structure is called a *reaction intermediate*. It is a reactive species and will not survive very long with the bromide ion in the vicinity. The *carbocation* is an *electrophile* since it is positively charged. The *bromide* ion is a *nucleophile* since it is negatively charged. Therefore, the bromide ion uses one of its lone pairs of electrons to form a new o bond to the carbocation and the final product is formed.

The addition of HBr to the alkene is an electrophilic addition since the first step of the mechanism involves the addition of the electrophilic hydrogen to the alkene. The second step involves a nucleophilic addition of the bromide ion to the carbocation intermediate, but *it is the first step which defines this reaction.*

The mechanism is shown in the below figure, in which the π electrons of the alkene provided the electrons for a new bond between the right hand carbon and hydrogen. They could equally well have been used to form a bond between the left hand carbon and hydrogen.

Carbocation intermediate

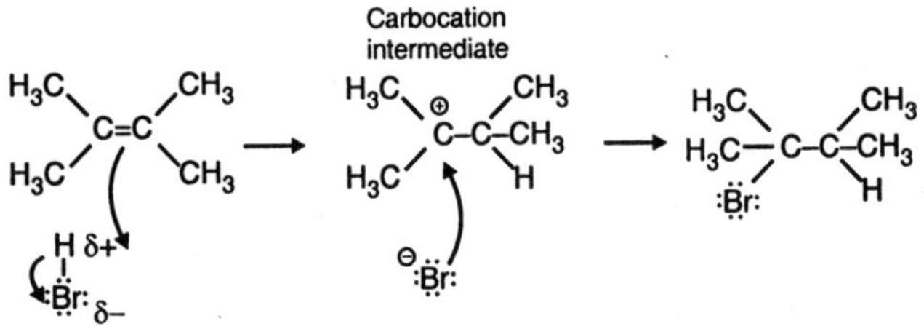

Fig. Mechanism of electrophilic addition of HBr to 2,3-dimethyl-2-butene.

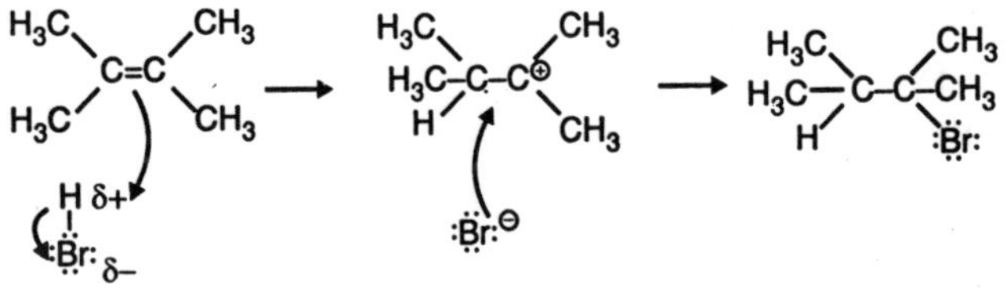

Fig. 'Alternative' mechanism for electrophilic addition of HBr to 2,3-dimethyl-2-butene.

In case of a symmetrical alkene, the product is the same and so it does not matter which end of the double bond is used for the new bond to hydrogen. The chances are equal of the hydrogenating to one side or the other.

The electrophilic additions of H–Cl and H–I follow the same mechanism to produce alkyl chlorides and alkyl iodides.

Addition of Halogens

The addition reaction of an alkene with a halogen like bromine or chlorine gives a vicinal dihalide. The halogen molecule is split and the halogens are added to each end of the double bond. Vicinal dibromides are quite useful in the purification or protection of alkenes since the bromine atoms can be removed under different reaction conditions to restore the alkene. Vicinal dibromides can also be converted to alkynes.

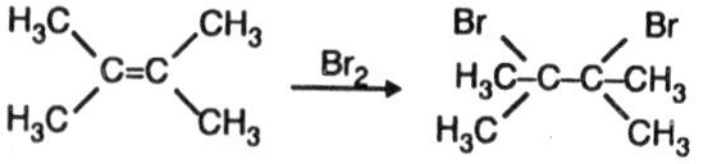

Fig. Electrophilic addition of bromine to 2,3-dimethyl-2-butene.

The mechanism followed in this reaction is similar to that discussed for alkenes with HBr. However, the first stage of the mechanism involves the nucleophilic alkene reacting with an electrophilic centre, and yet there is no obvious electrophilic centre in bromine. The bond between the two bromine atoms is a covalent a bond with both electrons equally shared between the bromine atoms.

When the bromine molecule approaches end-on to the alkene double bond and an electrophilic centre is included. Since the alkene double bond is electron rich, it repels the electrons in the bromine molecule and this results in a polarisation of the Br–Br bond in such a way that the nearer bromine becomes electron deficient (electrophilic). In this way, when an electrophilic centre has been generated, the mechanism is the same as before.

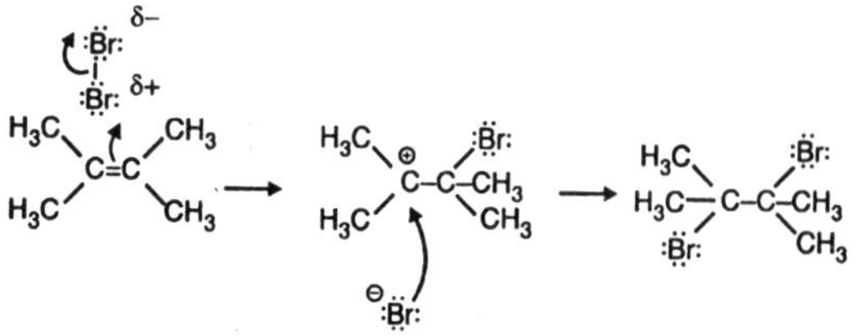

Fig. Mechanism for the electrophilic addition of Br_2 with 2.3-dimethyl-2-butene.

The carbocation intermediate can be stabilised by neighbouring alkyl groups through *inductive* and *hyperconjugation effects.* However, it can also be stabilised by sharing the positive charge with the bromine atom and a second carbon atom.

The positively charged carbon is an electrophilic centre. The bromine is a weak nucleophilic centre. A neutral halogen does not normally act as a nucleophile, but in this case the halogen is held close to the carbocation making reaction more likely. Once the lone pair of electrons on bromine is used to form a bond to the carbocation, a bromonium ion is formed in which the bromine gains a positive charge.

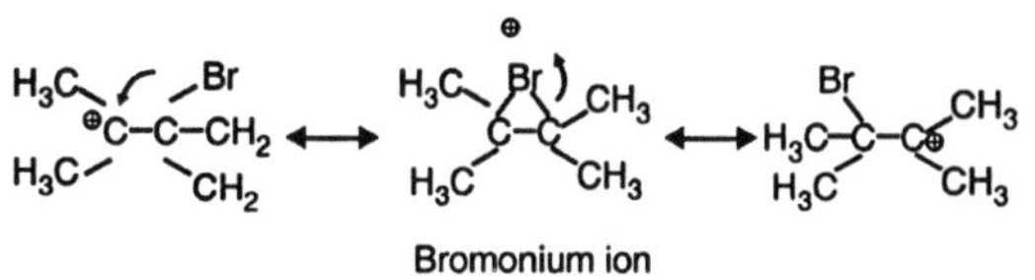

Bromonium ion

Fig. Formation of the bromonium ion.

The mechanism can go in reverse to regenerate the original carbocation. Alternatively, the other carbon-bromine bond can break with both electrons moving onto the bromine. This gives a second carbocation where the other carbon bears the positive charge. Thus, the positive charge is shared between three different atoms and is further stabilised.

Evidence for the existence of the bromonium ion is provided from the observation that bromine adds to cyclic alkenes (e.g. cyclopentene) in an anti-stereochemistry. Thus, each bromine adds to opposite faces of the alkene to produce only the *trans* isomer. None of the *cis* isomer is formed. If the intermediate was a carbocation, a mixture of *cis* and *trans* isomers would be expected as the second bromine could add form either side. With a bromonium ion, the second bromine must approach from the opposite side.

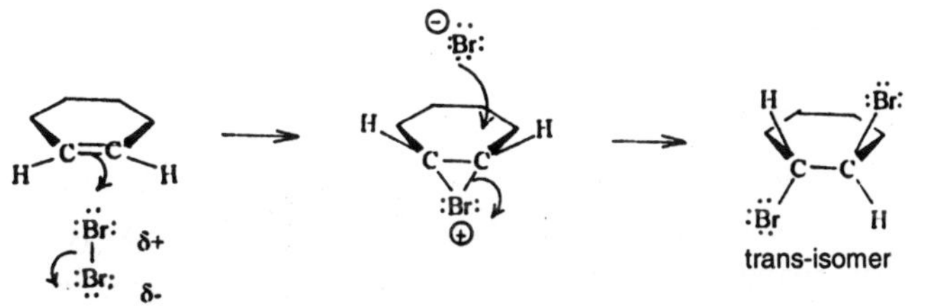

trans-isomer

Fig. Anti-stereochemistry of bromine addition to a cyclic alkene.

The reaction of an alkene with a halogen like bromine and chlorine generally gives a vicinal dihalide. However, if the reaction is carried out in water as solvent, the product obtained is a halohydrin where the halogen adds to one end of the double bond and a hydroxyl group from water adds to the other.

Fig. Formation of a bromohydrin from 2,3-dimethyl-2-butene.

In this reaction, the first stage of the mechanism proceeds as normal, but then water acts as a nucleophile and 'intercepts' the carbocation intermediate. Because water is the solvent, there are far more molecules of it present compared to the number of bromide ions generated from the first stage of the mechanism.

Fig. Mechanism of bromohydrin formation.

Water makes use of its a lone pair of electrons on oxygen to form a bond to the carbocation. Because of this, the oxygen effectively 'loses' an electron and gains a positive charge. This charge is lost and the oxygen regains its second lone pair when one of the O–H bonds breaks and both electrons move onto the oxygen.

Addition of H_2O

Alkenes get converted to alcohols by treatment with aqueous acid (sulphuric or phosphoric acid). This electrophilic addition reaction *involves the addition of water across the double bond.* The hydrogen adds to one carbon while a hydroxyl group adds to the other carbon.

Fig. Synthesis of an alcohol from an alkene.

Alkenes to Alcohols

Sometimes the reaction conditions used in this reaction are too harsh since heating is involved and rearrangement reactions can occur. A milder method that gives better results is to treat the alkene with mercuric acetate [$Hg(OAc)_2$] then sodium borohydride. The reaction involves electrophilic addition of the mercury reagent to form an intermediate mercuronium ion. This reacts with water to give an organomercury intermediate. Reduction with sodium borohydride replaces the mercury substituents with hydrogen and gives the final product.

Alkenes can also be converted to alcohols by hydroboration.

Fig. Synthesis of an alcohol from an alkene using mercuric acetate.

Alkenes to Ethers

A similar reaction to the mercuric acetate/sodium borohydride synthesis of alcohols allows the conversion of alkenes to ethers. In this case, mercuric trifluoracetate is used:

Fig. Synthesis of an ether from an alkene using mercuric trifluoroacetate.

Alkenes to Arylalkanes

The reaction of an aromatic ring such as benzene with an alkene under acid conditions results in the formation of an arylalkane. As far as the alkene is concerned this is another example of electrophilic addition involving the addition of a proton to one end of the double bond and the addition of the aromatic ring to the other. As far as the aromatic ring is concerned this is an example of an electrophilic substitution reaction called the *Friedel-Crafts alkylation*.

$$R_2C{=}CR_2 \xrightarrow[H^+]{\text{Benzene}} H{-}CR_2{-}CR_2{-}Ph$$

Fig. Synthesis of arylalkanes from alkenes.

Unsymmetrical Alkenes: Electrophilic Addition

Addition of Hydrogen Halides

The reaction of a symmetrical alkene with hydrogen bromide produces the same product irrespective of whether the hydrogen of HBr is added to one end of the double bond or the other. However, this is not the case with unsymmetrical alkenes. In this case, two different products are possible. These are not formed to an equal extent and the *more substituted alkyl halide (II) is preferred.* The reaction proceeds according to Markovnikov rule with hydrogen ending up on the least substituted position and the halogen ending up on the most substituted position. *Markovnikovs rule* states that 'in the addition of HX to an alkene, the hydrogen atom adds to the carbon atom that already has the greater number of hydrogen atoms'. This produces the more substituted alkyl halide.

$$(H_3C)_2C{=}CH_2 \xrightarrow{HBr} (H_3C)_2C(H){-}CH(R)(Br)\ \text{(I)} \quad (H_3C)_2C(Br){-}CH_2(H)\ \text{(II)}$$

Hydrogen added to left hand end of double bond

Hydrogen added to right hand end of double bond

Fig. Electrophilic addition of HBr to an unsymmetrical alkene.

Carbocation Stabilities

This reaction can be understood by assuming that in it the carbocation intermediate is formed which leads to product II, which is more stable than the carbocation intermediate leading to product I. It is possible to predict the more stable carbocation by counting the number of alkyl groups attached to the positive centre. The more stable carbocation on the right has three alkyl substituents attached to the positively charged carbon whereas the less stable carbocation on the left only has one such alkyl substituent. The stability of carbocation is 3° > 2° > 1°.

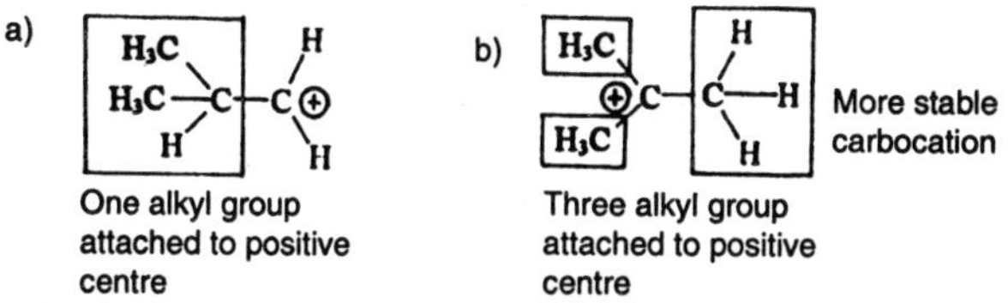

a) One alkyl group attached to positive centre

b) Three alkyl group attached to positive centre

More stable carbocation

Fig. (a) Carbocation leading to product I, (b) carbocation leading to product II.

However, Markovnikov's rule is not always hold true. For example, the reaction of $CF_3CH = CH_2$ with HBr gives $CF_3CH_2CH_2Br$ rather than $CF_3CHBrCH_3$ Here, *the presence of electron-withdrawing fluorine* substituents has a destabilising influence on the two possible intermediate carbocations. *The destabilising effect will be greater for the more substituted carbocation since the carbocation is closer to the fluorine substituents and so the favoured carbocation is the least substituted one in this case.*

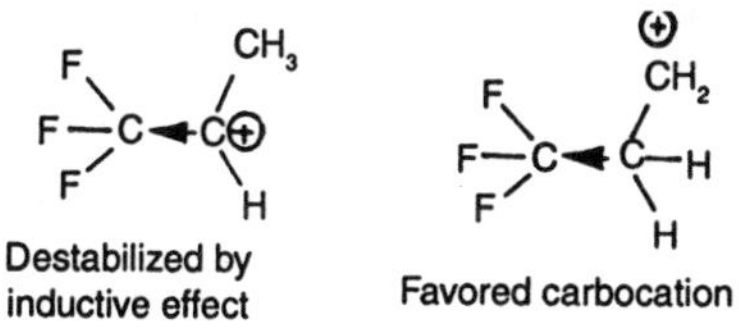

Destabilized by inductive effect

Favored carbocation

Fig. Comparison of carbocations.

Addition of Halogens

There is no possibility of different products when a halogen such as bromine or chlorine is added to an unsymmetrical alkene. However, in case water is used as a solvent, the halogen is attached to the least substituted carbon and the hydroxyl group is attached to the more substituted carbon.

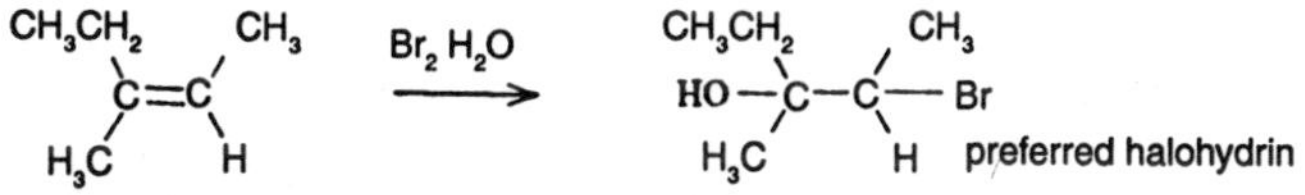

Br_2, H_2O

preferred halohydrin

Fig. Reaction of 3-methyl-2-pentene with bromine and water.

This can be explained by assuming that the bromonium ion is not symmetrical and that although the positive charge is shared between the bromine and the two carbon atoms, the *positive charge is greater on the more substituted carbon compared with the less substituted carbon.*

Addition of Water

With unsymmetrical alkenes, the more substituted alcohol is preferred product:

$$(H_3C)_2C{=}CH_2 \xrightarrow[S_2HO_4]{H_2O} HO{-}C(CH_3)_2{-}CH_3 \quad \left(H{-}C(CH_3)_2{-}CH_2{-}OH \right)$$

3° alcohol preferred product

1° alcohol

Fig. Reaction of 2-methyl-1 -propane with aqueous sulphuric acid.

The same holds true for the organomercuric synthesis of alcohols:

$$(H_3C)_2C{=}CH_2 \xrightarrow[H_2O/THF]{Hg(OAC)_2} \left[(H_3C)_2\overset{\oplus}{C}{-}CH_2{-}Hg(OAc) \right] \xrightarrow{H_2O} HO{-}C(CH_3)_2{-}CH_2{-}Hg(OAc) \xrightarrow{NaBH_4} HO{-}C(CH_3)_2{-}CH_3$$

Fig. Organomercuric synthesis of alcohols.

Reaction and Mechanism

Formation of Enolates

Acidic C–H Protons

Most acidic protons are attached to heteroatoms like halogen, oxygen, and nitrogen. Protons attached to carbon are not normally acidic but there are exceptions. One such exception occurs with aldehydes or ketones when there is a CHR_2, CH_2R or CH_3 group next to the carbonyl group. The protons indicated are acidic and are attached to the a (alpha) carbon. They are therefore called as *α protons.*

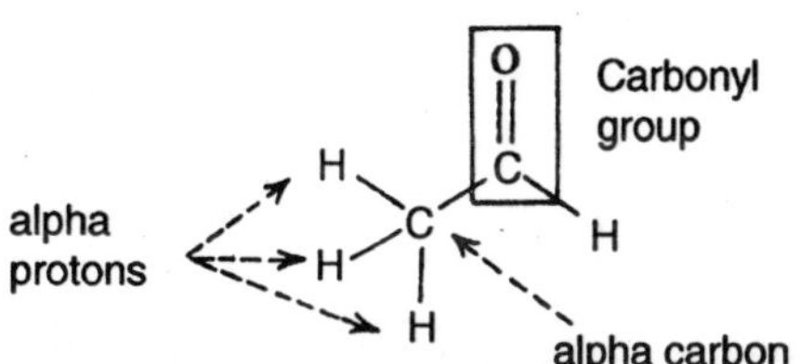

Fig. Acidic α protons.

Treatment with a base results in loss of one of the acidic α protons:

Fig. Loss of an a proton and formation of a carbanion.

A lone pair on the hydroxide oxygen forms a new bond to an α proton. Simultaneously, the C–H bond breaks. Both electrons of that bond end up on the carbon atom and give it a lone pair of electrons and a negative charge (a carbanion). However, carbanions are generally very reactive, unstable species that are not easily formed. Therefore, some form of stabilisation is involved here.

Stabilisation

As carbon is not electronegative so it cannot stabilise the charge. However, stabilisation is possible through resonance. The lone pair of electrons on the carbanion form a new π bond to the carbonyl carbon. As this bond is formed, the weak π bond of the carbonyl group breaks and both these electrons move onto the oxygen.

This results in the negative charge ending up on the electronegative oxygen where it is more stable. This mechanism is exactly the same as the one for the carboxylate ion. However, whereas both resonance structures are equally stable in the carboxylate ion but this is not the case here.

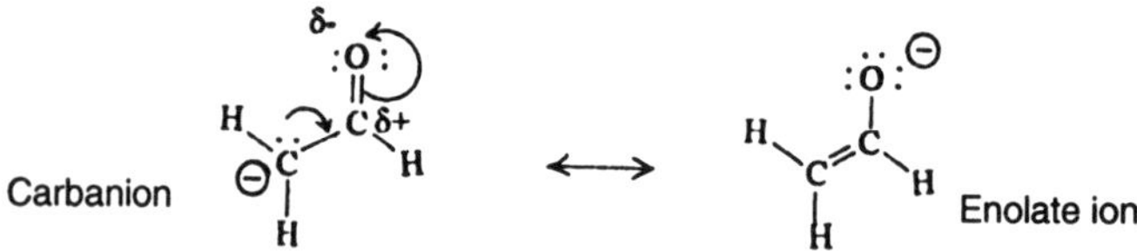

Fig. Resonance interaction between carbanion and enolate ion.

The resonance structure having the charge on the oxygen atom (an enolate ion) is more stable than the original carbanion resonance structure. Therefore, the enolate ion will predominate over the carbanion.

Mechanism

Because the enolate ion is the preferred resonance structure so a better mechanism for the acid base reaction shows the enolate ion being formed simultaneously as the acidic proton is lost. As the hydroxide ion forms its bond to the acidic proton, the C–H bond breaks, and the electrons in that bond form a π bond to the carbonyl carbon atom. Simultaneously, the carbonyl π bond breaks in such a way that both electrons move onto the oxygen. The electronegative oxygen is responsible for making the α proton acidic.

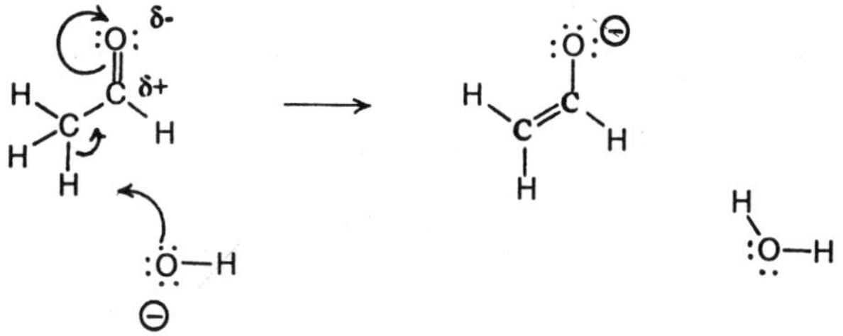

Fig. Mechanism for the formation of the enolate ion.

Enolate Ion

Resonance structures represent the extreme possibilities for a particular molecule and the true structure is really a hybrid of both. The 'hybrid' structure shows that the negative charge is 'smeared' or delocalised between three sp^2 hybridised atoms. Since these atoms are sp^2 hybridised, they are planar and have a $2p$ orbital that can interact with its neighbours to form one molecular orbital, thus spreading the charge between the three atoms. Keeping this in mind, we can state which of the methyl hydrogens is most likely to be lost in the formation of an enolate ion. The hydrogen circled is the one which will be lost since the σ C–H bond is correctly orientated to interact with the π orbital of the carbonyl bond. The orbital diagram illustrates this interaction. A Newman diagram can also be drawn by looking along the C–C bond to indicate the relative orientation of the α hydrogen which will be lost. In this example, there is no difficulty in the proton being in the correct orientation since there is free rotation around the C–C single bond. However, in cyclic systems, the hydrogen atoms are locked in space and the relative stereochemistry becomes important if the α proton is to be acidic.

Fig. Resonance structures and 'hybrid' structure for the enolate ion.

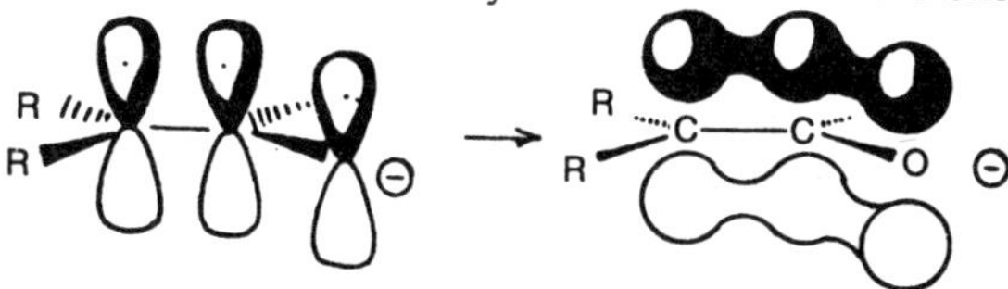

Fig. Interaction of 2p orbitals to form a molecular orbital.

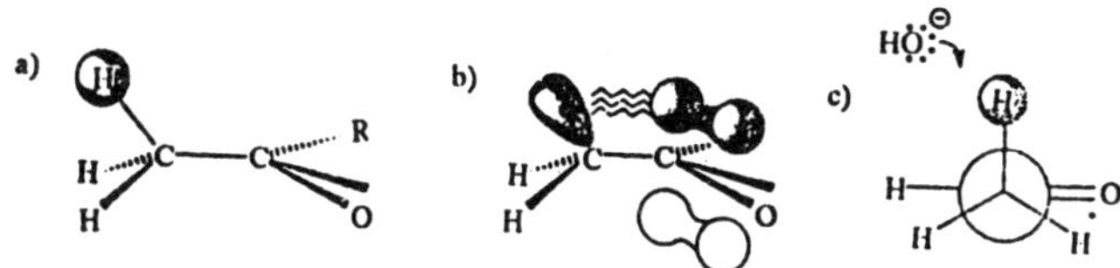

Fig. (a) a proton; (b) orbital diagram illustrating orbital interactions; (c) Newman projection.

Enolate ions formed from, ketones or aldehydes are extremely important in the synthesis of more complex organic molecules. The ease with which an enolate ion is formed is related to the acidity of the α proton. The pK_a of propane (acetone) is ≈ 19.3 that means that it is α stronger acid compared to ethane (pK_a ≈ 60) and a much weaker acid than acetic acid (pK_a 4.7), i.e. strong bases like sodium hydride, sodium amide, and lithium diisopropylamide $LiN(i–C_3H_7)_2$ are needed to form an enolate ion.

However, the acidity of the a proton gets increased if it is flanked by two carbonyl groups rather than one, for example, 1, 3-diketones (β-diketones) or 1,3-diesters (β-keto esters). This is due to the fact that the negative charge of the enolate ion can be stabilised by both carbonyl groups which results in three resonance structures. For example, the pK_a of 2, 4-pentanedione is 9.

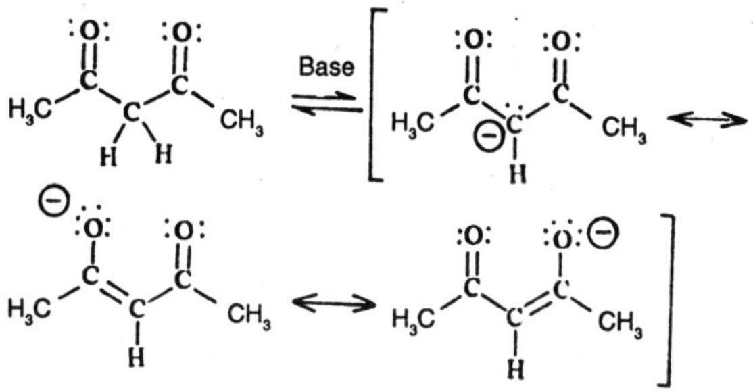

Fig. Resonance structures for the conjugate base of a 1, 3-dikctone.

Reaction between Acid and Base

Bronsted-Lowry Acids and Bases

Definition: According to Bronsted-Lowry concept an acid is a molecule that can donate a proton and a base is a molecule that can accept that proton.

An example of a simple acid/base reaction is the reaction of ammonia with water. In it, water loses a proton and acts as an acid. Ammonia accepts the proton and acts as the base.

$$H_2\ddot{O} + \ddot{N}H_3 \rightleftharpoons \overset{\ominus}{\ddot{O}}H + \overset{\oplus}{N}H_4$$

Fig. Reaction of ammonia with water.

In it, the ammonia uses its lone pair of electrons to form a new bond to the proton and so it is acting as a nucleophile, thus, the water is acting as an electrophile.

As the nitrogen uses its lone pair of electrons to form the new bond, the bond between hydrogen and oxygen must break because hydrogen can form only one bond. The electrons making up the O–H bond will move onto oxygen to produce a third lone pair of electrons, thus giving the oxygen a negative charge. Since the nitrogen atom or ammonia has used its lone pair of electrons to form a new bond, it now has to share the electrons with hydrogen and so nitrogen gains a positive charge.

Fig. Mechanism for the reaction of ammonia with water.

Bronsted-Lowry Acids

According to this concept an acid is a molecule that contains an acidic hydrogen. In order to be acidic, the hydrogen must be slightly positive or electrophilic. This is possible if hydrogen is attached to an electronegative atom like a halogen, oxygen, or nitrogen.

The following mineral acids and functional groups contain hydrogen's that are potentially acidic:

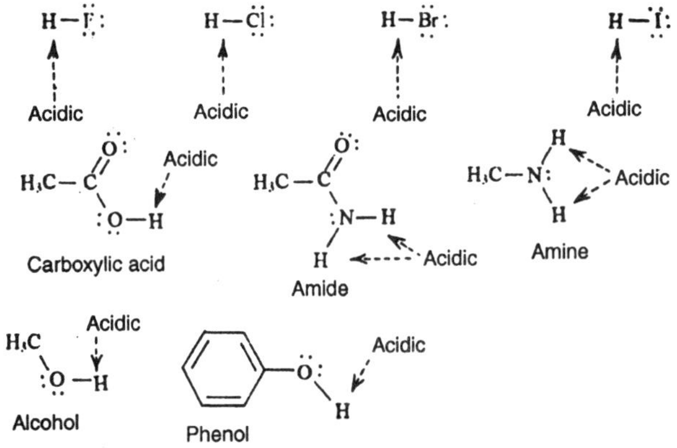

Fig. Acidic protons in mineral acids and common functional groups.

Hydrogens attached to carbon are not normally acidic. However, in some special cases hydrogens attached to carbon are acidic.

Bronsted-Lowry Bases

According to Bronsted-Lowry concept a base is a molecule that can form a bond to a proton. They may be negatively charged ions with a lone pair of electrons:

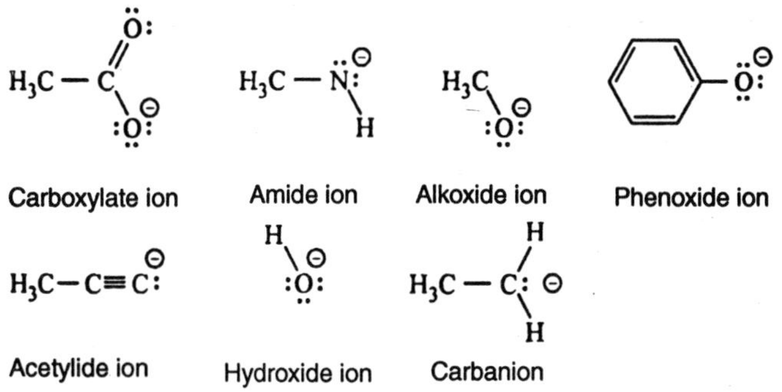

Fig. Examples of Bronsted-Lowry bases.

Neutral molecules can also act as bases if they contain an oxygen or nitrogen atom. The most common examples are amines. However, water, ethers and alcohols are also capable of acting as bases:

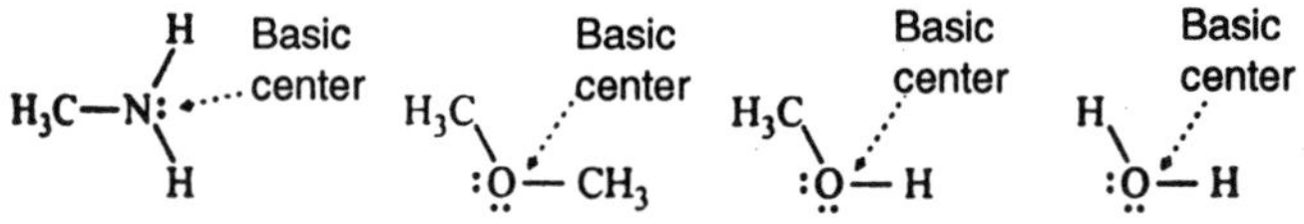

Fig. Examples of neutral Bronsted-Lowry bases.

Strength of Acid

Electronegativity

The acidic protons of various molecules depends on various factors, such as the electronegativity of the atom to which they are attached. For example, if we consider hydrofluoric acid, ethanoic acid, and methylamine. Hydrofluoric acid has the most acidic proton because the hydrogen is attached to a strongly electronegative fluorine. The fluorine strongly polarises the H-F bond such that the hydrogen becomes highly electron deficient and can be easily lost. Once the proton is lost, the fluoride ion can stabilise the resulting negative charge.

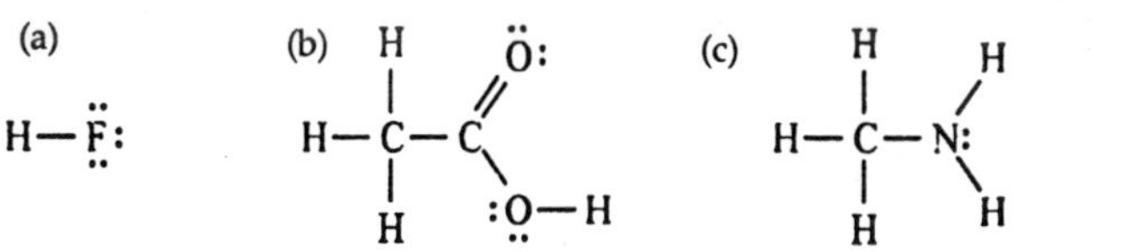

Fig. (a) Hydrofluoric acid; (b) ethanoic acid; (c) methylamine.

The acidic protons on methylamine are attached to nitrogen that is less electronegative than fluorine. Therefore, the N–H bonds are less polarised, and the protons are less electron deficient. If one of the protons is lost, the nitrogen is left with a negative charge, which it cannot stabilise as efficiently as a halide ion.

Thus, methylamine is a much weaker acid than hydrogen fluoride. Ethanoic acid is more acidic than methylamine but less acidic than hydrofluoric acid. This is because the electronegativity of oxygen lies between that of a halogen and that of a nitrogen atom.

These differences in acid strength can be shown if the three molecules above are placed in water. Mineral acids like HF, HCl, HBr, and HI are strong acids and dissociate or ionise completely when dissolved in water:

$$H{-}Cl + H{-}O{-}H \longrightarrow Cl^{\ominus} + H{-}\overset{\oplus}{O}H_2$$

Fig. Ionisation of hydrochloric acid.

Ethanoic acid (acetic acid) partially dissociates in water and an equilibrium is set up between the carboxylic acid (called the *free acid*) and the carboxylate ion. An acid that only partially ionises in this way is called a *weak acid*.

$$H_3C{-}C(=O){-}O{-}H + H_2O \rightleftharpoons H_3C{-}C(=O){-}O^{\ominus} + H_3O^{\oplus}$$

Fig. Partial ionisation of ethanoic acid.

When methylamine is dissolved in water, none of the acidic protons are lost at all and so the amine behaves as a weak base instead of an acid, and is in equilibrium with its protonated form:

$$H_3C{-}NH_2 + H_2O \rightleftharpoons H_3C{-}\overset{\oplus}{N}H_3 + HO^{\ominus}$$

Fig. Equilibrium acid-base reaction of methylamine with water.

Methylamine can act as an acid when it is treated as a strong base like butyl lithium.

The hydrogen atoms attached to carbon are not generally acidic because carbon atoms are not electronegative. There are some exceptions to this rule.

$$H_3C-NH_2 + H_2C^{\ominus}CH_2CH_2CH_3 \xrightarrow{Li^{\oplus}} H_3C-\ddot{N}H^{\ominus} + H_3C\text{-}CH_2CH_2CH_3$$

Fig. Methylamine acting as an acid with a strong base (butyl lithium).

pK_a

Acids can be called as being weak or strong comparing by their pK_a values. Dissolving acetic acid in water, results in an equilibrium between the carboxylic acid and the carboxylate ion:

$$H_3C-C(=O)-OH + H_2O \rightleftharpoons H_3C-C(=O)-O^{\ominus} + H_3O^{\oplus}$$

Fig. Equilibrium acid-base reaction of ethanoic acid with water.

Ethanoic acid on the left hand of the equation is called the *free acid,* whereas the carboxylate ion formed on the right hand side is called its *conjugate base.* The extent of ionisation or dissociation is given by the equilibrium constant (K_{eq}):

$$K_{eq} = \frac{[\text{Products}]}{[\text{Reactants}]} = \frac{\left[CH_3CO_2^-\right]\left[H_3O^+\right]}{[CH_3CO_2][H_2O]}$$

K_{eq} is generally measured in a dilute aqueous solution of the acid and therefore the concentration of water is high and assumed to be constant. Hence, we can rewrite the equilibrium equation in a simpler form where K_a is the acidity constant and includes the concentration of pure water (55.5M).

$$K_a = K_{eq}[H_2O] = \frac{\left[CH_3CO_2^-\right]\left[H_3O^+\right]}{[CH_3CO_2H]}$$

The acidity constant can also be a measure of dissociation and of how much acidic a particular acid is. The stronger be acid, is more ionised and therefore the greater the concentration of products in the above equation. Hence, a strong acid has a high K_a value. The K_a values for the following ethanoic acid are in brackets and these show that the strongest acid in the series in trichloroacetic acid.

Cl_3CCO_2H (23200 × 10^{-5}) > Cl_2CHCO_2H(5530 × 10^{-5}) > $ClCH_2CO_2H$ (136 × 10^{-5})·> CH_3CO_2H (1.75 × 10^{-5}).

K_a values are less commonly used and it is more usual to measure the acidic strength as a pK_a value rather than K_a. The pK_a is the negative logarithm of K_a ($pK_a = -\log_{10} K_a$ and results in more manageable numbers. The pK_a values for each of the above ethanoic acids is shown in brackets below. *The strongest acid (trichloroacetic acid) has the lowest pK_a value.*

Cl_3CCO_2H (0.63) < Cl_2CHCO_2H (1.26) < $ClCH_2CO_2H$ (2.87) < CH_3CO_2H(4.76)

Thus, the *stronger* the acid, the *higher* the value of K_a, and the *lower* the value of pK_a. An amine like ethylamine ($CH_3CH_2NH_2$) is a very weak acid (pK_a = 40) compared to ethanol (pK_a = 16). This is because of the relative electronegativities of oxygen and nitrogen. However, the electronegativity of neighbouring atoms is not the only influence on acidic strength. For example, the pK_a values of ethanoic acid (4.76), ethanol (16), and phenol (10) show that ethanoic acid is more acidic than phenol, and that phenol is more acidic than ethanol. The difference in acidity is quite marked, yet hydrogen is attached to oxygen in all three structures.

Similarly, the ethanoic acids Cl_3CCO_2H (0.63), Cl_2CHCO_2H (1.26), $ClCH_2CO_2H$ (2.87), and CH_3CO_2H (4.76) have significantly different pK_a values and yet the acidic hydrogen is attached to an oxygen in each of these structures. Therefore, factors other than electronegativity also play a role in determining acidic strength.

Inductive Effect

Permanent displacement of electrons along a certain chain when some atom or group of atoms with different electronegativity than carbon is attached to carbon chain is called *inductive effect*. For illustration consider a carbon chain which has a chlorine atom attached to one end.

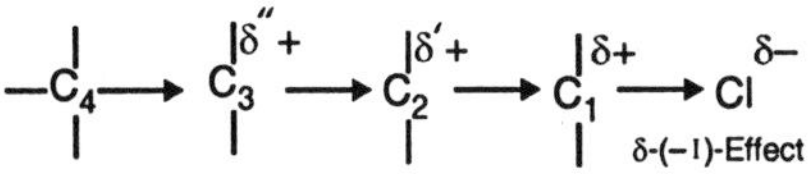

$$-\overset{|}{\underset{|}{C_4}} \longrightarrow \overset{|\delta''+}{\underset{|}{C_3}} \longrightarrow \overset{|\delta'+}{\underset{|}{C_2}} \longrightarrow \overset{|\delta+}{\underset{|}{C_1}} \longrightarrow \overset{\delta-}{Cl}$$

δ-(−I)-Effect

Since chlorine atom has higher electronegativity than carbon so the shared pair of electrons between C_1 and chlorine atom lies closer to chlorine atom as a result of which chlorine atom acquires a small negative charge and carbon atom C_1 acquires small +ve charge. As C_1 is now somewhat positively charged it attracts towards itself the shared pair of electrons between C_1 and C_2. This results in a small positive charge on C_2 also, but this charge is less than on C_1. In this way electrons are displaced towards chlorine in the carbon chain. This type of displacement of electrons along a carbon chain is known as *Electron Withdrawing Inductive Effect* or *(–I) Effect*.

If the electronegativity of the atom or group of atoms attached to the carbon chain is less than the electronegativity of the carbon then the displacement of electrons takes place away from group along the C-chain and this effect is called the *Electron Releasing Inductive Effect* or *(+I)-Effect*. This inductive effect can be represented as shown below:

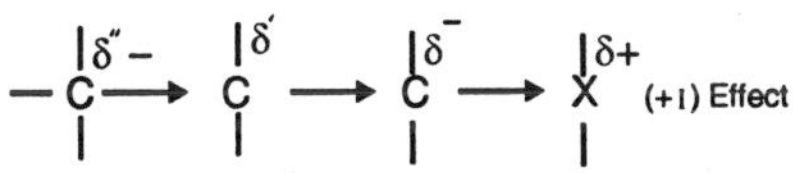

$$-\overset{|\delta''-}{\underset{|}{C}} \longrightarrow \overset{|\delta'}{\underset{|}{C}} \longrightarrow \overset{|\delta^-}{\underset{|}{C}} \longrightarrow \overset{|\delta+}{\underset{|}{X}} \quad \text{(+I) Effect}$$

Inductive effect is a permanent effect and decreases rapidly as the distance from the source increases. Some of the groups which show Inductive effect are shown below:

– *I effect:* $-NO_2 > -F > -Cl > -Br > -I > -OCH_3 > -C_6H_5$

+ *I effect:* $-C(CH_3)_3 > -C(CH_3)_2 > -C_2H_5 > -CH_3$

Any effect that stabilises the negative charge of the conjugate base will increase the acid strength. Substituents can stabilise a negative charge and they do so by an inductive effect. This can be illustrated by comparing the pK_a values of the alcohols CF_3CH_2OH and CH_3CH_2OH (12.4 and 16, respectively) where CF_3CH_2OH is more acidic than CH_3CH_2OH. This means that the anion $CF_3CH_2O^-$ is more stable than $CH_3CH_2O^-$:

a) $F_3C-CH_2-\ddot{O}:^{\ominus}$ b) $H_3C-CH_2-\ddot{O}:^{\ominus}$

Fig. (a) 2,2,2-Trifluorocthoxy; (b) ethoxy ion.

Fluorine atoms are strongly electronegative and therefore each C–F bond is strongly polarised. In such a way that the carbon bearing the fluorine atoms becomes strongly electropositive. As this carbon atom is now electron deficient, so it will 'demand' a greater share of the electrons in the neighbouring C–C bond. Due to this the electrons are being withdrawn from the neighbouring carbon, making it electron deficient too. This inductive effect will continue to be felt through the various bonds of the structure. It will decrease through the bonds but it is still significant enough to be felt at the negatively charged oxygen. As the inductive effect is electron withdrawing it will decrease the negative charge on the oxygen and help to stabilise it. Therefore, the original fluorinated alcohol will lose its proton more readily and will be a stronger acid.

The inductive effect can also explain the relative acidities of the chlorinated ethanoic acids Cl_3CCO_2H (0.63), Cl_2CHCO_2 (1.26), $ClCH_2,CO_2H$ (2.87), and CH_3CO_2H (4.76). Trichloroethanoic acid is the strongest acid as its conjugate base (the carboxylate ion) is stabilised by the inductive effect created by three electronegative chlorine atoms. As the number of chlorine atoms decrease, the inductive effect also decreases.

Inductive effect can also explain the difference between the acid strengths of ethylamine ($pK_a \sim 40$) and ammonia ($pK_a \sim 33$).

The pK_a values show that ammonia is a stronger acid than ethylamine. In this case, the inductive effect is electron donating. The alkyl group of ethylamine increases the negative charge of the conjugate base and so destabilises it thus, making ethylamine a weaker acid than ammonia:

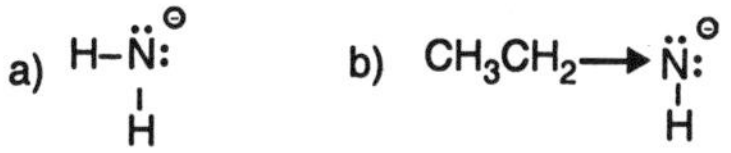

a) $H-\ddot{N}:^{\ominus}$ with H below N   b) $CH_3CH_2 \rightarrow \ddot{N}:^{\ominus}$ with H below N

Fig. Conjugate bases (a) ammonia and (b) ethylamine.

Resonance

Sometimes it is not possible to assign a single electronic structure to a molecule that may account for all its properties. In such a case the molecule is represented by two or more electronic structures but none of these represents the actual structure of the molecule. The actual structure of the molecule lies somewhere in between these structures but can not be expressed on paper. Such a molecule is said to exhibit resonance. The various structures assigned to the molecule are called *contributing structures* or *canonical structures* where as the intermediate structures is called the *resonance hybrid*. For example benzene (C_6H_6) may be assigned the following two structures:

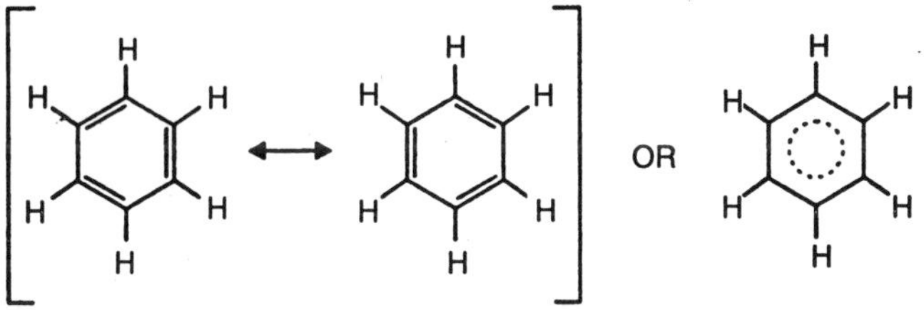

[Two Kekulé structures of benzene, each ring carbon bonded to H, joined by ↔ inside brackets] OR [benzene ring with dotted inner circle, each carbon bonded to H]

Any of these structures alone cannot account for all the properties of benzene. According to these structures there should be three single bonds (1.54Å) and three double bonds (1.34Å) between carbon atoms. But actually it has been found that all the six carbon-carbon bonds in benzene have same bond length (1.39Å). The actual structure of benzene is resonance hybrid of the above two structures and may be represented as under:

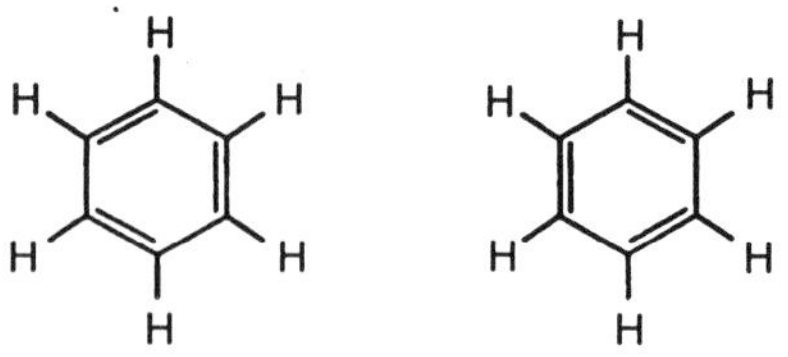

I II

The resonance hybrid is more stable than any of the contributing structures and the difference between the energy of the most stable contributing structure and resonance hybrid is called *resonance energy.*

Carboxylic acids and carboxylate ions also exhibit resonance.

Carboxylic acid

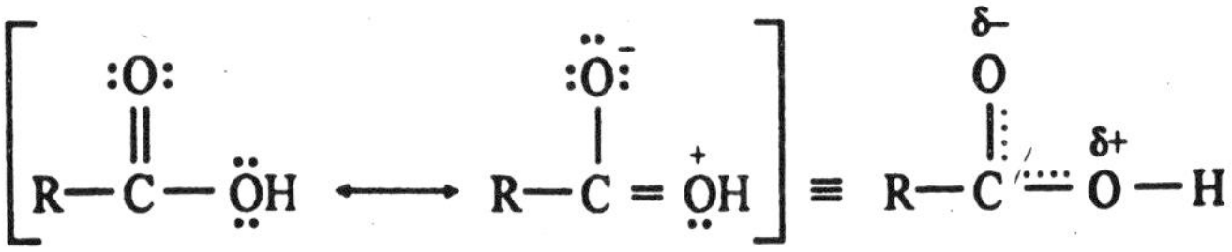

$$\left[R-\overset{\overset{:O:}{\|}}{C}-\ddot{O}H \longleftrightarrow R-\overset{\overset{:\ddot{O}:^-}{|}}{C}=\overset{+}{\ddot{O}}H \right] \equiv R-\overset{\overset{\delta-}{O}}{\vdots}C \cdots \overset{\delta+}{O}-H$$

Contributing structures Hybrid structure

Carboxylate Ion

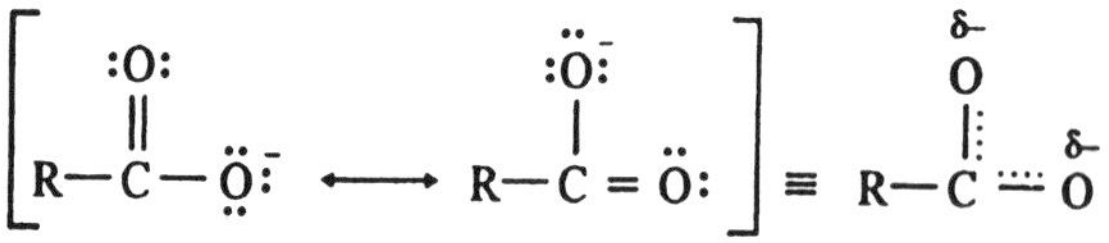

$$\left[R-\overset{\overset{:O:}{\|}}{C}-\ddot{O}:^- \longleftrightarrow R-\overset{\overset{:\ddot{O}:^-}{|}}{C}=\ddot{O}: \right] \equiv R-\overset{\overset{\delta-}{O}}{\vdots}C \cdots \overset{\delta-}{O}$$

Contributing structures Hybrid structures

Some important features of resonance are:

1. The various contributing structures may differently in electronic arrangement but should have same arrangement of atoms.
2. The number of unpaired electrons should be same in all the contributing structures.
3. All contributing structures should have almost same energy.

4. The more stable contributing structure makes more contribution.
5. The bond distances of hybrid structure are intermediate of those of resonating forms.

Resonance Effect

It may be noted that in conjugated systems (having alternate single and double bonds) resonance causes displacement of electrons from one part of the system to another part creating centres of high and low electron density. This is called *resonance effect*.

The negative charge on some conjugate bases can be stabilised by resonance. Resonance involves the movement of valence electrons around a structure, which results in the sharing of charge between different atoms. This process is called *delocalisation*. The effects of resonance can be shown by comparing the acidities of ethanoic acid (pK_a 4.76), phenol (pK_a 10.0) and ethanol (pK_a 12.4). The pK_a values show that ethanoic acid is a stronger acid than phenol, and that phenol is a stronger acid than ethanol.

The varying acidic strengths of ethanoic acid, phenol and ethanol can be explained by considering the relative stabilities of their conjugate bases:

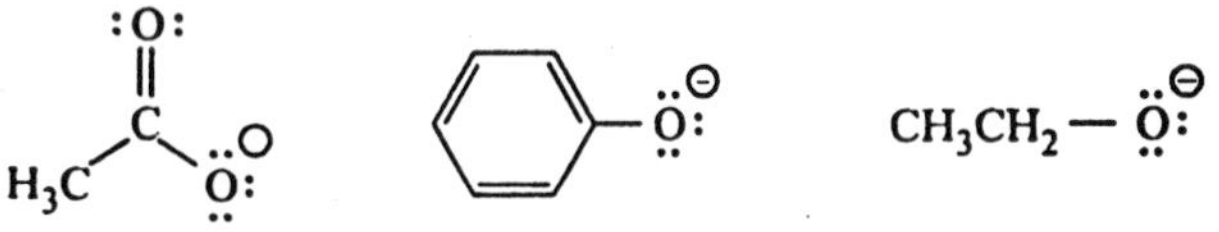

$H_3C-C(=\ddot{O}:)-\ddot{O}:^{\ominus}$ (a) $C_6H_5-\ddot{O}:^{\ominus}$ (b) $CH_3CH_2-\ddot{O}:^{\ominus}$ (c)

Fig. Conjugate bases of (a) ethanoic acid; (b) phenol; (c) ethanol.

The charge of the carboxylate ion is on an oxygen atom, and because oxygen is electronegative so the charge is stabilised. However, the charge can be shared with the other oxygen leading to delocalisation of the charge. This arises by a resonance interaction between a lone pair of electrons on the negatively charged oxygen and the π electrons of the carbonyl group. A lone pair of electrons on the 'bottom' oxygen forms a new π bond to the neighbouring carbon. Simultaneously, the weak π bond of the carbonyl group breaks. This is essential or else the carbonyl carbon would end up with five bonds and that is not allowed. Both electrons in the original π bond now end up on the 'top' oxygen that means that this oxygen ends up with three lone pairs and gains a negative charge. Please note that the π bond and the charge have effectively 'swapped places'. Both the structures involved are referred to as resonance structures and are easily interconvertible.

The negative charge is now shared or delocalised equally between both oxygens and is stabilised. Hence, ethanoic acid is a stronger acid than one would expect based on the electronegativity of oxygen alone.

$H_3C-C(=\ddot{O}:)-\ddot{O}:^{\ominus} \longleftrightarrow H_3C-C(-\ddot{O}:^{\ominus})=\ddot{O}:$

Fig. Resonance interaction for the carboxylate ion.

Phenol is less acidic than ethanoic acid but is more acidic than ethanol. Once again the resonance concept can explain the differences. The conjugate base of phenol is known as the *phenolate ion*. In this case, the resonance process can be carried out several times to place the negative charge on four separate atoms, i.e. the oxygen atom and three of the aromatic carbon atoms. Since the negative charge can be spread over four atoms might suggest that all the phenolate anion should be more stable than the carboxylate anion, since the charge is spread over more atoms.

However, with the phenolate ion, three of the resonance structures place the charge on a carbon atom that is much less electronegative than an oxygen atom. These resonance structures will therefore be far less important than the resonance structure having the charge on oxygen. Because of this, delocalisation is weaker for the phenolate ion than for the ethanoate ion. However, a certain amount of delocalisation still occurs that is why a phenolate ion is more stable than an ethoxide ion.

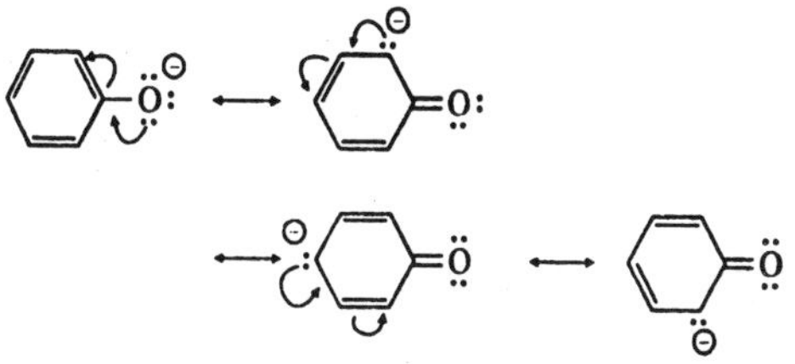

Fig. Resonance interactions for the phenolate ion.

In case of ethanol, the conjugate base is the ethoxide ion that cannot be stabilised by delocalising the charge, because resonance is not possible. There is no π bond available to participate in resonance. Thus, the negative charge is localised on the oxygen. Moreover, the inductive donating effect of the neighbouring alkyl group (ethyl) increases the charge and destabilises it. This makes the ethoxide ion the least stable (or most reactive) of the three anions that we have studied. Due to this, ethanol is the weakest acid.

$$CH_3CH_2 \longrightarrow O^{\ominus}$$

Fig. Destabilising inductive effect of the ethoxide ion.

Amines and Amides

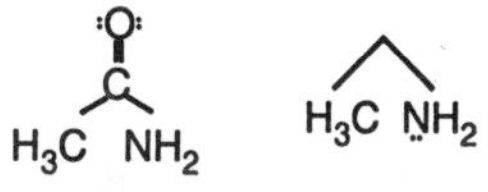

(a) (b)

Fig. (a) Ethanamide; (b) ethylamine.

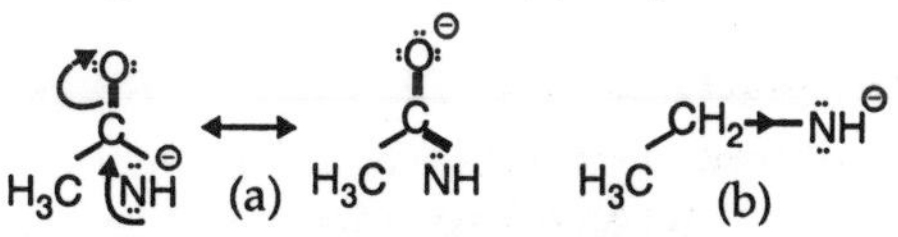

Fig. (a) Resonance stabilisation for the conjugate base of ethanamide: (b) inductive destabilisation for the conjugate bases of ethylamine.

Amines and amides are very weak acids and they only react with very strong bases. The pK_a values for ethanamide and ethylamine are 15 and 40, respectively, which means that ethanamide has the more acidic proton. This can be explained by making use of resonance and inductive effects.

Strength of Base

Electronegativity

Electronegativity influences the basic strength of the compound. If we compare the fluoride ion, hydroxide ion, amide ion and the methyl carbanion, then the order of basicity is as shown in the following figure.

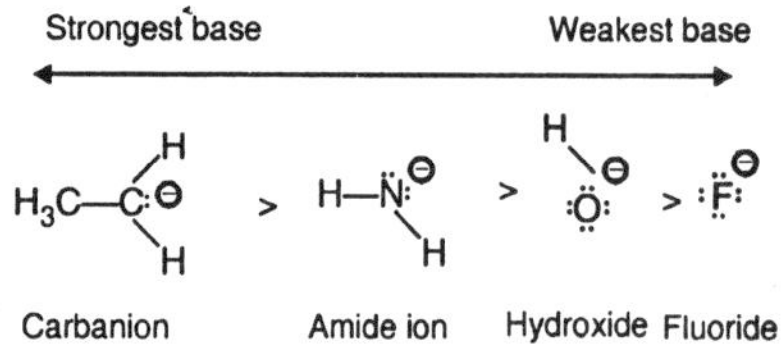

Strongest base ⟷ Weakest base

$H_3C{-}\ddot{C}H_2^{\ominus} > H_2\ddot{N}^{\ominus} > H\ddot{O}^{\ominus} > :\ddot{F}^{\ominus}$

Carbanion, Amide ion, Hydroxide, Fluoride

Fig. Comparison of basic strength.

The strongest base is the carbanion as this has the negative charge situated on the least electronegative atom, i.e. the carbon atom. The weakest base is the fluoride ion which has the negative charge situated on the most electronegative atom, i.e. the fluorine atom. Strongly electronegative atoms like fluorine are able to stabilise a negative charge making the ion less reactive and less basic. The order of basicity of the anions formed from alkanes, amines, and alcohols follows a similar order because of the same reason:

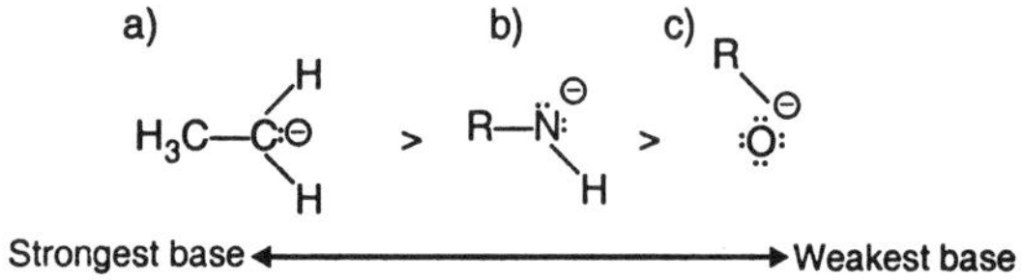

a) $H_3C{-}\ddot{C}H_2^{\ominus}$ > b) $R{-}\ddot{N}H^{\ominus}$ > c) $R\ddot{O}^{\ominus}$

Strongest base ⟷ Weakest base

Fig. Comparison of basic strengths: (a) a Carbanion; (b) an amide ion; (c) an alkoxide ion.

Electronegativity can also explain the order of basicity for neutral molecules like amines, alcohols, and alkyl halides:

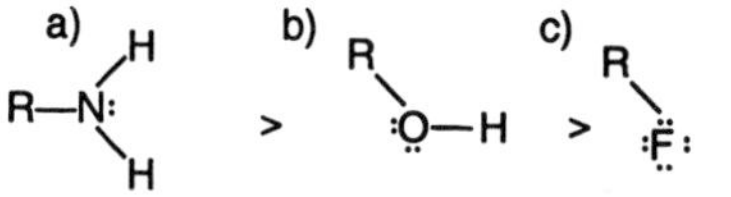

Fig. Comparison of basic strengths: (a) an amine; (b) an alcohol; (c) an alkyl fluoride.

These neutral molecules are much weaker bases than their corresponding anions, but the order of basicity is still the same and can be explained by considering the relative stability of the cations that are formed when these molecules bind a proton:

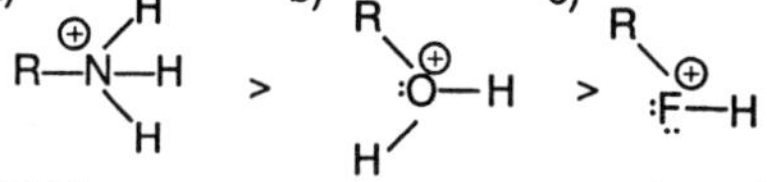

Fig. Relative stability of the carbons formed form (a) an amine; (b) an alcohol: (c) an alkyl fluoride.

A nitrogen can stabilise a positive charge better than a fluorine atom because the former is less electronegative. Electronegative atoms prefer to have a negative charge rather than a positive charge. Fluorine is so electronegative that its basicity is negligible. Therefore, amines act as weak bases in aqueous solution and are partially ionised. Alcohols only act as weak bases in acidic solution. Alkyl halides are essentially non-basic even in acidic solutions.

pK_b Values

pK_b value is a measure of basic strength of a compound. When methylamine is dissolved in water, the following equilibrium is set up:

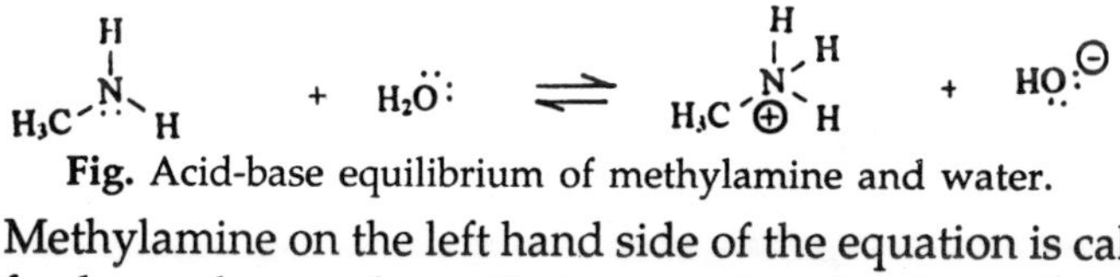

$$H_3C-\ddot{N}H_2 + H_2\ddot{O}: \rightleftharpoons H_3C-\overset{\oplus}{N}H_3 + HO:^{\ominus}$$

Fig. Acid-base equilibrium of methylamine and water.

Methylamine on the left hand side of the equation is called the *free base*, whereas the methyl ammonium ion formed on the right hand side is called the *conjugate acid*. The extent of ionisation or dissociation in the equilibrium reaction is defined by the equilibrium constant (K_{eq}):

$$K_{eq} = \frac{[\text{Products}]}{[\text{Reactants}]} = \frac{[CH_3NH_3^+][HO^-]}{[CH_3NH_2][H_2O]}$$

$$K_b = K_{eq}[H_2O] = \frac{[CH_3NH_3^+][HO^-]}{[CH_3NH_2]}$$

K_{eq} is generally measured in a dilute aqueous solution of the base and so the concentration of water is high and assumed to be constant. Therefore, we can rewrite the equilibrium equation in a simpler form where K_b is the basicity constant and includes the concentration of pure water (55.5M). pK_b is the negative logarithm of K_b and is used as a measure of basic strength ($pK_b = -Log_{10}K_b$).

A large pK_b indicates a weak base. For example, the pK_b values of ammonia and methylamine are 4.74 and 3.36, respectively, which indicates that ammonia is a weaker base than methylamine.

pK_b and pK_a are related by the equation $pK_a + pK_b = 14$. Therefore, if we know the pK_a of an acid, the pK_b of its conjugate base can be calculated and vice versa.

Inductive Effects

Inductive effects affect the strength of a charged base by influencing the negative charge. For example, an electron-withdrawing group helps to stabilise a negative charge, which results in a weaker base. An electron-donating group will destabilise a negative charge, which results in a stronger base. Amongst Cl_3CCO_2H, Cl_2CHCO_2H, $ClCH_2CO_2H$, and CH_3CO_2H, trichloroacetic acid is a strong acid as its conjugate base(the carboxylate ion) is stabilised by the three electronegative chlorine groups.

$$Cl_3C-C(=\ddot{O}:)-\ddot{O}H \rightleftharpoons Cl_3C \leftarrow C(=\ddot{O}:)-\ddot{O}:^{\ominus}$$

Strong acid $pK_a = 0.63$ Weak Conjugate base (stabilized)

Fig. Inductive effect on the conjugate base of trichloroacetic acid.

The chlorine atoms possesses electron-withdrawing effect that helps to stabilise it. If the negative charge is stabilised, it makes the conjugate base less reactive and a weaker base. We know that the conjugate base of a strong acid is weak, whereas the conjugate base of a weak acid is strong. Therefore, the order of basicity for the ethanoate ions $Cl_3CCO_2^-$, $Cl_2CHCO_2^-$, $ClCH_2CO_2^-$, and $CH_3CO_2^-$ is the opposite to the order of acidity for the corresponding carboxylic acids, i.e. the ethanoate ion is the strongest base, while the trichlorinated ethanoate ion is the weakest base.

Inductive effects can also influence the basic strength of neutral molecules (e.g. amines). The pK_b for ammonia is 4.74, which compares with pK_b values for methylamine, ethylamine, and propylamine of 3.36, 3.25 and 3.33, respectively.

The alkylamines are stronger bases than ammonia due to the inductive effect of an alkyl group on the alkyl ammonium ion (RNH_3^-). Alkyl groups donate electrons towards a neighbouring positive centre gets partially dispersed over the alkyl group. If the ion is stabilised, the equilibrium of the acid-base reaction will shift to the ion, that means that the amine is more basic. The larger the alkyl group, the more significant this effect.

$$R—\ddot{N}H_2 + H_2\ddot{O} + \rightleftharpoons R\rightarrow \overset{\oplus}{N}H_3 + H\ddot{O}:^{\ominus}$$

Inductive effect

Fig. Inductive effects of an alkyl group on the alkyl ammonium ion.

If one alkyl group can influence the basicity of an amine, then further alkyl groups should have an even greater inductive effect. Therefore, one might expect secondary and tertiary amine is to be stronger bases than primary amines. In fact, this is not necessarily the case. There is no easy relationship between basicity and the number of alkyl groups attached to nitrogen. Although the inductive effect of more alkyl groups is certainly greater, this effect is counterbalanced by a solvation effect.

Solvation Effects

After the formation of an alkyl ammonium ion, it is solvated by water molecules. This process involves hydrogen bonding between the oxygen atom of water and any N–H. group present in the alkyl ammonium ion. Water solvation is a stabilising factor that is as important as the inductive effect of the alkyl substituents and the more hydrogen bonds that are possible, the greater the stabilisation.

Solvation is stronger for the alkyl ammonium ion formed from a primary amine than for the alkyl ammonium ion formed from a tertiary amine. This is due to the fact that the former ion has three N–H hydrogens available for H-bonding, compared with only one such N–H hydrogen the latter.

Because of this there is more solvent stabilisation experienced for the alkyl ammonium ion of a primary amine compared to that experienced by the alkyl ammonium ion of a tertiary amine. This means that tertiary amines are generally weaker bases than primary or secondary amines.

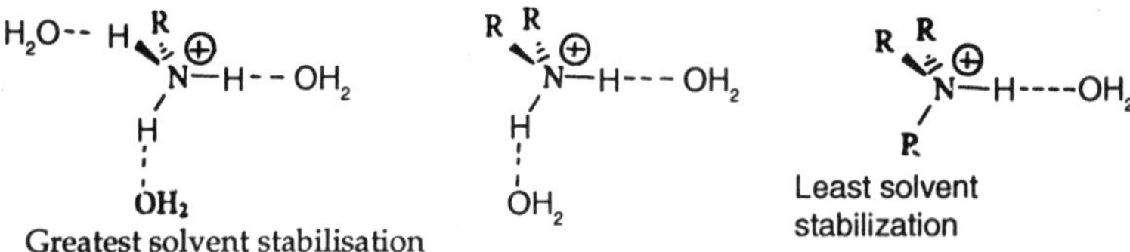

Fig. Solvent effect off alkyl ammonium ions from primary, secondary, and tertiary amines.

Resonance

We have learnt that resonance can stabilise a negative charge by delocalising it over two or more atoms. Resonance explains why a carboxylate ion is more stable than an alkoxide ion. The negative charge in the former can be delocalised between two oxygens whereas the negative charge on the former is localised on the oxygen. We used this to explain why a carboxylic acid is a stronger acid than an alcohol. We can use the same argument in reverse to explain the difference in basicities between a carboxylate ion and an alkoxide ion. Because the latter is less stable, it is more reactive and is therefore a stronger base.

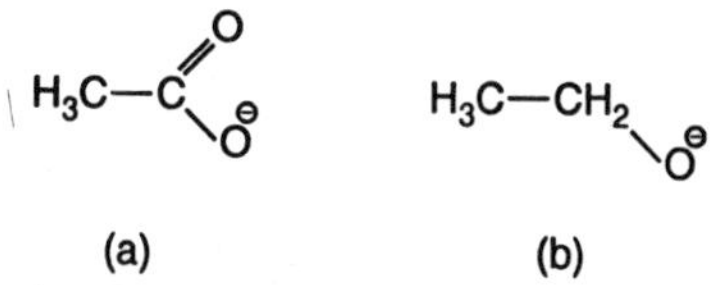

(a) (b)

Fig. (a) Carboxylate ion; (b) alkoxide ion.

Resonance effects can also explain why aromatic amines (arylamines) are weaker bases than alkylamines. The lone pair of electrons on nitrogen can interact with the π system of the aromatic ring resulting in the possibility of three zwitterionic resonance structures. (A zwitterion is a neutral molecule containing a positive and a negative charge). Since nitrogen's lone pair of electrons is involved in this interaction, it is less available to form a bond to a proton and so the amine is less basic.

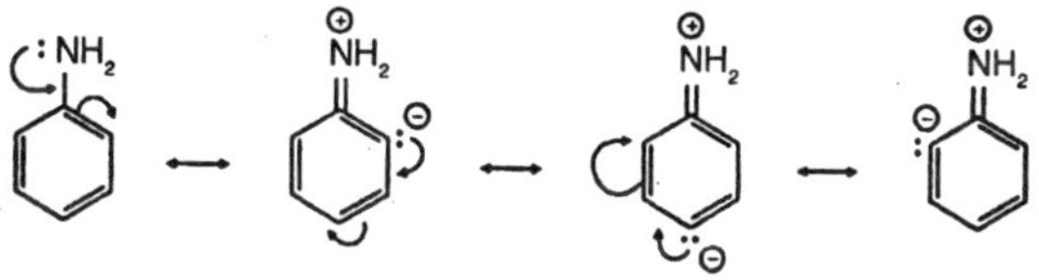

Fig. Resonance structures for aniline.

Amines and Amides

Amines are weak bases. They form water soluble slats in acidic solutions and in aqueous solution they are in equilibrium with their conjugate acid.

Amines are the basic because they have a lone pair of electrons that can form a bond to a proton. Amides also have a nitrogen with a lone pair of electrons, but unlike amines they are not basic. This is because a resonance occurs within the amide structure that involves the nitrogen lone pair.

The driving force behind this resonance is the electronegative oxygen of the neighbouring carbonyl group that is 'hungry' for electrons.

a)

$$R{-}\ddot{N}H_2 + HCl \longrightarrow R{-}\overset{\oplus}{N}H_3 \quad :\ddot{Cl}:^{\ominus}$$

Free base Soluble salt

b)

$$R{-}\ddot{N}H_2 + H{-}\ddot{O}{-}H \rightleftharpoons R{-}\overset{\oplus}{N}H_3 \quad H{-}\ddot{O}:^{\ominus}$$

Free base Conjugate acid

Fig. (a) Salt formation; (b) acid-base equilibrium.

The lone pair of electrons on nitrogen forms a π bond to the neighbouring carbon atom. As this occurs, the π bond of the carbonyl group breaks and both electrons move onto the oxygen to give it a total of three lone pairs and a negative charge. Because the nitrogen's lone pair is involved in this resonance, it is unavailable to bind to a proton and therefore amides are not basic.

$$H_3C{-}C(=O){-}\ddot{N}H_2 \longleftrightarrow H_3C{-}C(O^{\ominus})={\overset{\oplus}{N}}H_2$$

Fig. Resonance interaction of an amide.

Acids and Bases of Lewis

Lewis Acids

Lewis acids are ions or electron deficient molecules having an unfilled valence shell. They are known as *acids* because they can accept a lone pair of electrons from another molecule to fill their valence shell. Lewis acids include all the Bronsted-Lowry acids as well as ions (e.g. H^+, Mg^{2+}), and neutral species such as BF_3 and $AlCl_3$.

Both Al and B are in Group 3 of the periodic table and have three valence electrons in their outer shell. These elements can form three bonds. However, there is still room for a fourth bond. For example in BF_3, boron is surrounded by six electrons (three bonds containing two electrons each). However, boron's valence shell can accommodate eight electrons and so a fourth bond is possible if the fourth group can provide both electrons for the new bond. Since both boron and aluminium are in Group 3 of the periodic table, they are electropositive and will react with electron-rich molecules so as to obtain this fourth bond. Many transition metal compounds can also act like Lewis acids (e.g. $TiCl_4$ and $SnCl_4$).

Lewis Bases

A Lewis base is a molecule that can donate a lone pair of electrons to fill the valence shell of a Lewis acid. The base can be a negatively charged group such as a halide, or a neutral molecule like water, an amine, or an ether, as long as there is an atom present with a lone pair of electrons (i.e. O, N or a halogen).

All the Bronsted-Lowry bases can also be defined as Lewis bases. The crucial feature is the presence of a lone pair of electrons that is available for bonding. Therefore, all negatively charge ions and all functional groups containing a nitrogen, oxygen, or halogen atom can act as Lewis bases.

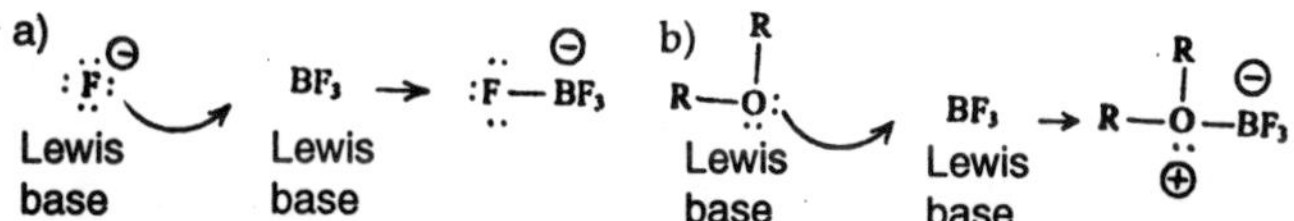

Fig. Reactions between Lewis acids and Lewis bases.

The Reactions

Organic reactions can be classified into following four types:

(a) Substitution Reactions

(b) Addition Reactions

(c) Elimination Reactions

(d) Rearrangement Reactions

All reactions involve the bond cleavage and the bond formation.

Bond Formation: Basically, most reactions involve electron-rich molecules forming bonds to electron deficient molecules (i.e. nucleophiles forming bonds to electrophiles). The bond will be formed particularly between the nucleophilic centre of the nucleophile and the electrophilic centre of the electrophile.

Classification of Reactions: We can also classify reactions as:

(a) acid/base reactions

(b) functional group transformations

(c) carbon-carbon bond formations

The reaction of type (a) are relatively simple and involves the reaction of an acid with a base to give a salt. The reaction of type (b) are one functional group can be converted into another. Normally these reactions are relatively straightforward and proceed in high yield. The reactions of type (c) are extremely important to organic chemistry as these are the reactions that allow the chemist to construct complex molecules from simple starting materials. In general, these reactions are the most difficult and temperamental to carry out. Some of these reactions are so important that they are named after the scientists who developed them (e.g. Grignard and Aldol reactions).

These reactions can also be classified by grouping together, depending on the process or mechanism involved. This is particularly useful since specific functional groups will undergo certain types of reaction category. Table given below serves as a summary of the types of reactions which functional groups normally undergo.

Table: Different categories of reaction undergone by functional groups

Reaction Category	Functional Group
Electrophillic addition	Alkenes and alkynes
Electrophilic Substitution	Aromatic
Nucleophilic addition	Aldehydes and ketones
Nucleophilic Substitution	Carboxylic acid derivatives
	Alkyl halides

Elimination	Alcohols and alkyl halides
Reduction	Alkenes, alkynes, aromatic, aldehydes, ketones, nitriles, carboxylic acids, and carboxylic acid derivatives
Oxidation	Alkenes, alcohols, aldehydes
Acid/base reactions	Carboxylic acids, phenols, amines

(a) ***Substitution Reactions:*** These reactions involve the replacement of an atom or group from the organic molecule by some other atom or group without changing the remaining part of the molecule. The product formed as a result of replacement is called *substitution product,* e.g.:

(i) $\underset{\text{Ethanol}}{CH_3CH_2OH} \xrightarrow{PCl_5} \underset{\text{Ethyl chloride}}{CH_3CH_2Cl} + POCl_3 + HCl$

(ii) $\underset{\text{Iodoalkane}}{R\text{—}I} + KOH\,(aq) \longrightarrow \underset{\text{(alcohol)}}{R\text{—}OH} + KI$

(iii) $\underset{\text{Benzene}}{C_6H_5H} + \underset{\text{(Conc.)}}{HNO_3} \xrightarrow{\text{Conc. } H_2SO_4} \underset{\text{Nitorbenzene}}{C_6H_5NO_2} + H_2O$

(b) ***Addition Reactions:*** These reactions are generally given by the organic molecule containing multiple bonds. They involve combination of two molecules to form a single molecule. In general in these reactions one p-bond is cleaved and two sigma bonds are formed. The product formed is known as ***addition product*** or ***adduct***. Some examples are:

(i) $\underset{\text{Ethene}}{CH_2 = CH_2} + HBr \rightarrow \underset{\text{Ethyl bromide}}{CH_3CH_2Br}$

(ii) $\underset{\text{Ethene}}{CH \equiv CH + 2H_2} \xrightarrow[575\text{ K}]{Ni} \underset{\text{Ethane}}{CH_3\text{—}CH_3}$

(iii) $\underset{\text{Acetaldehyde}}{CH_3C(H) = O} + HCN \longrightarrow \underset{\text{Cynohydrin}}{CH_3CH(OH)CN}$

(c) ***Elimination Reactions:*** These reactions involve the removal of two or more atoms/groups from the organic molecule under suitable conditions to form a product with multiple bond. Elimination can be considered as reverse of addition. Some examples are:

(i) $\underset{\text{Ethanol}}{\overset{H\quad\ \ OH}{H_2C{-}CH_2}} \xrightarrow{H_2SO_4} \underset{\text{Ethene}}{H_2C = CH_2} + H_2O$

(ii) $\underset{\text{Ethyl bromide}}{\overset{H\quad\ \ Br}{H_2C{-}CH_2}} \xrightarrow{KOH(alc.)} \underset{\text{Ethene}}{H_2C = CH_2} + HBr$

(d) ***Rearrangement Reactions:*** These reactions involve the rearrangement of atoms within the molecule under suitable conditions to form the product with different properties. Some examples are:

(i) $\underset{\text{Ammonium cyanate}}{NH_4CNO} \xrightarrow{\text{Heat}} \underset{\text{Urea}}{(H_2N)_2C = O}$

(ii) $\underset{\text{n-Butane}}{CH_3CH_2CH_2CH_3} \xrightarrow[\text{Heat}]{AlCl_3} \underset{\text{Isobutane}}{CH_3{-}\overset{CH_3}{\overset{|}{CH}}{-}CH_3}$

The Mechanisms

Definition: A clear understanding of electrophilic and nucleophilic centres permits us to predict where reactions might occur but not what sort of reaction will occur. To understand and predict the outcome of reactions, it is essential to understand what goes on at the electronic level. This process is a *mechanism.*

A mechanism tells us as to how a reaction occurs. It explains how molecules react together to give the final product. The mechanism tells us how bonds are formed and how bonds are broken and in what order. It explains what is happening to the valence electrons in the molecule as it is the movement of these electrons that result in a reaction. Consider the reaction between a hydroxide ion and a proton to form water.

The hydroxide ion is a nucleophile and the proton is an electrophile. A reaction occurs between the nucleophilic centre (the oxygen) and the electrophilic centre (the hydrogen) and water is formed. A new bond is formed between the oxygen of the hydroxide ion and the proton.

The mechanism of this reaction suggests that a lone pair of electrons from oxygen is used to form a bond to the proton. In this way, the oxygen effectively 'loses' one electron and the proton effectively gains one electron. Because of this, the oxygen loses its negative charge and the proton loses its positive charge.

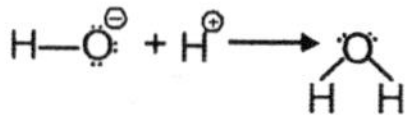

$$H{-}\ddot{O}^{\ominus} + H^{\oplus} \longrightarrow H_2O$$

Fig. Reaction of hydroxide ion and a proton form water.

Curly Arrows: To understand what happens to the valence electrons during a reaction mechanism there is a diagrammatic way making use of curly arrows. For example, the above mechanism can be explained by using a curly arrow to show what happens to the lone pair of electrons. In this case, the arrow starts from a lone pair of electrons on the oxygen (the source of the two electrons) and points to where the centre of the new bond will be formed.

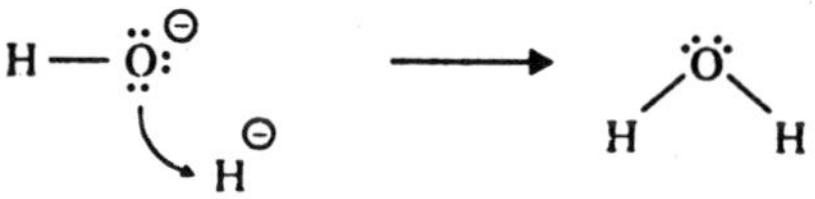

$$H{-}\ddot{O}:^{\ominus} \quad H^{\ominus} \longrightarrow H{-}\ddot{O}{-}H$$

Fig. Mechanism for the reaction of a hydroxide ion with a proton.

Sometimes the arrow is written directly to the proton. Formally, this is incorrect. Arrows should only be drawn directly to an atom if the electrons are going to end up to that atom as a lone pair of electrons.

$$H{-}\ddot{O}:^{\ominus} \quad H^{\oplus} \longrightarrow H{-}\ddot{O}{-}H$$

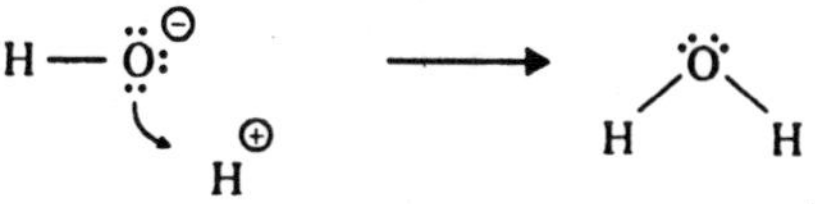

Fig. Incorrect way of drawing a curly arrow.

The following rules are useful when drawing arrows:

(i) Curly arrows show the movement of electrons, not atoms.

(ii) Curly arrows start from the source of two electrons (i.e. a lone pair of electrons on an atom or the middle of a bond which is about to be broken).

(iii) Curly arrows point to an atom if the electrons are going to end up as a lone pair on that atom.

(iv) Curly arrows point to where a new bond will be formed if the electrons are being used to form a new bond.

Figure given below is a demonstration of how arrows should be drawn. One of the lone pairs of electrons on the hydroxide ion is used to form a bond to the acidic proton of the carboxylic acid. The curly arrow representing this starts from a lone pair of electrons and points to the space between the two atoms to show that a bond is being formed.

At the same time as this new bond is being formed, the O–H bond of the carboxylic acid must break. This is because the hydrogen atom can form only one bond. The electrons in this bond end up on the carboxylate oxygen as a third lone pair of electrons. The arrow representing this starts from the centre of the bond being broken and points directly to the atom where the electrons will end up as a lone pair.

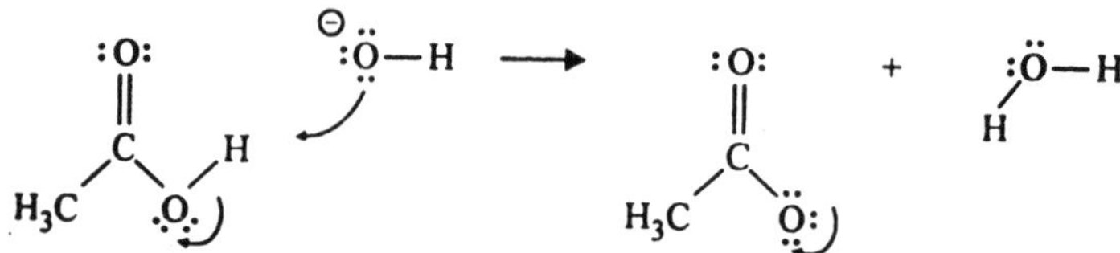

Fig. Mechanism for the reaction of a hydroxide ion with ethanoic acid.

In the process, the negatively charged oxygen of the hydroxide ion ends up as a neutral oxygen in water, because one of the oxygen's lone pairs is used to form the new bond. Both electrons are now shared between two atoms and so the oxygen effectively loses one electron and its negative charge. The oxygen in the carboxylate ion (which was originally neutral in the carboxylic acid) becomes negatively charged since it now has three lone pairs of electrons and has effectively gained an extra electron.

Half Curly Arrows: Sometimes reactions take place that involve the movement of single electrons rather than pairs of electrons. Such reactions are called *radical reactions*. For example, a chlorine molecule can be split into two chlorine radicals on treatment with light. One of the original bonding electrons ends up on one chlorine radical and the second bonding electrons ends up on the other chlorine radical. The movement of these single electrons can be illustrated by using half curry arrows rather than full curly arrows:

$$:\ddot{\underset{\cdot\cdot}{Cl}}-\ddot{\underset{\cdot\cdot}{Cl}}: \xrightarrow{\text{Light}} :\ddot{\underset{\cdot\cdot}{Cl}}\cdot \qquad \cdot\ddot{\underset{\cdot\cdot}{Cl}}:$$

Fig. Use of half curly arrows in a mechanism (homolytic cleavage).

This form of bond breaking is a homolytic cleavage. The radical atoms obtained are neutral but highly reactive species as they have an unpaired valence electron.

There are some important radical reaction in organic chemistry, but the majority of organic reactions involves the heterolytic cleavage of covalent bonds where electrons move together as a pair:

$$H-\ddot{\underset{\cdot\cdot}{Cl}}: \longrightarrow \overset{\oplus}{H} \qquad \cdot\ddot{\underset{\cdot\cdot}{Cl}}:^{\ominus}$$

Fig. Heterolytic cleavage of a bond.

Free Radicals: These are the neutral species having an unpaired electron, e.g. Cl, Br, OR, R CH_3.

Significance of Nucleophilics

Reactions in Enolate Ion

Enolate ions are formed by treating aldehydes or ketones with a base. An α-proton must be present. Enolate ions can undergo a number of important reactions including alkylation and the Aldol reactions.

Alkylation

When an enolate ion is treated with an alkyl halide it results in a reaction called *alkylation* (Fig.E). The overall reaction involves the replacement of an α-proton with an alkyl group. The nucleophilic and electrophilic centres of the enolate ion and methyl iodide are shown (Fig.F). The enolate ion has its negative charge shared between the oxygen atom and the carbon atom because of resonance and so both of these atoms are nucleophilic centres. Iodomethane has a polar C—I bond where the iodine is a weak nucleophilic centre and the carbon is a good electrophilic centre.

Fig.E. Alkylation of a ketone.

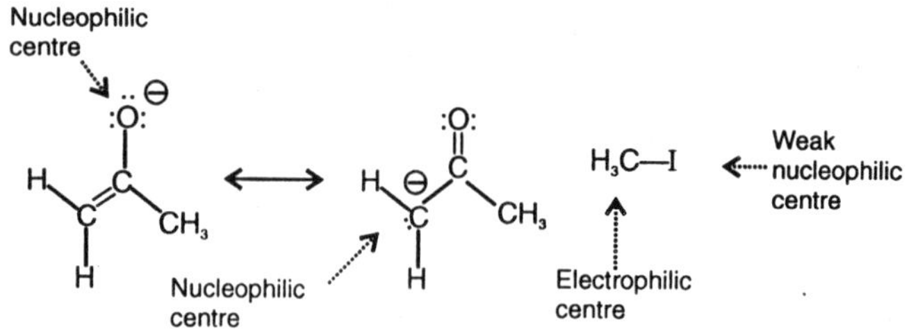

Nucleophilic centre

Weak nucleophilic centre

Nucleophilic centre

Electrophilic centre

Fig.F. Nucleophilic and electrophilic centres.

The possible reaction between these molecules involves the nucleophilic oxygen using one of its lone pairs of electrons to form a new bond to the electrophilic carbon on iodomethane (Following fig.). At the same time, the C—I bond and both electrons move onto iodine to give it a fourth lone pair of electrons and a negative charge.

This reaction is possible, but the actual product obtained is more like to arise from the reaction of the alternative carbanion structures reacting with methyl iodide (Following fig.). This is a more useful reaction as it involves the formation of a carbon-carbon bond and allows the construction of more complex carbon skeletons.

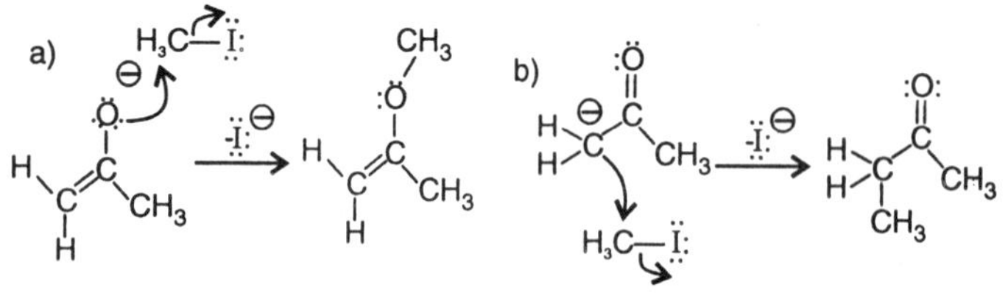

Fig. (a) O-Alkylation; (b) C-Alkylation.

An alternative mechanism to that shown in above fig., but which gives the same product, starts with the enolate ion. The enolate ion is more stable than the carbanion because the charge is on the electronegative oxygen and so it is more likely that the reaction mechanism will occur in this manner (Following fig.). This is a very useful reaction in organic synthesis. However, there are limitations to the type of alkyl halide which can be used in the reaction. The reaction is S_{N2}

with respect to the alkyl halide and so the reaction is suitable with primary alkyl, primary benzylic, and primary allylic halides. The enolate ion is a strong base and if it is reacted with secondary and tertiary halides, elimination of the alkyl halide occurs to give an alkene.

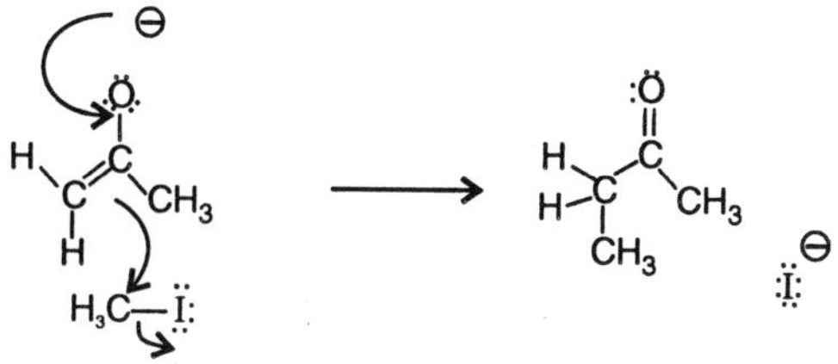

Fig. Mechanism for C-alkylation of the enolate ion.

α-Alkylation occurs easily with ketones, but not so easily for aldehydes because the latter tend to undergo Aldol condensations instead.

The α-protons of a ketone like propanone are only weakly acidic and so a powerful base (e.g. lithium diisopropylamide) is required to generate the enolate ion needed for the alkylation. An alternative method of preparing the same product by using a milder base is to start with ethyl acetoacetate (a β-keto ester) (Fig.G). The α-protons in this structure are more acidic because they are flanked by two carbonyl groups.

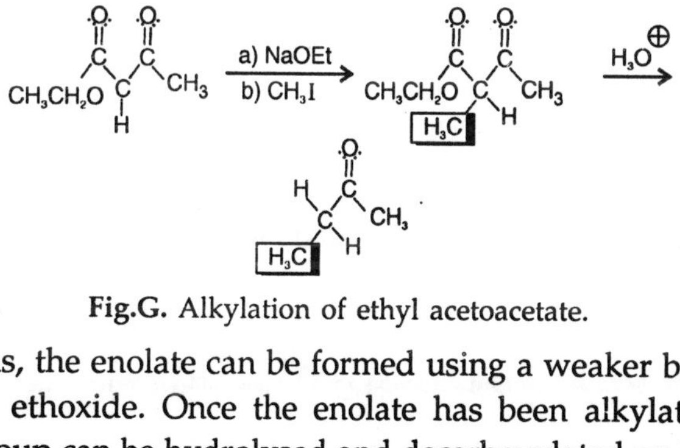

Fig.G. Alkylation of ethyl acetoacetate.

Thus, the enolate can be formed using a weaker base like sodium ethoxide. Once the enolate has been alkylated, the ester group can be hydrolysed and decarboxylated on heating with aqueous hydrochloric acid.

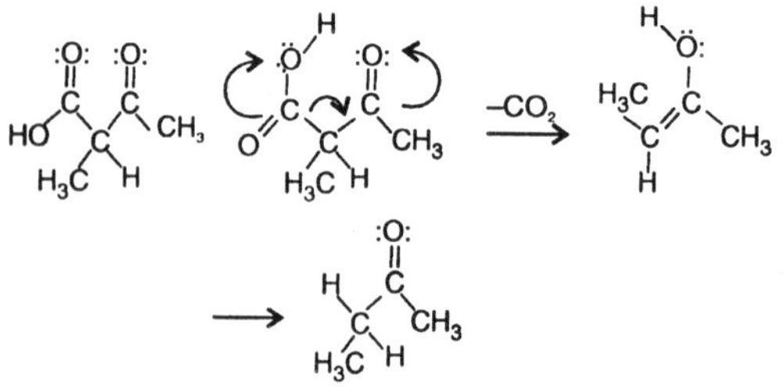

Fig.H. Decarboxylation mechanism.

The decarboxylation mechanism involves the β-keto group and would not take place if this group was absent (Fig.H). Carbon dioxide is lost and the enol tautomer is formed. This can then form the keto tautomers by the normal keto-enol tautomerism.

It is possible for two different alkylations to be done on ethyl acetoacetate as there is more than one α-proton present:

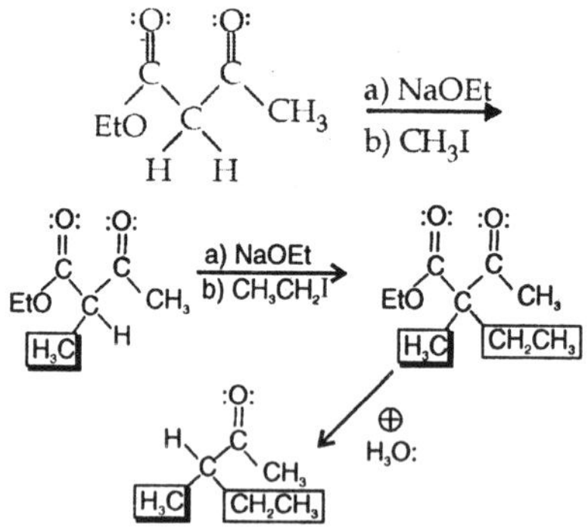

Fig. Double alkylation of ethylacetoacetate.

β-keto esters like ethylacetoacetate are also useful in solving a problem involved in the alkylation of unsymmetrical ketones.

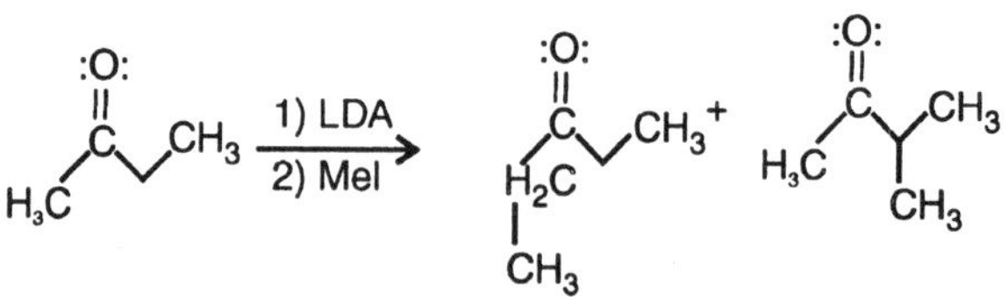

Fig.I Alkylation of butanone.

For example alkylating butanone with methyl iodide leads to two different products as there are α-protons on either side of the carbonyl group (Fig.I). One of these products is obtained specifically by using a β-keto ester to make the target alkylation site more acidic (Fig.J).

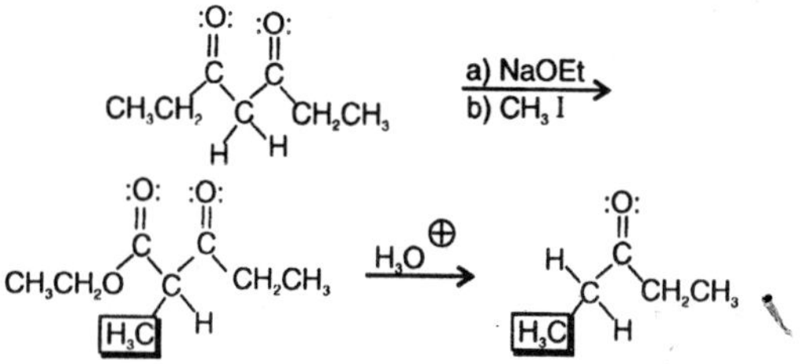

Fig.J. Use of a β-keto ester to direct alkylation.

The alternative alkylation product could be obtained by using a different β-keto ester:

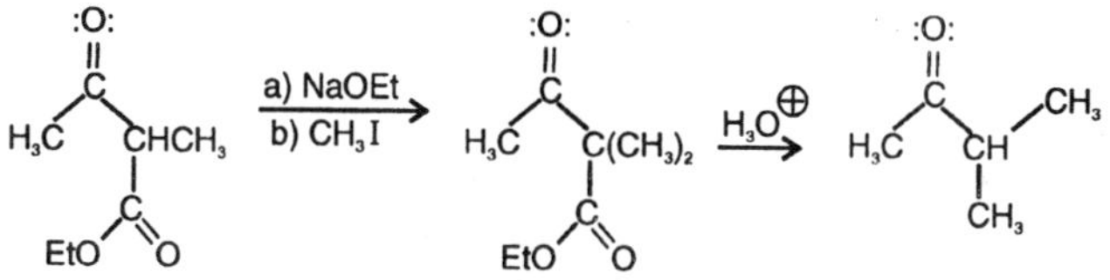

Fig. Use of a β-keto ester to direct alkylation.

Aldol Reaction

Enolate ions can also react with aldehydes and ketones by nucleophilic addition. The enolate ion acts as the nucleophile while the aldehyde or ketone acts as an electrophile. As the enolate ion is formed from a carbonyl compound itself, and can then react with a carbonyl compound, it is possible for an aldehyde or ketone to react with itself. To illustrate this we look at the reaction of acetaldehyde with sodium hydroxide. Under these conditions, two molecules of acetaldehyde are linked together to form β-hydroxyaldehyde.

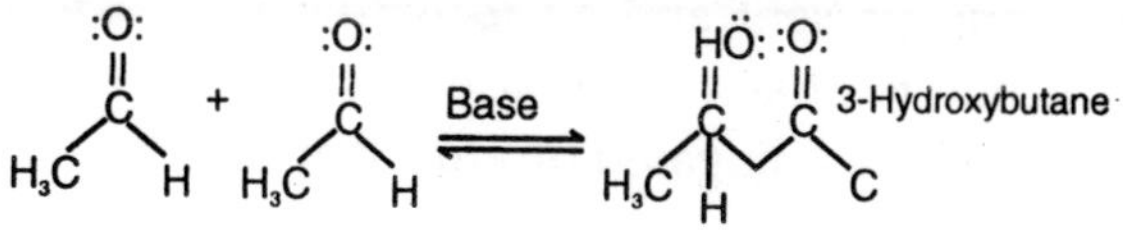

Fig. Aldol reaction.

In this reaction, two separate reactions are going on, i.e. the formation of an enolate ion from one molecule of acetaldehyde, and the addition of that enolate to a second molecule of acetaldehyde. The mechanism starts with the formation of the enolate ion, however, it must be noted that not all of the acetaldehyde is converted to the enolate ion and so we still have molecules of acetaldehyde present in the same solution as the enolate ions. Since acetaldehyde is susceptible to nucleophilic attack, the next stage in the mechanism is the nucleophilic attack of the enolate ion on acetaldehyde (Following fig.). The enolate ion has two nucleophilic centres, i.e. the carbon and the oxygen, however, the preferred reaction is at the carbon atom. The first step is nucleophilic addition of the aldehyde to form a charged intermediate. The second step involves protonation of the charged oxygen. As a dilute solution of sodium hydroxide is used in this reaction so water is available to supply the necessary proton. [It would be wrong to show a free proton (H^+) because the solution is alkaline.]

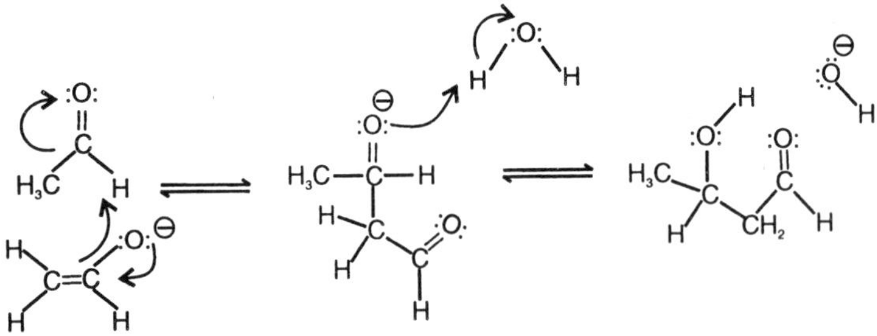

Fig. Mechanism of the Aldol reaction

If the above reaction is carried out with heating, then a different product is obtained (Fig.K). This arises from elimination of a molecule of water from the *Aldol reaction product*. The two reasons for occurrence of this reaction are: First, the product still has an acidic proton (i.e. there is still a carbonyl group present and an α-hydrogen next to it). This proton is prone to attack from base. Second, the dehydration process results in a conjugated product that results in increased stability.

The mechanism of dehydration is shown below (Fig.L). First of all, the acidic proton is removed and a new enolate ion is formed. The electrons in the enolate ion can then move in such a fashion that the hydroxyl group is expelled to give the final product, i.e. an α, β-unsaturated aldehyde. In this example, it is possible to change the conditions such that one gets the Aldol reaction product or the α, β-unsaturated aldehyde, but in some cases only the α, β-unsaturated carbonyl product is obtained, particularly when extended conjugation is possible.

The Aldol reaction is best carried out with aldehydes. Some ketones will undergo an Aldol reaction, but an equilibrium is set up between the products and starting materials and it is essential to remove the product as it is being formed so as take the reaction to completion.

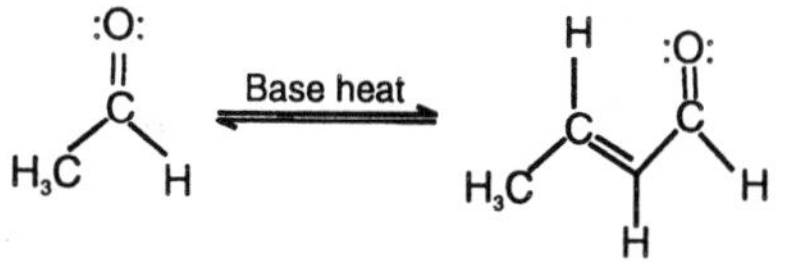

Base heat

Fig.K. Formation of 2-butenal.

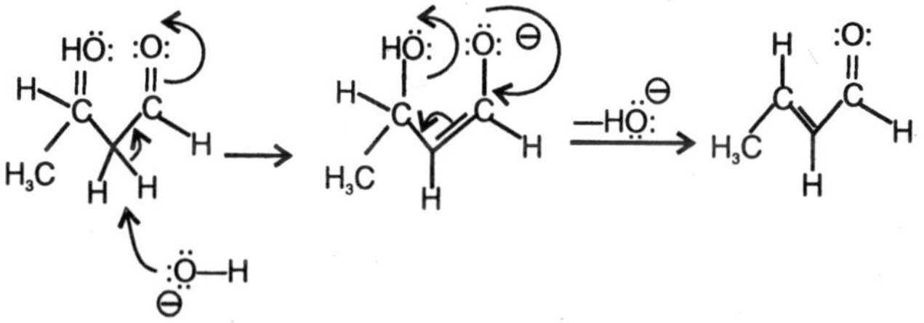

Fig.L. Mechanism of dehydration.

Crossed Aldol Reaction

An Aldol reaction is used to link two molecules of the same aldehyde or ketone, however, it is also possible to link two different carbonyl compounds. This is called a *crossed Aldol reaction*. For example, benzaldehyde and ethanal can be linked in the presence of sodium hydroxide (Following fig.). In this, ethanal reacts with sodium hydroxide to form the enolate ion which then reacts with benzaldehyde. Elimination of water takes place easily to give an extended conjugated system involving the aromatic ring, the double bond, and the carbonyl group.

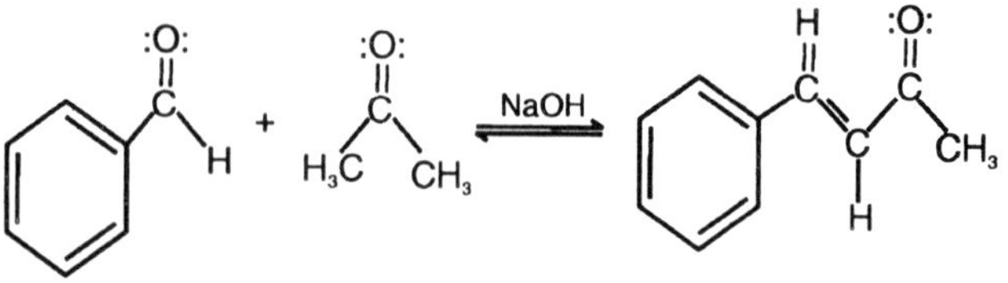

Fig. Crossed Aldol reaction.

This reaction proceeds well because the benzaldehyde has no α-protons and cannot form an enolate ion. Therefore, there is no chance of benzaldehyde undergoing self-condensation. It can only act as the electrophile for another enolate ion. However, what is to stop the ethanal undergoing an Aldol addition with itself as already described.

This reaction can be limited by only having benzaldehyde and sodium hydroxide initially present in the reaction flask. Since benzaldehyde has no α-protons, no reaction can occur. A small quantity of ethanal can then be added. Reaction with excess sodium hydroxide changes most of the ethanal into its enolate ion.

There will only be a very small amount of 'free' ethanal left compared to benzaldehyde and so the enolate ion is more likely to react with benzaldehyde. Once the reaction has taken place, the next small addition of ethanal can occur and the process is repeated.

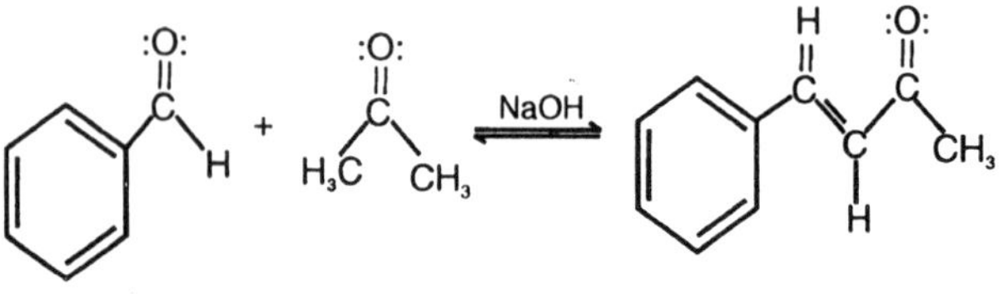

Fig. Claisen-Schmidt reaction.

Ketones and aldehydes can also be linked by the same method, i.e. a reaction called the *Claisen-Schmidt reaction*. The most successful reactions are those where the aldehyde does not have an α-proton.

α-Halogenation

Definition

Aldehydes and ketones react with chlorine, bromine or iodine in acidic solution, this is called *halogenation* at the α-carbon:

Fig. α-Halogenation.

Mechanism

Because acid conditions are used so this process does not involve an enolate ion. Instead, the reaction occurs through the enol tautomers of the carbonyl compound. The enol tautomer acts as a nucleophile with a halogen by the mechanism shown below. In the final step, the solvent acts as a base to remove the proton.

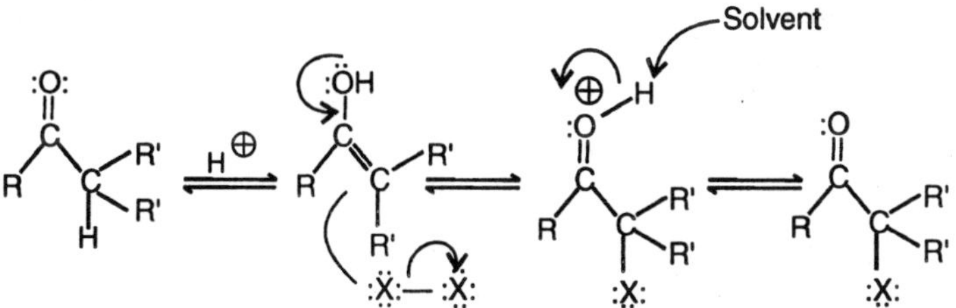

Fig. Mechanism of α-halogenation.

Iodoform Reaction

α-Halogenation can also be done in the presence of base. The reaction proceeds through an enolate ion that is then halogenated (Following fig.).

However, it is difficult to stop the reaction at mono-halogenation because the product formed is generally more acidic than the starting ketone because of the electron-withdrawing effect of the halogen. Due to this, another enolate ion is quickly formed leading to further halogenation.

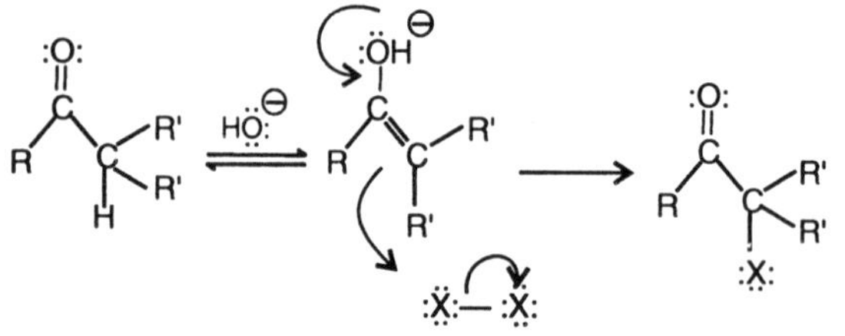

Fig. α-Halogenation in the presence of base.

This tendency towards multiple halogenation is the basis for the *iodoform test* which is used to identify methyl ketones. The ketone to be tested is treated with excess iodine and base and if a yellow precipitate is formed, a methyl ketone is indicated. Under these conditions, methyl ketones undergo α-halogenation three times (Following fig.).

The product obtained is then susceptible to nucleophilic substitution whereby the hydroxide ion substitutes the tri-iodomethyl ($^{-}CI_3$) carbanion — a good leaving group because of the three electron — withdrawing iodine atoms. Tri-iodomethane is then formed as the yellow precipitate.

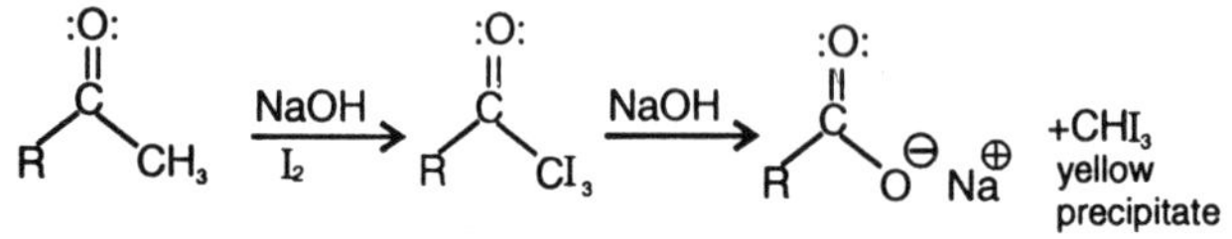

Fig. The iodoform reaction.

Reduction and Oxidation

Reduction to Alcohols

Aldehydes and ketones when reduced yield alcohols with a hydride ion that is provided by reducing reagents like sodium borohydride or lithium borohydride. Primary alcohols are obtained from aldehydes and secondary alcohols from ketones.

Reduction to Alkanes

Aldehydes and ketones on complete reduction give alkanes by three different methods that are complementary to each other. The *Woff-Kishner reduction* is done under basic conditions and is suitable for compounds that might be sensitive to acid (Following fig.). The reaction involves the nucleophilic addition by hydrazine followed by elimination of water to form a hydrazone. The mechanism is the same as that described for the synthesis of 2,4-dinitro-phenylhydrazones.

$$R(C{=}O)R' \xrightarrow[NaOH]{NH_2NH_2} \left[R(C{=}N{-}NH_2)R' \right] \xrightarrow{-N_2(g)} RCH_2R' \qquad R' = H \text{ or alkyl}$$

hydrazone

Fig. Wolf-Kishner reduction.

But the simple hydrazone formed under these reaction conditions spontaneously decomposes with the loss of nitrogen gas.

The *Clemmenson reduction* (Following fig.) gives a similar product but is done under acid conditions and so this is a suitable method for compounds which are unstable to basic conditions.

$$R(C{=}O)R' \xrightarrow[HCl]{Zn(Hg)} RCH_2R' \qquad R' = H \text{ or alkyl}$$

Fig. Clemmenson reduction.

Compounds that are sensitive to both acid and base can be reduced under neutral conditions by forming the thioacetal or thioketal, then reducing with Raney nickel:

$$R(C{=}O)R' \xrightarrow{HSCH_2CH_2SH} R(C(SCH_2CH_2S))R' \xrightarrow[(H_2)]{\text{Raney nickel}} RCH_2R' + CH_3\,CH_3 + NiS$$

Fig. Reduction via a cyclic thioketal.

Aromatic aldehydes and ketones can also be deoxygenated with hydrogen over a palladium charcoal catalyst. The reaction occurs because the aromatic ring activates the carbonyl group towards reduction. *Aliphatic aldehydes and ketones are not reduced* in this.

Oxidation

Ketones are resistant to oxidation whereas aldehydes are easily oxidised. When an aldehyde is treated with an oxidising agent a carboxylic acid is obtained [Following fig. (a)]. Some compounds may be sensitive to the acid conditions used in this reaction and their oxidation can be carried out by using a basic solution of silver oxide [Following fig. (b)].

a) $RCHO \xrightarrow[H_2SO_4]{CrO_3} RCOOH$

b) $RCHO \xrightarrow[NH_3/H_2O/EtOH]{Ag_2O} RCOOH$

Fig. (a) Oxidation of an aldehyde to form a carboxylic acid, (b) Oxidation of an aldehyde using silver oxide.

Both reactions involve the nucleophilic addition of water to form a 1,1-diol or hydrate which is then oxidised in the same way as an alcohol:

$$RCHO \underset{}{\overset{H_2O}{\rightleftharpoons}} [RCH(OH)_2] \xrightarrow{\text{Oxidation}} RCOOH$$

Fig. 1,1-Diol intermediate.

α, β-Unsaturated Aldehydes and Ketones

Definition: α, β-Unsaturated aldehydes and ketones are aldehydes and ketones that are conjugated with a double bond. The α-position is defined as the carbon atom next to the carbonyl group, while the β-position is the carbon atom two bonds removed:

Fig. (a) α,β-unsaturated aldehyde; (b) α,β-unsaturated ketone.

Nucleophilic and Electrophilic Centres

The carbonyl group of α,β-unsaturated aldehydes and ketones are made up nucleophilic oxygen and an electrophilic carbon. However, α,β-unsaturated aldehydes and ketones also have another electrophilic carbon—the β-carbon. This is because of the influence of the electronegative oxygen that can result in the resonance shown below. Since two electrophilic centres are present, there are two places where a nucleophile can react. In both situations, an addition reaction occurs. If the nucleophile reacts with the carbonyl carbon, this is a normal nucleophilic addition to an aldehyde or ketone and is known as a *1,2-nucleophilic addition.* If the nucleophile adds to the β-carbon, this is called a *1,4-nucleophilic addition* or a *conjugate addition.*

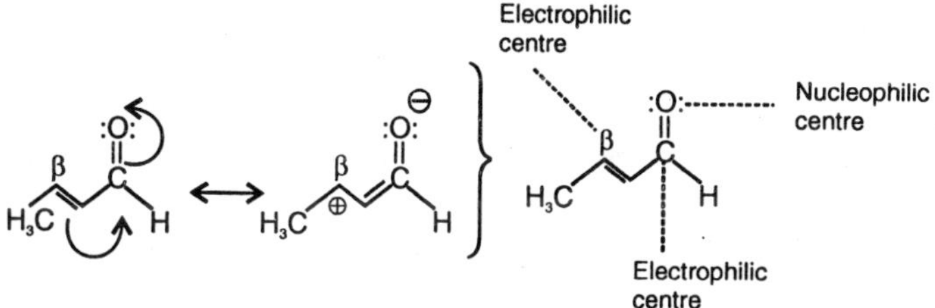

Fig. Nucleophilic and electrophilic centres.

1,2-Addition

The mechanism of 1,2-nucleophilic addition is the same as already described. It is found that Grignard reagents and organolithium reagents will react with α,β-unsaturated aldehydes and ketones in this way and do not attack the β-position:

Fig. 1,2-Nucleophilic addition.

1,4-Addition

The mechanism for 1,4-addition involves two stages (Following fig.). In the first stage, the nucleophile uses a lone pair of electrons to form a bond to the β-carbon. Simultaneously, the C=C π bond breaks and both electrons are used to form a new π bond to the carbonyl carbon. This in turn forces the carbonyl π bond to break with both of the electrons involved moving onto the oxygen as a third lone pair of electrons.

The resulting intermediate is an enolate ion. Aqueous acid is now added to the reaction mixture. The carbonyl π bond is reformed, which forces open the C=C π bond. These electrons are now used to form a σ bond to a proton at the α carbon.

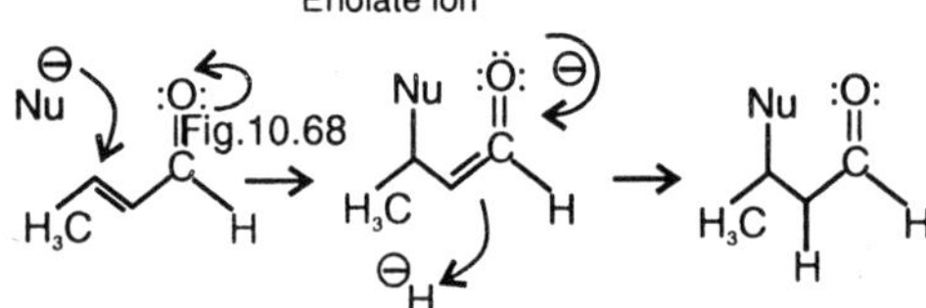

Fig. Mechanism of 1,4-nucleophilic addition.

Conjugate addition reactions can be done with amines, or a cyanide ion. Alkyl groups can also be added to the β-position by using orsanocuprate reagents (Following fig.). A large number of organocuprate reagents can be prepared allowing the addition of primary, secondary and tertiary alkyl groups, aryl groups, and alkenyl groups.

:O: C H_3C H

$Li^+(Me_2C\bar{u})$

H_3C :O: C H_3C H

Fig. Alkylation with organocuprate reagents.

Reduction

The reduction of α,β-unsaturated ketones to allylic alcohols can be easily done using lithium aluminium hydride under carefully controlled conditions (Following fig.). With sodium borohydride, some reduction of the alkene also occurs.

$$H_3C{-}CH{=}CH{-}C({=}O){-}H \xrightarrow[\text{b) } H_3O^{\oplus}]{\text{a) } LiAlH_4} H_3C{-}CH{=}CH{-}CH(OH){-}R$$

$$H_3C{-}CH{=}CH{-}C({=}O){-}R \xrightarrow[\text{b) } H_3O^{\oplus}]{\text{a) } NaBH_4} H_3C{-}CH{=}CH{-}CH(OH){-}R \quad H_3C{-}CH_2{-}CH_2{-}CH(OH){-}R$$

Nucleophilic Additions

Aldehydes and Ketones

Functional Group Transformations: Functional group transformations help us in the conversion of a functional group to an aldehyde or a ketone without affecting the carbon skeleton of the molecule. Aldehydes can be synthesised by the oxidation of primary alcohols, or by the reduction of esters, acid chlorides, or nitriles. Since nitriles can be obtained from alkyl halides, this a way of adding an aldehyde unit (CHO) to an alkyl halide:

$$\underset{\text{Alkyl halide}}{R{-}X} \xrightarrow{KCN} \underset{\text{Nitrite}}{R{-}C{\equiv}N} \xrightarrow[2.H_3C\ \oplus]{\text{1. DIBAH, Toluene}} R{-}C({=}H){-}H$$

Fig. Synthesis of an aldehyde from an alkyl halide with 1C chain extension.

Ketones can be synthesised by the oxidation of secondary alcohols. Methyl ketones can be synthesised from terminal alkynes.

C–C Bond Formation

Reactions that result in the formation of ketones by carbon-carbon bond formation are extremely important as they can be used to synthesise complex carbon skeleton from simple starting materials. Ketones can be synthesised from the reaction of acid chlorides with organocuprate reagents, or from the reaction of nitriles with a Grignard or organolithium reagent. Aromatic ketones can be synthesised by the Friedel-Crafts acylation of an aromatic ring.

C–C Bond Cleavage

Aldehydes and ketones can be obtained from the ozonolysis of suitably substituted alkenes.

Charateristics of Aldehydes and Ketones

Carbonyl Group: Both aldehydes and ketones have a carbonyl group (C=O). The substituents attached to the carbonyl group determine whether it is an aldehyde or a ketone, and whether it is aliphatic or aromatic.

The geometry of the carbonyl group is planar with bond angles of 120° (Following fig.). The carbon and oxygen atoms of the carbonyl group are sp^2 hybridised and the double bond between the atoms consists of a strong σ bond and a weaker π bond.

The carbonyl bond is shorter than a C–O single bond (1.22 Å vs. 1.43 Å) and is also stronger since two bonds are present as opposed to one (732 kJ mol^{-1} vs. 385 kJ mol^{-1}). The carbonyl group is more reactive than a C–O single bond because of the relatively weak π bond.

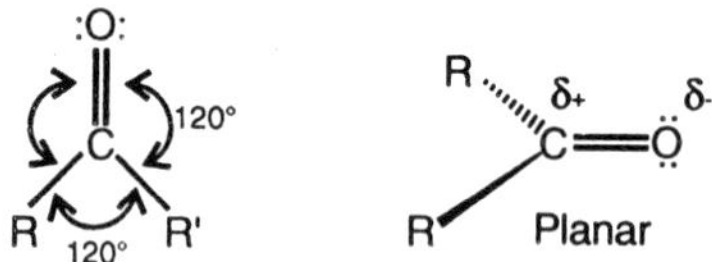

Fig. Geometry of the carbonyl group.

The carbonyl group is polarised in such a way that the oxygen is slightly negative and the carbon is slightly positive. Both the polarity of the carbonyl group and the presence of the weak π bond explain much of the chemistry and the physical properties of aldehydes and ketones. The polarity of the bond also means that the carbonyl group has a resultant dipole moment.

Properties

Because of polar nature of the carbonyl group, aldehydes and ketones have higher boiling points than alkanes of similar molecular weight. However, hydrogen bonding is not possible between carbonyl groups and so aldehydes and ketones have lower bonding points than alcohols or carboxylic acids.

Low molecular weight aldehydes and ketones (e.g. formaldehyde and acetone) are soluble in water. This is because the oxygen of the carbonyl group can participate in intermolecular hydrogen bonding with water molecules:

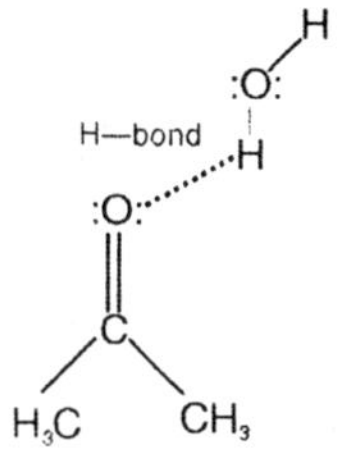

Fig. Intermolecular hydrogen bonding of a ketone with water.

As molecular weight increases, the hydrophobic character of the attached alkyl chains begins to outweigh the water solubility of the carbonyl group and the result is that large molecular weight aldehydes and ketones are insoluble in water. Aromatic ketones and aldehydes are insoluble in water because of the hydrophobic aromatic ring.

Nucleophilic and Electrophilic Centres

Because of the polarity of the carbonyl group, aldehydes and ketones have a nucleophilic oxygen centre and an electrophilic carbon centre as shown for propanol (Following fig.). Therefore, nucleophiles react with aldehydes and ketones at the carbon centre, and electrophiles react at the oxygen centre.

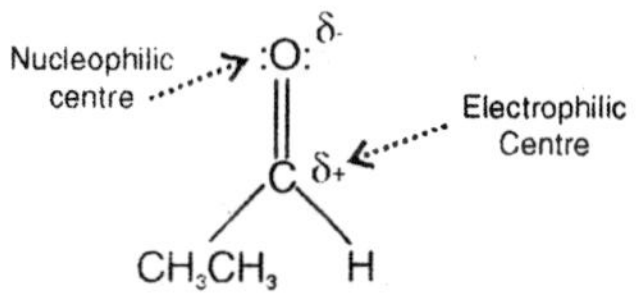

Fig. Nucleophilic and electrophilic centres of the carbonyl group.

Keto-enol Tautomerism

Ketones that have hydrogen atoms on their α-carbon (the carbon next to the carbonyl group) are in rapid equilibrium with an isomeric structure known as an *enol* in which the α-hydrogen ends up on the oxygen instead of the carbon. The two isomeric forms are known as *tautomers* and the process of equilibrium is known as *tautomerism* (Following fig.). Generally the equilibrium favours the keto tautomer and the enol tautomer may only be present in very small quantities.

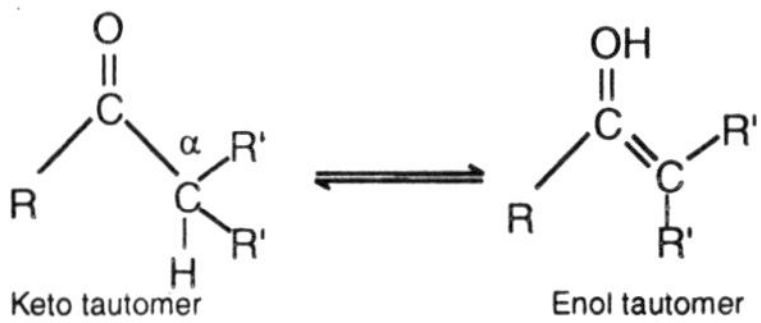

Keto tautomer

Enol tautomer

Fig. Keto-enol tautomerism.

The tautomerism mechanism is catalysed by acid or base. When catalysed by acid (Following fig.), the carbonyl group acts a nucleophiles with the oxygen using a lone pair of electrons to form a bonds to a proton. This results in the carbonyl oxygen gaining a positive charge that activates the carbonyl group to attack by weak nucleophiles (Step 1). The weak nucleophiles is a water molecule that removes the α-proton from the ketone, resulting in the formation of a new C=C double bond and cleavage of the carbonyl π bond. The enol tautomer is formed thus neutralising the unfavourable positive charge on the oxygen (Step 2).

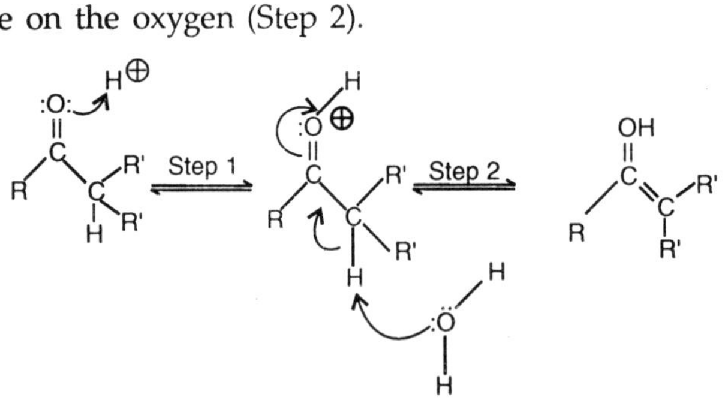

Step 1

Step 2

Fig. Acid-catalysed mechanism for keto-enol tautomerism.

Under basic condition (Following fig.), an enolate ion is formed, which then reacts with water to form the enol.

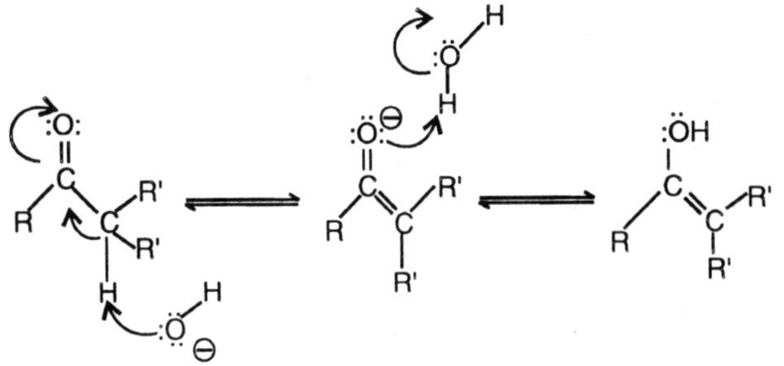

Fig. Base-catalysed mechanism for keto-enol tautomerism.

Nucleophilics in Operation

Definition

The nucleophilic addition involves the addition of a nucleophilic to a molecule. This is a distinctive reaction for ketones and aldehydes and the nucleophile will add to electrophilic carbon atom of the carbonyl group. The nucleophile can be a negatively charged ion like cyanide or hydride, or it can be a neutral molecule like water or alcohol.

Overview

Generally, addition of charged nucleophiles results in the formation of a charged intermediate (Following fig.). The reaction stops at this stage and acid has to be added to complete the reaction.

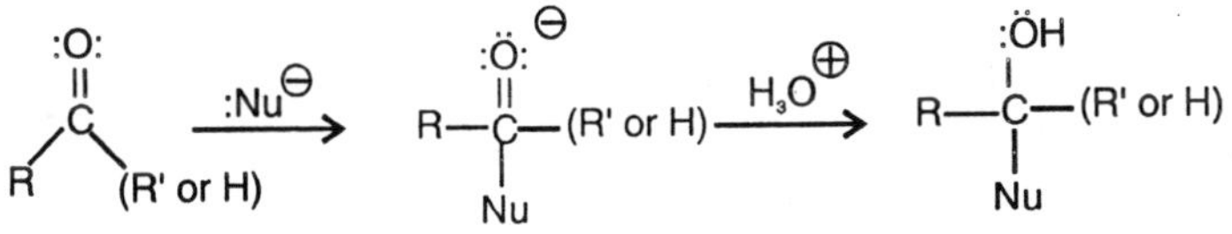

Fig. Nucleophilic addition to a carbonyl group.

Neutral nucleophiles where nitrogen or oxygen is the nucleophilic centre are relatively weak nucleophiles, and an acid catalyst is generally needed. After nucleophilic addition has taken place, further reactions may occur leading to structures such as imines, enamines, acetals, and ketals:

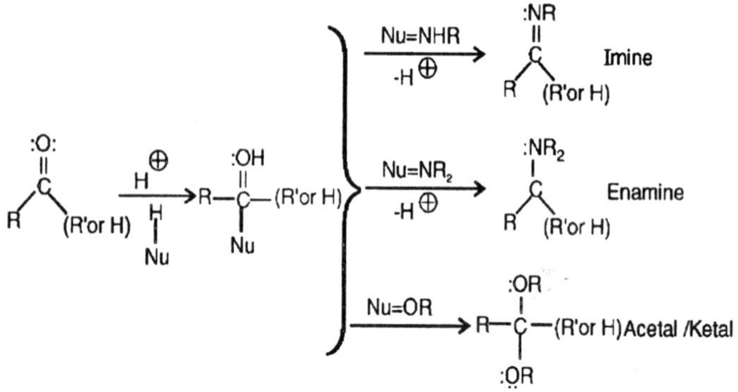

Fig. Synthesis of imines, enamines, acetals, and ketals.

Bibliography

Adams, H. R.: *Veterinary Pharmacology and Therapeutics*, Iowa State University Press, Iowa, 1995.

Adams, R. D.: *Principles of Neurology*, McGraw Hill, New York, 1997.

Arnold, H. L.: *Andrews Diseases of the Skin*, Saunders, Philadelphia, 1990.

Barclay, A. N.: *The Leucocyte Antigen Factsbook*, Academic Press, San Diego, 1997.

Behrman, R. E.: *Nelson Textbook of Pediatrics*, Saunders, Philadelphia, 2004.

Belshe, R. B.: *Textbook of Human Virology*, Mosby, St. Louis, 1991.

Birks, J. W.: *Chemiluminescence and Photochemical Reaction Detection in Chromatography*, VCH Publishers, New York, 1989.

Bradley, W. G.: *Neurology in Clinical Practice*, Butterworth Heinemann, Boston, 2000.

BrainInfo: *Seattle*: *Neurosciences Division*, National Primates Research Center, University of Washington, 2000.

Bruice, P. Y. : *Organic Chemistry*, Prentice Hall, Upper Saddle River, New Jursey, 2002.

Brunton, L.: *Goodman and Gilman's the Pharmacological Basis of Therapeutics*, McGraw Hill, New York, 2006.

Burr, J. G.: *Chemi and Bioluminescence,* Marcel Dekker, New York, 1985.

Campbell, A. K.: *Chemiluminescence: Principles and Applications in Biology and Medicine,* VCH, Ellis Horwood Ltd., New York, 1988.

Carey, F. A.: *Organic Chemistry,* McGraw Hill, London, 1989.

Carlisle, J. G.: *Red Tide in California,* Published by State of California, California, 1968.

Castenholz, C. W.: *Bergey's Manual of Systematic Bacteriology,* Springer, New York, 2001.

Catterall, W. A.: *IUPHAR Compendium of Voltage Gated ion Channels,* IUPHAR Media, Leeds, 2002.

Champion, R. H.: *Textbook in Dermatology,* Blackwell, Oxford, 1992.

Cilento, G.: *Chemical and Biological Generation of Excited States,* Academic Press, New York, 1982.

Clemente, C. D.: *Gray's Anatomy,* Philadelphia: Lea & Febiger, 1985.

Dale, D. C.: *Scientific American Medicine,* Scientific American, New York, 2004.

Day, W. T.: *Bacterial Luminescence in the Deep-sea Anglerfish Oneirodes Acanthias,* Natural History Museum, Los Angles, 1974.

DeLuca, M. A.: *Bioluminescence and Chemiluminescence,* Academic Press, Orlando, 1978.

DeVita, V. T.: *Cancer: Principles and Practice of Oncology,* Lippincott, Philadelphia, 2005.

Dewick. P. M.: *Medicinal Natural Products,* Wiley, West Sussex, 2002.

Donaghy, M.: *Brain's Diseases of the Nervous System,* Oxford, New York, 2001.

Dorland's: *Medical Dictionary*, Philadelphia, Saunders, 2003.

Dyke, K. and Castranova, V.: *Cellular Chemiluminescence*, CRC Press, Boca Raton, 1987.

Evans, R. and Barnett, N.: *Luminescence: Overview*, Academic Press, Orlando, 1995.

Fauquet, C. M.: *Virus Taxonomy: Classification and Nomenclature of Viruses*, Academic Press, San Diego, 2005.

Fishman, A. P.: *Fishman's Pulmonary Diseases and Disorders*, McGraw Hill, New York, 1998.

Fitzgerald, K. A.: *The Cytokine Factsbook*, Academic Press, San Diego, 2001.

Fontijn, A.: *Gas-phase Chemiluminescence and Chemi-ionization*, Elsevier, New York, 1985.

Fritz, A.: *International Classification of Diseases for Oncology (ICD O)*, World Health Organization, Geneva, 2000.

Gennaro, A. R.: *Remington: The Science and Practice of Pharmacy*, Lippincott Williams and Wilkins, Baltimore, 2000.

Goldman, L.: *Cecil Textbook of Medicine*, Saunders, Philadelphia, 2004.

Gosling, P.: *Dictionary of Parasitology*, CRC Press, New York, 2005.

Gundermann, K. D.: *Chemiluminescence in Organic Chemistry*, Springer Verlag, Berlin, 1987.

Harvey, E. N.: *Bioluminescence*, Academic Press, New York, 1952.

————————: *Living Light*, Hafner Publishers, New York, 1965.

Haubrich, W. S.: *Bockus Gastroenterology*, Saunders, Philadelphia, 1995.

Herring, P. J.: *Bioluminescence in Action*, Academic Press, New York, 1978.

Hunter, G. W.: *Hunter's Tropical Medicine and Emerging Infectious Diseases*, Saunders, Philadelphia, 2000.

Hurley, H. J. and Moschella, S. L.: *Dermatology*, Saunders, Philadelphia, 1992.

Ibelgaufts, H.: *Dictionary of Cytokines*, VCH Publishers, New York, 1995.

Jablonski, S.: *Dictionary of Medical Acronyms and Abbreviations*, Hanley and Belfus, Philadelphia, 2001.

Johnson, F. H.: *Luminescence, Narcosis and Life in the Deep Sea*, Vantage Press, New York, 1988.

Johnston; D. and Bodner, G. M.: *A Multicolored Luminescence Demonstration*, Journal of Chemical Education, 1992.

Junqueira, L. C.: *Basic Histology*, McGraw Hill, New York, 1998.

Kahn, C. M.: *Merck Veterinary Manual*, Merck, New Jursey, 2005.

Kasper, D. L.: *Harrison's Principles of Internal Medicine*, McGraw Hill, New York, 2005.

Kenneth, W. L.: *Macroscale and Microscale Organic Experiments*, Houghton Mifflin, Boston, 1994.

Kirk, P. M.: *Ainsworth and Bisby's Dictionary of the Fungi*, CAB International, England, 2001.

Klein, H. A.: *Bioluminescence*, Lippincott Publishers, Philadelphia, 1965.

Konev, S. V.: *Fluorescence and Phosphorescence of Proteins and Nucleic Acids*, Plenum Press, New York, 1967.

Kumar, V.: *Robbins and Cotran, Pathological Basis of Disease*, Elsevier Saunders, Philadelphia, 2005.

Larock, R. C.: *Comprehensive Organic Transformations*, Houghton Mifflin, Boston, 1994.

Larsen, P. R.: *Williams Textbook of Endocrinology*, Saunders, Philadelphia, 2003.

Lee, G. R.: *Wintrobe's Clinical Hematology,* Lea and Febiger, Philadelphia, 1999.

Lewis, R. J.: *Chemical Dictionary,* John Wiley, New York, 2001.

Magalini, S. C.: *Dictionary of Medical Syndromes,* Lippincott, Philadelphia, 1997.

March, J.: *Advanced Organic Chemistry,* Houghton Mifflin, Boston, 1994.

Markell, E. K.: *Medical Parasitology,* Saunders, Philadelphia, 1992.

McElroy, W. D.: *Bioluminescence and Chemiluminescence,* Academic Press, Orlando, 1986.

McGown, L. B.: *Phosphorescence, and Chemiluminescence Spectrometry,* Analytical Chemistry, 1992.

Menkes, J. H.: *Textbook of Child Neurology,* Williams and Wilkins, Baltimore, 1995.

Moore, K. L.: *Clinically Oriented Anatomy,* Williams and Wilkins, Baltimore, 2006.

Nealson, K. H.: *Bioluminescence, Current Perspectives,* Burgess Publication, Minneapolis, 1981.

Nieman, T.: *Chemiluminescence: Techniques, Liquid Phase Chemiluminescence,* Academic Press, Orlando, 1995.

————————: *Chemiluminescence: Theory and Instrumentation, Overview,* Academic Press, Orlando, 1995.

O'Day, W. T.: *Luminescent Silhouetting in Stomiatoid Fishes,* Published by Natural History Museum; Los Angles, 1973.

Paparella, M.: *Otolaryngology,* Saunders, Philadelphia, 1991.

Parent, A.: *Carpenters Human Neuroanatomy,* Williams and Wilkins, Media, 1996.

Paula, B. Y.: *Solutions Manual to Accompany Organic Chemistry,* Prentice Hall, Upper Saddle River, New Jursey, 2004.

Phipson, T. L.: *Phosphorescence or the Emission of Light by Minerals, Plants, and Animals*, L. Reeve and Co., London, 1862.

Price, D. and Worsfold, P. J.: *Flow Injection Assays with Chemiluminescence and Bioluminescence Detection: A Review*, Journal of Bioluminescence and Chemiluminescence, 1993.

Pringle, M. J.: *Analytical Applications of Chemiluminescence*, Academic Press, New York, 1993.

Reeder, D. M.: *Mammal Species of the World*, Johns Hopkins University Press, Baltimore, 2005.

Scott, T. A.: *Concise Encyclopedia of Biology*, W. De Gruyter, Berlin, 1996.

Scriver, C. R.: *The Metabolic and Molecular Basis of Inherited Disease*, McGraw Hill, New York, 2001.

Segen, J. C.: *Dictionary of Modern Medicine*, McGraw Hill, New York, 2006.

Shakhashiri, B. Z.: *Chemical Demonstrations*, University of Wisconsin Press, Madison, 1983.

Sloane, S. B.: *Medical Abbreviations and Eponyms*, Saunders, Philadelphia, 1997.

Smith, M. : *Organic Chemistry*, Oxford, New York, 2001.

Solomons, T. W.: *Organic Chemistry*, Wiley, New York, 2000,

Staley, J. T. and Krieg, N. R.: *Bergey's Manual of Systematic Bacteriology*, Springer, New York, 2005.

Staley, J. T.: *Bergey's Manual of Systematic Bacteriology*, Williams and Wilkins, Baltimore, 1989.

Stedman: *Medical Dictionary, Illustrated*, Lippincott Williams and Wilkins, Philadelphia, 2000.

Sundberg, R. J. and Carey, F. A.: *Advanced Organic Chemistry: Reactions and Synthesis*, Houghton Mifflin, Boston, 1994.

Testen, M. E. and L.G. Williams: *Chemiluminescence*, University of Wisconsin Press, Madison, 1983.

Townshend, A. *Flame Chemiluminescence Analysis by Molecular Emission Cavity Detection*, Wiley, New York, 1994.

Turoverov, K.: *Luminescence of Biopolymers and Cells*, Plenum Press, New York, 1969.

Van, D. K.: *Bioluminescence and Chemiluminescence, Instruments and Applications*, CRC Press, Boca Raton, 1985.

Walsh, P. C.: *Campbell's Urology*, Saunders, Philadelphia, 1997.

Warner, I. M and McGown, L. B.: *Molecular Fluorescence, Phosphorescence, and Chemiluminescence Spectrometry*, Analytical Chemistry, 1990.

Williams, S. T.: *Bergey's Manual of Systematic Bacteriology*, Williams and Wilkins, Baltimore, 1989.

Williamson, K. L.: *Organic Experiments*, Houghton Mifflin, Boston, 1998.

Yanoff, M.: *Ophthalmology*, Mosby, St. Louis, 2004.

Yates, M. V.: *Dictionary of Environmental Microbiology*, Academic Press, San Diego, 2003.

Zipes, O. P.: *Braunwald's Heart Disease*, Elsevier Saunders, Philadelphia, 2005.

Index

A

Acid Anhydrides, 66, 143, 148, 149, 155, 156, 178, 180, 185, 186, 187.
Acid Chlorides, 30, 66, 126, 143, 148, 149, 153, 155, 156, 178, 180, 183, 185, 186, 187, 290.
Acidity, 142, 143, 246, 251, 252, 263.
Alcohols, 8, 17, 28, 33, 34, 35, 36, 37, 39, 71, 108, 112, 115, 122, 128, 130, 131, 132, 133, 134, 135, 136, 137, 138, 139, 140, 141, 142, 143, 144, 155, 156, 157, 162, 169, 170, 171, 179, 180, 188, 191, 193, 212, 216, 239, 240, 242, 249, 254, 260, 261, 269, 285, 289, 290, 291, 305.
Alkoxides, 112.
Ammonia, 9, 31, 106, 117, 118, 122, 127, 134, 166, 167, 199, 230, 247, 248, 254, 255, 262, 263.

C

Carbohydrates, 1, 16.
Carbon, 1, 2, 3, 4, 5, 6, 7, 8, 9, 10, 11, 12, 13, 14, 16, 21, 24, 25, 32, 33, 34, 37, 46, 54, 57, 59, 62, 63, 64, 65, 66, 68, 69, 70, 71, 72, 73, 74, 75, 76, 77, 78, 79, 80, 97, 99, 100, 101, 106, 109, 115, 116, 119, 121, 126, 127, 128, 130, 131, 132, 139, 141, 158, 159, 162, 163, 164, 165, 168, 169, 172, 173, 174, 175, 176, 177, 178, 179, 181, 182, 183, 185, 186, 194, 195, 196, 197, 201, 204, 207, 208, 211, 231, 232, 233, 235,

237, 239, 241, 242, 243, 244, 245, 248, 251, 253, 254, 255, 258, 260, 266, 268, 275, 276, 278, 280, 283, 287, 288, 290, 291, 292, 294, 295, 296, 297, 298, 299, 300, 301, 303.

Carboxylic Acid, 23, 25, 30, 31, 36, 39, 44, 97, 119, 127, 144, 150, 151, 154, 160, 164, 178, 179, 180, 181, 182, 184, 185, 186, 188, 215, 216, 250, 251, 256, 264, 268, 269, 272, 286, 287.

Cyanoborohydride, 118.

E

Energy, 63, 67, 68, 69, 71, 72, 75, 77, 78, 79, 80, 173, 189, 190, 208, 219, 220, 256.

Ethanoic Acid, 31, 160, 249, 250, 251, 252, 257, 258, 272.

H

Halogen Atom, 46, 117, 162, 227, 228, 234, 267.

Hydrofluoric Acid, 249, 250.

Hydroperoxide, 108, 194.

I

Isomerism, 20, 21.

M

Metal, 1, 42, 52, 88, 113, 134, 170, 189, 198, 199, 206, 267.

Methanol, 17, 31, 38, 105, 130, 133, 152, 177, 183, 303.

Molecular, 67, 68, 76, 77, 112, 115, 122, 130, 133, 163, 179, 180, 218, 245, 246, 291, 292.

N

Nitrogen, 2, 21, 31, 32, 34, 36, 65, 70, 71, 80, 84, 88, 99, 119, 120, 121, 123, 125, 126, 127, 166, 186, 214, 225, 228, 243, 247, 248, 249, 250, 252, 261, 263, 265, 266, 267, 285, 294, 298, 299, 300, 301, 304.

Nucleophiles, 54, 99, 112, 113, 116, 125, 126, 131, 132, 137, 138, 165, 166, 169, 171, 172, 180, 181, 183, 184, 186, 188, 227, 268, 292, 293, 294, 295, 296, 298, 302, 305.

Nucleophilic Chloride Ion, 138.

O

Orbitals, 54, 62, 63, 65, 67, 68, 69, 70, 71, 72

73, 74, 75, 76, 77, 78, 79, 80, 81, 112, 120, 121, 176, 201, 204, 231, 232, 246.

Organic Compounds, 10, 16, 21, 66, 180.

Oxidation, 42, 85, 86, 112, 113, 140, 141, 142, 145, 189, 190, 191, 192, 194, 195, 216, 269, 285, 286, 287, 290.

Oxygen, 2, 6, 8, 9, 16, 21, 29, 36, 37, 51, 65, 70, 71, 75, 108, 112, 114, 115, 116, 123, 129, 130, 131, 135, 136, 138, 139, 141, 142, 151, 152, 156, 179, 182, 183, 186, 187, 191, 194, 212, 213, 214, 223, 224, 225, 228, 239, 243, 244, 245, 247, 248, 249, 250, 252, 254, 257, 258, 259, 264, 265, 266, 267, 270, 271, 272, 275, 276, 277, 280, 287, 288, 291, 292, 293, 294, 295, 299, 300, 302, 303, 304.

P

Phenols, 36, 128, 129, 130, 131, 132, 133, 142, 143, 144, 145, 215, 269.

Polymer, 12, 47.

S

Sulphur, 2, 9, 21, 54, 112, 114, 116, 158, 213, 214, 302, 306.

T

Temperature, 8, 38, 189.

Thiolate ions, 112, 113.

W

Water, 6, 8, 13, 38, 39, 46, 48, 52, 53, 99, 100, 119, 122, 123, 125, 127, 130, 131, 132, 133, 134, 135, 136, 137, 138, 147, 149, 150, 151, 152, 158, 163, 166, 167, 168, 169, 170, 178, 179, 180, 190, 195, 206, 212, 213, 230, 232, 238, 239, 242, 247, 248, 249, 250, 251, 261, 262, 264, 265, 267, 270, 271, 272, 280, 281, 282, 285, 287, 291, 292, 293, 294, 300, 301, 302, 304, 305.

Z

Zinc, 112, 190.

□□□